T0179809

The Fundamentals of Heavy Tails

Heavy tails – extreme events or values more common than expected – emerge everywhere: the economy, natural events, and social and information networks are just a few examples. Yet after decades of progress, they are still treated as mysterious, surprising, and even controversial, primarily because the necessary mathematical models and statistical methods are not widely known.

This book, for the first time, provides a rigorous introduction to heavy-tailed distributions accessible to anyone who knows elementary probability. It tackles and tames the zoo of terminology for models and properties, demystifying topics such as the generalized central limit theorem and regular variation. It tracks the natural emergence of heavy-tailed distributions from a wide variety of general processes, building intuition. And it reveals the controversy surrounding heavy tails to be the result of flawed statistics, then equips readers to identify and estimate with confidence. Over 100 exercises complete this engaging package.

JAYAKRISHNAN NAIR is Associate Professor in Electrical Engineering at IIT Bombay. His research focuses on modeling, performance evaluation, and design issues in online learning environments, communication networks, queueing systems, and smart power grids. He is the recipient of best paper awards at IFIP Performance (2010 and 2020) and ACM e-Energy (2020).

ADAM WIERMAN is Professor of Computing and Mathematical Sciences at the California Institute of Technology (Caltech). His research develops tools in machine learning, optimization, control, and economics with the goal of making the networked systems that govern our world sustainable and resilient. He is best known for his work spearheading the design of algorithms for sustainable data centers and he is the recipient of numerous awards including the ACM Sigmetrics Rising Star award, the ACM Sigmetrics Test of Time award, the IEEE Communication Society William Bennet Prize, and multiple teaching and best paper awards.

BERT ZWART is group leader at CWI Amsterdam and Professor of Mathematics at Eindhoven University of Technology. He has expertise in stochastic operations research, queueing theory, and large deviations, and in the context of heavy tails, he has focused on sample path properties, designing Monte Carlo methods and applications to computer-communication and energy networks. He was area editor of *Operations Research*, the flagship journal of his profession, from 2009 to 2017, and was the recipient of the INFORMS Applied Probability Society Erlang prize, awarded every two years to an outstanding young applied probabilist.

This series of high-quality upper-division textbooks and expository monographs covers all aspects of stochastic applicable mathematics. The topics range from pure and applied statistics to probability theory, operations research, optimization, and mathematical programming. The books contain clear presentations of new developments in the field and also of the state of the art in classical methods. While emphasizing rigorous treatment of theoretical methods, the books also contain applications and discussions of new techniques made possible by advances in computational practice.

A complete list of books in the series can be found at www.cambridge.org/statistics..
Recent titles include the following:

The Fundamentals of Heavy Tails

Properties, Emergence, and Estimation

Jayakrishnan Nair

Indian Institute of Technology, Bombay

Adam Wierman

California Institute of Technology

Bert Zwart

CWI Amsterdam and Eindhoven University of Technology

CAMBRIDGE
UNIVERSITY PRESS

CAMBRIDGE
UNIVERSITY PRESS

University Printing House, Cambridge CB2 8BS, United Kingdom

One Liberty Plaza, 20th Floor, New York, NY 10006, USA

477 Williamstown Road, Port Melbourne, VIC 3207, Australia

314–321, 3rd Floor, Plot 3, Splendor Forum, Jasola District Centre, New Delhi – 110025, India

103 Penang Road, #05-06/07, Visioncrest Commercial, Singapore 238467

Cambridge University Press is part of the University of Cambridge.

It furthers the University's mission by disseminating knowledge in the pursuit of
education, learning, and research at the highest international levels of excellence.

www.cambridge.org
Information on this title: www.cambridge.org/9781316511732
DOI: 10.1017/9781009053730

First published 2022

Printed in the United Kingdom by TJ Books Limited, Padstow Cornwall

A catalogue record for this publication is available from the British Library.

ISBN 978-1-316-51173-2 Hardback

Contents

Preface

Heavy tails are a continual source of excitement and confusion as they are repeatedly "discovered" in new contexts. This is true across the social sciences, the physical sciences, and especially the information sciences, where heavy-tails pop up seemingly everywhere – from degree distributions in the web and social networks to file sizes and interarrival times of compute jobs. However, despite nearly a century of work on heavy tails, they are still treated as *mysterious*, due to their counterintuitive properties; *surprising*, since the central limit theorem trains us to expect the "Normal" distribution; and even *controversial*, since intuitive statistical tools can lead to false discoveries.

The goal of this book is to show that heavy-tailed distributions need not be mysterious and should not be surprising or controversial either. The book strives to demystify heavy-tailed distributions by showing how to reason formally about their counterintuitive properties by demonstrating that the emergence of heavy-tailed phenomena should be expected (not surprising) and by revealing that most of the controversy surrounding heavy-tails is the result of bad statistics and can be avoided by using the proper tools. It proceeds by offering a coherent collection of bite-sized introductions to the big questions:

How can we understand the seemingly counterintuitive properties of heavy-tailed distributions?
Why are heavy-tailed distributions so common?
How can heavy-tailed distributions be estimated from data?

These questions are important not just to mathematicians but also to computer scientists, economists, social scientists, astronomers, and beyond. Heavy-tailed distributions truly come up everywhere in the world around us, and understanding their consequences is crucial to an understanding of the world. Despite this, many students learn probability and statistics without being taught how to think carefully about the heavy tails. Our mission is to make this practice a thing of the past.

This book aims to provide an accessible introduction to the fundamentals of heavy tails. It covers mathematically deep concepts that are typically taught in graduate-level courses, such as the generalized central limit theorem, extreme value theory, and regular variation; but it does so using only elementary mathematical tools in order to make these topics accessible to anyone who has had an introductory probability course. To make this possible we have explicitly chosen, at times, not to prove results in full generality. Instead, we have worked to rigorously present the core idea in as simple a manner as possible, pushing aside technicalities that are necessary to extend the analysis to the most general cases. However, we always provide pointers to other books and papers where proofs of the most general results can be found.

How to Use This Book

The book is designed for teaching as well as for independent study by anyone who wants to develop the intuition and mathematical tools for reasoning about heavy-tailed distributions. In teaching from the book, we have found that each chapter corresponds to approximately two 80-minute lectures (though we do not cover all the material in each chapter during lectures) and that it is exciting to include a project component in the course where students explore heavy tails in their areas of interest. Due to the interdisciplinary nature of heavy tails, we have found that students tend to come from widely varying backgrounds and so the projects can be extremely diverse.

The first chapter of the book introduces the class of heavy-tailed distributions formally, and discusses a few examples of common heavy-tailed distributions. These examples illustrate how the behavior of heavy-tailed distributions contrasts with that of light-tailed distributions, but do not yet build intuition for these differences or explain why heavy-tailed distributions are so common in the world around us. The chapter also begins to introduce the controversy that often surrounds heavy-tailed distributions because intuitive statistical approaches for identifying heavy tails in data are flawed.

The remainder of the book is organized to first provide intuition, both qualitative and mathematical, about the defining properties of heavy-tailed distributions (Part I: Properties), then explain why heavy-tailed distributions are so common in the world around us (Part II: Emergence), and finally develop the statistical tools for the estimation of heavy-tailed distributions (Part III: Estimation).

Given the mystique and excitement that surrounds the discovery of heavy-tailed phenomena, detection and estimation of heavy tails in data is an activity that is often (over)zealously pursued. You may be tempted to skip directly to Part III on estimation. However, the book is written so that the tools used in Part II are developed in Part I, and the tools used in Part III are developed in Parts I and II. Thus, we encourage readers to work through the book in order. That said, we have organized the material in each chapter so that there is a main body that presents the core ideas important for later chapters, followed by sections that present examples and/or variations of the main topic. These later sections can be viewed as enrichment opportunities that can be skipped as desired if the goal is to move quickly to Part III. The quickest path for acquiring the background needed before digging into Part III consists of Chapter 2 from Part I, then Chapters 5 and 7 from Part II.

Finally, we would like to emphasize that the aim of this book is to present the fundamentals of heavy-tailed distributions in a way that is both rigorous and accessible to readers who have taken an introductory undergraduate course in probability. Given that the theory of heavy tails uses advanced mathematical tools and is typically presented in a way that is accessible only to graduate students in mathematics, our target has required us to rethink and reprove many classical results in the area with the goal of providing a simple, intuitive presentation. This often means stating theorems that have more restrictive assumptions than the most general results known and presenting proofs that convey the key ideas but address some of the difficult technical details via either an assumption or a reference to a technical lemma in another source. To provide interested readers with references to the full generality of the results we discuss, each chapter ends with an "Additional Notes" section that includes references to more detail on the topics presented in the chapter. We encourage readers to follow up on the references in these sections. Additionally, many of the exercises at the ends of the chapters ask the reader to derive extensions or fill in technical details that we have left out of the main body of the chapters, so we encourage readers to work through the exercises.

Our goal for this book is that, through reading it, heavy-tailed distributions will be demystified for you. That their properties will be intuitive, not mysterious. That their emergence will be expected, not surprising. And, that you will have the proper statistical tools for studying heavy-tailed phenomena and so will help resolve (or avoid) controversies rather than feed them. Happy reading!

Acknowledgments

When first teaching an iteration of our course on heavy tails at Caltech in 2011 we had hopes of finishing a book on the topic within a couple years. Instead, the writing process continued off and on over a decade, during which time the authors added eight children to their families! The process of writing a book like this is joyous. You get to return to the core results of a field you love and think deeply about how to simplify them to their core; attempting to isolate a form of the results that may be more restrictive than the most general formulation, but that expose the essence of the intuition for the results using as simple an argument as possible. In many cases, this process led us to develop new intuition for classical results, which has fed back into our own research over the years.

Given this lengthy on-again/off-again process of writing this book, it is not surprising that this book owes a great deal of thanks to a great number of people. Our approach to thinking about heavy tails has been shaped in many ways by our research collaborators and mentors throughout our careers. The book draws on perspectives that we have developed from our conversations with colleagues such as Onno Boxma, Sem Borst, Mor Harchol-Balter, Alan Scheller-Wolf, Sergey Foss, Sid Resnick, Gennady Samorodnitsky, Thomas Mikosch, John Doyle, K. Mani Chandy, Steven Low, Mike Pinedo, Alessandro Zocca, and Guannan Qu, Peter Glynn, Ramesh Johari, and others.

The book grew out of courses taught at Caltech and IIT Bombay, and we owe a huge debt of thanks to the teaching assistants and students of those courses for their feedback and suggestions. We also thank the faculty who have integrated parts of this book into their courses at other schools all around the world. The comments and suggestions from their experience have been invaluable. More broadly, thanks goes to all the students in the Rigorous Systems Research Group (RSRG) at Caltech, many of whom read early versions of the chapters and shared thoughts and insights at formative stages of the book.

We are extremely appreciative of Longbo Huang and Chenye Wu, who hosted Adam Wierman for a summer of intensive writing at Tsinghua University, which was crucial to finally finishing the book. The conversations with Longbo and Chenye about the book and research more broadly were inspirational, and the input on the book from students at Tsinghua was irreplaceable.

As we have progressed in writing the book, we have kept a version of our draft available on our webpages and we want to thank the hundreds of readers who carefully studied those versions and provided us comments, found typos, and shared applications where they applied the ideas of the book. It was extremely motivating and stimulating to hear about the wide variety of places where the intuitions and techniques introduced in the book proved to be impactful.

Finally, we are grateful to the editorial team at Cambridge University Press, especially Diana Gillooly, for their patience with us as we prepared the book and for the advice and help they provided along the way.

1

Introduction

> The top 1% of the population controls 35% of the wealth. On Twitter, the top 2% of users send 60% of the messages. In the health care system, the treatment for the most expensive fifth of patients create four-fifths of the overall cost. These figures are always reported as shocking, as if the normal order of things has been disrupted, as if [it] is a surprise of the highest order. It's not. Or rather, it shouldn't be.
> – Clay Shirky, in response to the question "What scientific concept would improve everybody's cognitive toolkit?" [194]

Introductory probability courses often leave the impression that the Gaussian distribution is what we should expect to see in the world around us. It is referred to as the "Normal" distribution after all! As a result, statistics like the ones in the quote above tend to be treated as aberrations, since they would never happen if the world were Gaussian. The Gaussian distribution has a "scale," a typical value (the mean) around which individual measurements are centered and do not deviate from by too much. For example, if we consider human heights, which are approximately Gaussian, the average height of an adult male in the US is 5 feet 9 inches and most people's heights do not differ by more than 10 inches from this. In contrast, there are order-of-magnitude differences between individuals in terms of wealth, Twitter followers, health care costs, and so on.

However, order-of-magnitude differences like those just mentioned are not new and should not be surprising. Over a century ago, Italian economist Vilfredo Pareto discovered that the richest 20 percent of the population controlled 80 percent of the property in Italy. This is now termed the "Pareto Principle," aka the "80-20" rule and variations of this principle have shown up repeatedly in widely disparate areas in the time since Pareto's discovery. For example, in 2002 Microsoft reported that 80 percent of the errors in Windows are caused by 20 percent of the bugs [188], and similar versions of the Pareto principle apply (though not always with 80/20) to many aspects of business, for example, most of the profit is made from a small percentage of the customers and most of the sales are made by a small percentage of the sales team.

Statistics related to the Pareto principle make for compelling headlines, but they are typically an indication of something deeper. When we see such figures, it is likely that there is not a Gaussian distribution underlying them, but rather a heavy-tailed distribution is the reason for the "surprising" statistics. The most celebrated such distribution again carries Vilfredo Pareto's name: *the Pareto distribution*. Heavy-tailed distributions such as the Pareto distribution are just as prominent as (if not more so than) the Gaussian distribution and have been observed in hundreds of applications in physics, biology, computer science, the social sciences, and beyond over the past century. Some examples include the sizes of cities [92, 163], the file sizes in computer systems and networks [52, 146], the size of avalanches and earthquakes [109, 144], the length of protein sequences in genomes [130, 145], the size of meteorites [13, 162], the degree distribution of the web graph [36, 116], the returns of stocks [49, 94], the number of copies of books sold [14, 110], the number of households affected during blackouts in power grids [114], the frequency of word use in natural language [77, 227], and many more.

Given the breadth of areas where heavy-tailed phenomena have been observed, one might guess that, by now, observations of heavy-tailed phenomena in new areas are expected – that heavy tails are treated as *more normal than the Normal*. After all, Pareto's work has been widely known for more than a century. However, despite a century of experience, statistics related to the Pareto Principle and, more broadly, heavy-tailed distributions are still typically presented as surprising curiosities – anomalies that could not have been anticipated. Even in scientific communities, observations of heavy-tailed phenomena are often presented as mysteries to be explained rather than something to be expected a priori. In many cases, there is even a significant amount of controversy and debate that follows the identification of heavy-tailed phenomena in data.

Surprising? Mysterious? Controversial?

Given the century of mathematical and statistical work around heavy tails, it certainly should not be the case that heavy tails are surprising, mysterious, and controversial. In fact, there are many reasons why one should *expect* to see heavy-tailed distributions arise. Perhaps the main reason why they are still viewed as surprising is that the version of the central limit theorem taught in introductory probability courses gives the impression that the Gaussian will occur everywhere. However, this introductory version of the central limit theorem does not tell the whole story. There is a "generalized" version of the central limit theorem that states that either the Gaussian *or a heavy-tailed distribution* will emerge as the limit of sums of random variables. Unfortunately, the technical nature of this result means it rarely features in introductory courses, which leads to unnecessary surprises about the presence of heavy-tailed distributions. Going beyond sums of random variables, when random variables are combined in other natural ways (e.g., products or max/min) heavy tails are even more likely to emerge, whereas the Gaussian distribution is not.

So heavy-tailed phenomena should not be considered surprising. What about mysterious? The view of heavy tails as mysterious is, to some extent, a consequence of unfamiliarity. People are familiar with the Gaussian distribution because of its importance in introductory probability courses, and when something emerges that has qualitatively and quantitatively different properties it seems mysterious and counter-intuitive. The Pareto Principle is one illustration of the counterintuitive properties that make heavy-tailed distributions seem mysterious, but there are many others. For example, while the Gaussian distribution has a clear "scale" – most samples will be close to the mean – samples from heavy-tailed distributions frequently differ by orders of magnitude and may even be "scale free" (e.g., in the case of the Pareto distribution). Another example is that, while the moments (the mean, variance, etc.) of the Gaussian distribution are all finite, it is not uncommon to see data that fits a heavy-tailed distribution having an infinite variance, or even an infinite mean! For example, the degree distribution of many complex networks tends to have a tail that matches that of a Pareto with infinite variance (see, for example, [23]). This can potentially lead to mind-bending challenges when trying to apply statistical tools, which often depend on averages and variances.

The combination of surprise and mystery that surrounds heavy-tailed phenomena means that there is often considerable excitement that follows the discovery of data that fits a heavy-tailed distribution in a new field. Unfortunately, this excitement often sparks debate and controversy – often enough that an unfortunate pattern has emerged. A heavy-tailed phenomenon is discovered in a new field. The excitement over the discovery leads researchers to search for heavy tails in other parts of the field. Heavy tails are then discovered in many settings and are claimed to be a universal property. However, the initial excitement of discovery and lack of previous background in statistics related to heavy tails means that the first wave of research identifying heavy tails uses intuitive but flawed statistical tools. As a result, a controversy emerges – which settings where heavy tails have been observed really have heavy tails? Are they really universal? Over time, more careful statistical analyses are used, showing that some places really do exhibit heavy tails while others were false discoveries. By the end, a mature view of heavy tails emerges, but the whole process can take decades.

At this point, the pattern just described has been replicated in many areas, including computer science [68], biology [119], chemistry [160], ecology [10], and astronomy [216]. Maybe the most prominent example of this story is still ongoing in the area of *network science*. Near the turn of the century, the study of complex networks began to explode in popularity due to the growing importance of networks in our lives and the increasing ease of gathering data about large networks. Initial results in the area were widely celebrated and drove an enormous amount of research to look at the universality of scale-free networks. However, as the field matured and the statistical tools became more sophisticated, it became clear that many of the initial results were flawed. For example, claims that the internet graph [80] and the power network [24] are heavy-tailed were refuted [4, 222], among others. This led to a controversy in the area that continues to this day, 20 years later [37, 212].

Demystifying Heavy Tails

The goal of this book is to demystify heavy-tailed phenomena. Heavy tails are not anomalies – and their emergence should not be surprising or controversial either! Heavy tails are an unavoidable part of our lives, and viewing statistics like the ones that started this chapter as anomalies prevents us from thinking clearly about the world around us. Further, while properties of heavy-tailed phenomena like the Pareto Principle may initially make heavy-tailed distributions seem counterintuitive, they need not be. This book strives to provide tools and techniques that can make heavy tails as easy and intuitive to reason about as the Gaussian, to highlight when one should expect the emergence of heavy-tailed phenomena, and to help avoid controversy when identifying heavy tails in data.

Because of the ubiquitousness and seductive nature of heavy-tailed phenomena, they are a topic that has permeated wide ranging fields, from astronomy and physics, to biology and physiology, to social science and economics. However, despite their ubiquity, they are also, perhaps, one of the most misused and misunderstood mathematical areas, shrouded in both excitement and controversy. It is easy to get excited about heavy-tailed phenomena as you start to realize the important role they play in the world around us and become exposed to the beautiful and counterintuitive properties they possess. However, as you start to dig into the topic, it quickly becomes difficult. The mathematics that underlie the analysis of heavy-tailed distributions are technical and advanced, often requiring prerequisites of graduate-level probability and statistics courses. This is the reason why introductory probability courses typically do not present much, if any, material related to heavy-tailed distributions. If they are mentioned, they are typically used as examples illustrating that "strange" things can happen (e.g., distributions can have an infinite mean). Thus, a scientist or researcher in a field outside of mathematics who is interested in learning more about heavy tails may find it difficult, if not impossible, to learn from the classical texts on the topic.

It is exactly this difficulty that led us to write this book. In this book we hope to introduce the fundamentals of heavy-tailed distributions using only tools that one learns in an introductory probability course. The book intentionally does not spend much time on describing the settings where heavy tails arise – there are simply too many different areas to do justice to even a small subset of them. Instead, we assume that if you have found your way to this book, then heavy tails are important to you. Given that, our goal is to provide an introduction to how to think about heavy tails both intuitively and mathematically.

The book is divided into three parts, which focus on three foundational guiding questions.

- **Part I: Properties.** *What leads to the counterintuitive properties of heavy-tailed phenomena?*
- **Part II: Emergence.** *Why do heavy-tailed phenomena occur so frequently in the world around us?*
- **Part III: Estimation.** *How can we identify and estimate heavy-tailed phenomena using data?*

In Part I of the book we provide insight into some of most mysterious and elegant properties of heavy-tailed distributions, connecting these properties to formal definitions of subclasses of heavy-tailed distributions. We focus on three foundational properties: "scale-invariance" (aka, scale-free), the "catastrophe principle," and "increasing residual life." We illustrate that these properties provide qualitatively different behaviors than what is seen under light-tailed distributions like the Gaussian, and provide intuition underlying the properties. The three chapters that make up Part I strive to demystify some of the particularly exotic properties of heavy-tailed distributions and to provide a clear view of how these properties interact with each other and with the broader class of heavy-tailed distributions.

In Part II of the book we explore simple laws that can "explain" the emergence of heavy-tailed distributions in the same way that the central limit theorem "explains" the prominence of the Gaussian distribution. We study three foundational stochastic processes in order to understand when one should expect the emergence of heavy-tailed distributions as opposed to light-tailed distributions. Our discussions in the three chapters that make up Part II highlight that heavy-tailed distributions should not be viewed as anomalies. In fact, heavy tails should not be surprising at all; in many cases they should be treated as something as natural as, if not more natural than, the emergence of the Gaussian distribution.

In Part III of this book we focus on the statistical tools used for the estimation of heavy-tailed phenomena. Unfortunately, there is no perfect recipe for "properly" detecting and estimating heavy-tailed distributions in data. Our treatment, therefore, seeks to highlight a handful of important approaches and to provide insight into when each approach is appropriate and when each may be misleading. Combined, the chapters that make up Part III highlight a crucial point: one must proceed carefully when estimating heavy-tailed phenomena in real-world data. It is naive to expect to estimate *exact* heavy-tailed distributions in data. Instead, a realistic goal is to estimate the *tail* of heavy-tailed phenomena. Even in doing this, one should not rely on a single method for estimation. Instead, it is a necessity to build confidence through the use of multiple, complementary estimation approaches.

1.1 Defining Heavy-Tailed Distributions

Before we tackle our guiding questions, we start with the basic question: *What is a heavy-tailed distribution?*

One of the reasons for the mystique that surrounds heavy-tailed distributions is that if you ask five people from different communities this question, you are likely to get five different answers. Depending on the community, the term heavy-tailed may be used interchangeably with terms like scale-free, power-law, fat-tailed, long-tailed, subexponential, self-similar, stable, and others. Further, the same names may mean different things to different communities!

Sometimes the term "heavy-tailed" is used to refer to a specific distribution such as the Pareto or the Zipf distribution. Other times, it is used to identify particular properties of a distribution, such as the fact that it is scale-free, has an infinite (or very large) variance, a decreasing failure rate, and so on. As a result, there is often a language barrier when discussing heavy-tailed distributions that stems from different associations with the same terms across communities.

Hopefully, reading this book will equip you to navigate the zoo of terminology related to heavy-tailed distributions. Each of the terms mentioned earlier does have a concrete, precise, established mathematical definition. It is just that these terms are often used carelessly, which leads to confusion. It will take us most of the book to get through the definitions of all the terms mentioned in the previous paragraph, but we start in this section by laying the foundation – defining the term "heavy-tailed" and discussing some of the most celebrated examples.

The term "heavy-tailed" is inherently relative – heavier than what? A Gaussian distribution has a heavier tail than a Uniform distribution, and an Exponential distribution has a heavier tail than a Gaussian distribution, but neither of these is considered "heavy-tailed." Thus, the key feature of the definition is the comparison point chosen.

The comparison point that is used to define the class of heavy-tailed distributions is the Exponential distribution. That is, a distribution is considered to be heavy-tailed if it has a heavier tail than any Exponential distribution. Formally, this is stated in terms of the cumulative distribution function (c.d.f.) F of a random variable X, that is, $F(x) = \Pr(X \leq x)$, and the complementary cumulative distribution function (c.c.d.f.) $\bar{F}$, that is, $\bar{F}(x) = 1 - F(x)$.

Definition 1.1 A distribution function F is said to be heavy-tailed if and only if, for all $\mu > 0$,

$$\limsup_{x \to \infty} \frac{1 - F(x)}{e^{-\mu x}} = \limsup_{x \to \infty} \frac{\bar{F}(x)}{e^{-\mu x}} = \infty.$$

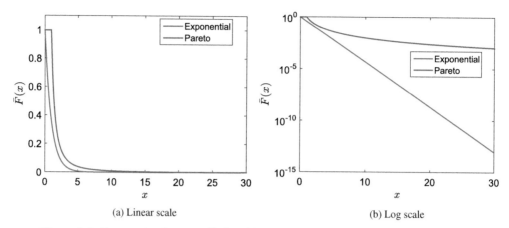

(a) Linear scale   (b) Log scale

Figure 1.1 Contrasting heavy-tailed and light-tailed distributions: The plots show the c.c.d.f. of the exponential distribution (with mean 1) and a heavy-tailed Pareto distribution (with minimal value $x_m = 1$, scale parameter $\alpha = 2$). While the contrast in tail behavior is difficult to discern on a linear scale (Fig. (a)), it is quite evident when the probabilities are plotted on a logarithmic scale (Fig. (b)).

Otherwise, F is light-tailed. A random variable X is said to be heavy-tailed (light-tailed) if its distribution function is heavy-tailed (light-tailed).

Note that the definition of heavy-tailed distributions given above applies to the *right* tail of the distribution, that is, it is concerned with the behavior of the probability of taking values larger than x as $x \to \infty$. In some applications, one might also be interested in the *left* tail. In such cases, the definition of heavy-tailed can be applied to both the right tail (without change) and the left tail (by considering the right tail of $-X$).

The definition of heavy-tailed is, in some sense, natural. It looks explicitly at the "tail" of the distribution (i.e., the c.c.d.f. $\bar{F}(x)$), and it is easy to see from the definition that the tails of distributions that are heavy-tailed are "heavier" (i.e., decay more slowly) than the tails of distributions that are light-tailed; see Figure 1.1.

The particular choice of the Exponential distribution as the boundary between heavy-tailed and light-tailed may, at first, seem arbitrary. In fact, without detailed study of the class of heavy-tailed distributions, it is difficult to justify this particular choice. But, as we will see throughout this book, the Exponential distribution serves to separate two classes of distributions that have qualitatively different behavioral properties and require fundamentally different mathematical tools to work with.

To begin to examine the distinction between heavy-tailed and light-tailed distributions, it turns out to be useful to consider two alternative, but equivalent, definitions of "heavy-tailed."

Lemma 1.2 *Consider a random variable X. The following statements are equivalent.*

(i)   *X is heavy-tailed.*
(ii) *The moment generating function $M(s) := \mathbb{E}\left[e^{sX}\right] = \infty$ for all $s > 0$.*
(iii) *$\liminf_{x \to \infty} -\frac{\log \Pr(X > x)}{x} = 0$.*

The proof of this lemma provides useful intuition about heavy-tailed distribution; however, before proving this result, let us interpret the two new, equivalent definitions of heavy-tailed that it provides.

First, consider *(ii)*, which states that a random variable is heavy-tailed if and only if its moment generating function $M(s) := \mathbb{E}\left[e^{sX}\right]$ is infinite for all $s > 0$. This definition highlights that heavy-tailed distributions require a different analytic approach than light-tailed distributions. For light-tailed distributions the moment generating function often provides an important tool for characterizing the distribution. It can be used to derive the moments of the distribution, but it also can be inverted to characterize the distribution itself. Further, it is a crucial tool for analysis because of the simplicity of handling convolutions via the moment generating function, for example, when deriving concentration inequalities such as Chernoff bounds. In contrast, the definition given by *(ii)* shows that such techniques are not applicable for heavy-tailed distributions.

Next, consider *(iii)*, which states that a random variable X is heavy-tailed if and only if the log of its tail, $\log \Pr\left(X > x\right)$, decays sublinearly. This again highlights that heavy-tailed distributions require a different analytic approach than light-tailed distributions. In particular, when studying the tail of light-tailed distributions it is common to use concentration inequalities such as Chernoff bounds, which inherently have an exponential decay. As a result, such bounds focus on determining the optimal decay rate, which is characterized by deriving a maximal μ such that $\Pr\left(X > x\right) \leq Ce^{-\mu x}$. However, the definition given by *(iii)* highlights that the maximum possible μ for heavy-tailed distributions is zero, and so fundamentally different analytic approaches must be used.

To build more intuition on the relationship between these three equivalent definitions of "heavy-tailed," as well as to get practice working with the definitions, it is useful to consider the proof of Lemma 1.2.

Proof of Lemma 1.2   To prove Lemma 1.2, we need to show the equivalence of each of the three definitions of heavy-tailed. We do this by showing that *(i)* implies *(ii)*, that *(ii)* implies *(iii)*, and finally that *(iii)* implies *(i)*.

$(i) \Rightarrow (ii)$. Suppose that X is heavy-tailed, with distribution F. By definition, this implies that for any $s > 0$, there exists a strictly increasing sequence $(x_k)_{k \geq 1}$ satisfying $\lim_{k \to \infty} x_k = \infty$, such that

$$\lim_{k \to \infty} e^{sx_k} \bar{F}(x_k) = \infty. \tag{1.1}$$

We can now bound $\mathbb{E}\left[e^{sX}\right]$ as follows.

$$\begin{aligned}
\mathbb{E}\left[e^{sX}\right] &= \int_0^\infty e^{sx} dF(x) \\
&\geq \int_{x_k}^\infty e^{sx} dF(x) \\
&\geq e^{sx_k} \bar{F}(x_k).
\end{aligned}$$

Since the above inequality holds for all k, it now follows from (1.1) that $\mathbb{E}\left[e^{sX}\right] = \infty$. Therefore, Condition *(i)* implies Condition *(ii)*.

$(ii) \Rightarrow (iii)$. Suppose that X satisfies Condition *(ii)*. For the purpose of obtaining a contradiction, let us assume that Condition *(iii)* does not hold. Since $-\frac{\log \Pr(X>x)}{x} \geq 0$, this means that

$$\liminf_{x\to\infty} -\frac{\log \Pr(X>x)}{x} > 0.$$

The above statement implies that there exist $\mu > 0$ and $x_0 > 0$ such that

$$-\frac{\log \Pr(X>x)}{x} \geq \mu \iff \Pr(X>x) \leq e^{-\mu x} \quad \forall x \geq x_0. \qquad (1.2)$$

Now, pick s such that $0 < s < \mu$. We may now bound the moment generating function of X at s as follows:

$$M(s) = \mathbb{E}\left[e^{sX}\right] = \int_0^\infty \Pr\left(e^{sX} > x\right) dx$$

$$= \int_0^{e^{sx_0}} \Pr\left(e^{sX} > x\right) dx + \int_{e^{sx_0}}^\infty \Pr\left(X > \frac{\log(x)}{s}\right) dx.$$

Here, we have used the following representation for the expectation of a nonnegative random variable Y: $\mathbb{E}[Y] = \int_0^\infty \Pr(Y > y)\, dy$. While the first term above can be bounded from above by e^{sx_0}, we may bound the second using (1.2), since $x \geq e^{sx_0}$ is equivalent to $\log(x)/s \geq x_0$.

$$M(s) \leq e^{sx_0} + \int_{e^{sx_0}}^\infty e^{-\mu \frac{\log(x)}{s}} dx$$

$$= e^{sx_0} + \int_{e^{sx_0}}^\infty x^{-\mu/s} dx.$$

Since $\mu/s > 1$, we have $\int_1^\infty x^{-\mu/s} dx < \infty$, which implies that $M(s) < \infty$, giving us a contradiction. Therefore, Condition *(ii)* implies Condition *(iii)*.

$(iii) \Rightarrow (i)$. Suppose that the random variable X, having distribution F, satisfies Condition *(iii)*. Thus, there exists a strictly increasing sequence $(x_k)_{k\geq 1}$ satisfying $\lim_{k\to\infty} x_k = \infty$, such that

$$\lim_{k\to\infty} -\frac{\log \bar{F}(x_k)}{x_k} = 0.$$

Given $\mu > 0$, this in turn implies that there exists $k_0 \in \mathbb{N}$ such that

$$-\frac{\log \bar{F}(x_k)}{x_k} < e^{-\frac{\mu}{2}} \quad \forall\, k > k_0$$

$$\iff \bar{F}(x_k) > e^{-\frac{\mu x_k}{2}} \quad \forall\, k > k_0$$

$$\iff \frac{\bar{F}(x_k)}{e^{-\mu x_k}} > e^{\frac{\mu x_k}{2}} \quad \forall\, k > k_0.$$

The last assertion above implies that $\lim_{k\to\infty} \frac{\bar{F}(x_k)}{e^{-\mu x_k}} = \infty$, which implies $\limsup_{x\to\infty} \frac{\bar{F}(x)}{e^{-\mu x}} = \infty$. Since this is true for any $\mu > 0$, we conclude that Condition *(iii)* implies Condition *(i)*.

$\square$

1.2 Examples of Heavy-Tailed Distributions

We now have three equivalent definitions of heavy-tailed distributions and, through the proof, we understand how these three definitions are related. But, even with these restatements, the definition of heavy-tailed is still opaque. It is difficult to get behavioral intuition about the properties of heavy-tailed distributions from any of the definitions. Further, it is very hard to see much about what makes heavy-tailed distributions have the mysterious properties that are associated with them using these definitions alone.

In part, this is due to the breadth of the definition of heavy-tailed. The important properties commonly associated with heavy-tailed distributions, such as scale invariance, infinite variance, the Pareto principle, etc., do not hold for all heavy-tailed distributions; they hold only for certain subclasses of heavy-tailed distributions.

As a result, it is important to build intuition for the class of heavy-tailed distributions by looking at specific examples. That is the goal of the remainder of this chapter. In particular, we focus in detail on the Pareto distribution, the Weibull distribution, and the LogNormal distribution with the goal of providing both the mathematical formalism for these distributions and some insight in their important properties and applications. Additionally, we briefly introduce some of the other important examples of heavy-tailed distributions that come up frequently in applications, including the Cauchy, Fréchet, Lévy, Burr, and Zipf distributions.

Perhaps the most important thing to keep in mind as you read these sections is the contrast between the properties of the heavy-tailed distributions that we discuss and the properties of light-tailed distributions, such as the Gaussian and Exponential distributions, with which you are likely more familiar. To set the stage, we summarize the important formulas for these two distributions next.

The Gaussian Distribution

The Gaussian distribution, also called the Normal distribution or the bell curve, is perhaps the most widely recognized distribution and is extremely important in statistics and beyond. It is defined using two parameters, the mean μ and the variance σ^2, and is expressed most conveniently through its probability density function (p.d.f.), $f(x)$, or its moment generating function (m.g.f.), $M(s)$. Given a random variable $Z \sim \text{Gaussian}(\mu, \sigma)$, we have

$$f_Z(x) = \frac{1}{\sigma\sqrt{2\pi}} e^{-\frac{(x-\mu)^2}{2\sigma^2}},$$
$$M_Z(s) = E[e^{sZ}] = e^{\mu s + \frac{1}{2}\sigma^2 s^2}.$$

Since $M_Z(s) < \infty$ for all $s > 0$, it follows that the Gaussian distribution is light-tailed. The light-tailedness of the Gaussian distribution can also be deduced directly by bounding its c.c.d.f. (see Exercise 2).

The particular Gaussian distribution with zero mean and unit variance ($\mu = 0$, $\sigma = 1$) is commonly referred to as the *standard Gaussian*.

The Exponential Distribution

The Exponential distribution is a widely known and broadly applicable distribution that serves as the light-tailed distribution on the boundary between light-tailed and heavy-tailed distributions. It is a nonnegative distribution defined in terms of one parameter: λ, which is referred to as the "rate" since the mean of the distribution is $1/\lambda$. Given a random variable $X \sim \text{Exponential}(\lambda)$, the p.d.f., c.c.d.f., and m.g.f., can be expressed as

$$f_X(x) = \lambda e^{-\lambda x} \qquad (x \geq 0),$$

$$\bar{F}(x) = e^{-\lambda x} \qquad (x \geq 0),$$

$$M_X(s) = \frac{1}{(1 - s/\lambda)} \qquad (s < \lambda).$$

Note that the tail of the Exponential distribution is heavier than that of the Gaussian because e^{-x} goes to zero more slowly than e^{-x^2}. Additionally, unlike the Gaussian, the moment generating function is not finite everywhere.

1.2.1 The Pareto Distribution

Vilfredo Pareto originally presented the Pareto distribution, and introduced the idea of the Pareto Principle, in the study of the allocation of wealth. But since then, it has been used as a model in numerous other settings, including the sizes of cities, the file sizes in computer systems and networks, the price returns of stocks, the size of meteorites, casualties and damages due to natural disasters, frequency of words, and many more. It is perhaps the most celebrated example of a heavy-tailed distribution, and as a result, the term Pareto is sometimes, unfortunately, used interchangeably with the term heavy-tailed.

Formally, a random variable X follows a Pareto(x_m, α) distribution if

$$\Pr(X \geq x) = \bar{F}(x) = \left(\frac{x}{x_m}\right)^{-\alpha}, \text{ for } \alpha > 0, x \geq x_m > 0.$$

Here, α is the shape parameter of the distribution and is also commonly referred to as the *tail index*, while x_m is the minimum value of the distribution, that is, $X \geq x_m$. Given the c.c.d.f. above, it is straightforward to differentiate and obtain the p.d.f.

$$f(x) = \frac{\alpha x_m^\alpha}{x^{\alpha+1}}, \quad x \geq x_m.$$

It is easy to see from the c.c.d.f. that the Pareto is heavy-tailed. In particular, using Definition 1.1, we can compute

$$\limsup_{x \to \infty} \frac{\bar{F}(x)}{e^{-\mu x}} = \limsup_{x \to \infty} \left(\frac{x_m}{x}\right)^\alpha e^{\mu x} = \infty, \tag{1.3}$$

since the exponential $e^{\mu x}$ grows more quickly than the polynomial x^α.

This highlights the key contrast between the Pareto distribution and common light-tailed distributions like the Gaussian and Exponential distributions: the Pareto tail decays *polynomially*, as $x^{-\alpha}$, instead of *exponentially* (as $e^{-\mu x}$) in the case of the Exponential, or *superexponentially* (as $e^{-x^2/2\sigma^2}$) in the case of the Gaussian. As a consequence, large values are much more likely to occur under a Pareto distribution than under a Gaussian or Exponential distribution. For example, you are much more likely to meet someone whose income is 10 times the average than someone whose height is 10 times the average.

This contrast is present visually too. Figure 1.2 shows that the tail of the Pareto is considerably heavier. The figure illustrates the p.d.f. and c.c.d.f. of the Pareto for different values of

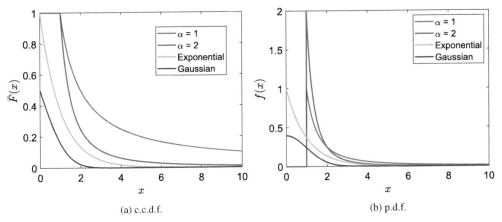

(a) c.c.d.f. (b) p.d.f.

Figure 1.2 Contrasting Pareto distribution with the Exponential and the Gaussian: The plots show (a) the c.c.d.f., and (b) the p.d.f., corresponding to Pareto distributions with $x_m = 1$ with different values of α, alongside the Exponential distribution (with unit mean) and the standard Gaussian.

the tail index α, which is typically the parameter of interest since it controls the degree of the polynomial decay of the p.d.f. and c.c.d.f., and thus determines the "weight" of the tail. As α decreases, the tail becomes heavier, while as $\alpha \to \infty$ the Pareto distribution approaches the Dirac delta function centered at x_m.

While Figure 1.2 already contrasts the Pareto, Gaussian, and Exponential distributions, we can better emphasize this contrast by presenting the figure in a different way, that is, by rescaling its axes. In particular, Figure 1.3 shows the same c.c.d.f.s but presents the data on a log-log scale, that is, with logarithmic horizontal and vertical axes. With this change, a remarkable pattern emerges – the Pareto c.c.d.f. becomes a straight line, while the Gaussian and Exponential distributions quickly drop off a cliff and disappear. This image viscerally highlights the heaviness of the Pareto's tail as compared to the tails of the Exponential and the Gaussian.

To understand why the Pareto is linear when viewed on a log-log scale, let us do a quick calculation. Letting $C_1 = x_m^\alpha$ we can write

$$\bar{F}(x) = \left(\frac{x}{x_m}\right)^{-\alpha} = C_1 x^{-\alpha}.$$

Taking logarithms of both sides then gives

$$\underbrace{\log \bar{F}(x)}_{'y'} = \underbrace{\log C_1}_{y\text{-intercept}} + \underbrace{(-\alpha)}_{\text{slope}} \underbrace{\log x}_{'x'},$$

which reveals that, on a log-log scale, the c.c.d.f. is simply a linear function with y-intercept $\log C_1$ and slope $-\alpha$. Not only that, the p.d.f. is also of the same form, that is, $f(x) = C_2 x^{-(\alpha+1)}$ where $C_2 = \alpha x_m^\alpha$ and so it also is linear in the log-log scaling.

This property – being (approximately) linear on a log-log scale – is important enough that it has received a few different names from different communities. Distributions of the form $\bar{F}(x) = C x^{-\alpha}$ for some constant C are referred to as *power law* distributions. A related set of distributions are *fat-tailed distributions*, which are distributions with $\bar{F}(x) \sim x^{-\alpha}$ as

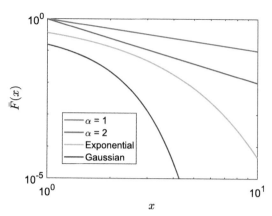

Figure 1.3 A clearer contrast between the Pareto distribution and the Exponential and Gaussian: The plots show the c.c.d.f.s corresponding to different Pareto distributions with $x_m = 1$ and different values of α, alongside the Exponential distribution (with unit mean) and the standard Gaussian, on a log-log scale. This scaling demonstrates clearly how the Pareto tail (linear on a log-log plot) is heavier than those of the Exponential and the Gaussian.

$x \to \infty$, where we use $a(x) \sim b(x)$ as $x \to \infty$ as shorthand for $\lim_{x \to \infty} a(x)/b(x) = 1$. Finally, the class of *regularly varying* distributions, which we introduce in Chapter 2, generalizes both power law and fat-tailed distributions and has strong connections to the concept of scale-invariance.

The fact that Pareto distributions, and more generally power law distributions, are approximately linear on a log-log plot has a number of important consequences. Maybe the most prominent one is that it provides an intuitive exploratory tool for identifying power laws in data. Specifically, when presented with data, one can look at the empirical p.d.f. and c.c.d.f. on a log-log scale and check whether they are approximately linear. If so, then there is the potential that the data comes from a power-law distribution. One can even go further and hope to estimate the tail index α using linear regression on the empirical p.d.f. and c.c.d.f. This is a common approach across fields, which we illustrate in Figure 1.4 using population data for US cities as per the 2010 census. Notice that the empirical c.c.d.f. (on a log-log scale) looks roughly linear for large populations. It is therefore tempting to postulate that the distribution of city populations (asymptotically) follows a power law, and further to estimate the tail index by fitting a least squares regression line to the empirical c.c.d.f. beyond, say 10^4 (since the tail "looks linear" beyond this point), as shown in Figure 1.4. However, as we discuss in Chapter 8, this approach is not statistically sound and may lead to incorrect conclusions. In fact, the temptation to make conclusions based on such naive analyses is one of the most common reasons for the controversy that often surrounds the identification of heavy-tailed phenomena.

Moments

One of the biggest contrasts between the Pareto distribution and light-tailed distributions such as the Gaussian and Exponential is the fact that the Pareto distribution can have infinite moments. In fact, for $X \sim \text{Pareto}(x_m, \alpha)$, $\mathbb{E}[X^n] = \infty$ if $n \geq \alpha$. More specifically, the mean of the Pareto distribution is

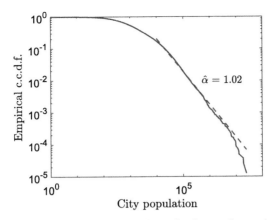

Figure 1.4 Visualizing data on a log-log plot: The figure shows the empirical c.c.d.f. of the populations of U.S. cities as per the 2010 census (data sourced from [2]). Note that on a log-log scale, the data beyond population 10^4 looks approximately linear. The least squares regression line on this data yields an estimate $\hat{\alpha} = 1.02$ of the power law exponent.

$$\mathbb{E}[X] = \begin{cases} \infty, & \alpha \leq 1; \\ \frac{\alpha x_m}{\alpha - 1}, & \alpha > 1. \end{cases}$$

The variance is

$$\mathrm{Var}[X] = \begin{cases} \infty, & \alpha \in (1, 2]; \\ \left(\frac{x_m}{\alpha - 1}\right)^2 \frac{\alpha}{\alpha - 2}, & \alpha > 2. \end{cases}$$

And, in general, the nth moment is

$$\mathbb{E}[X^n] = \begin{cases} \frac{\alpha x_m^n}{\alpha - n}, & n < \alpha; \\ \infty, & n \geq \alpha. \end{cases}$$

Importantly, it is not just a curiosity that the Pareto distribution can have infinite moments. In many cases where data has been modeled using the Pareto distribution, the distribution that is fit has infinite variance and/or mean. For example, file sizes in computer systems and networks [52] and the degree distributions of complex networks such as the web [68] appear to have infinite variance. Additionally, the logarithmic returns on stocks in finance tend to have finite variance, but infinite fourth moment [75], leading to values of α in the range $(2, 4)$.

The Pareto Principle

We began this chapter with a quote about the Pareto principle, so it is important to return to it now that we have formally introduced the Pareto distribution. The classical version of the Pareto principle is that the wealthiest 20 percent of the population holds 80 percent of the wealth. Mathematically, we can ask a more general question about what fraction of the wealth the largest P fraction of the population holds.

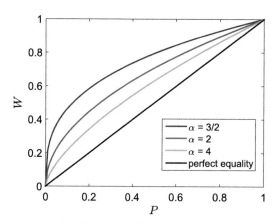

Figure 1.5 Lorenz curves for the Pareto distribution: Lorenz curves for different values of α. The smaller the value of α, the more pronounced the concentration of wealth within a small fraction of the population. The black line represents perfect equality, that is, the utopian scenario in which all individuals have exactly the same wealth.

To compute an analytic version of the Pareto principle, we consider the fraction of the population whose wealth exceeds x. Call this fraction $P(x)$, and then we can calculate $P(x)$ in the case of a Pareto distribution as follows:

$$P(x) = \int_x^\infty f(t)dt = \alpha x_m^\alpha \int_x^\infty t^{-(\alpha+1)}dt = \left(\frac{x}{x_m}\right)^{-\alpha}.$$

Then, the fraction of wealth that is in the hands of such people, which we denote by $W(x)$, is

$$W(x) = \frac{\int_x^\infty tf(t)dt}{\int_{x_m}^\infty tf(t)dt} = \frac{\alpha x_m^\alpha \int_x^\infty t^{-\alpha}dt}{\alpha x_m^\alpha \int_{x_m}^\infty t^{-\alpha}dt} = \left(\frac{x}{x_m}\right)^{-\alpha+1},$$

assuming that $\alpha > 1$. Combining the above equations then gives that, regardless of x, the fraction of wealth W owned by the richest P fraction of the population is

$$W = P^{(\alpha-1)/\alpha}.$$

We illustrate the curve of W as a function of P in Figure 1.5. It is always concave and increasing, and when α is close to 1, it indicates that wealth is concentrated in a very small fraction of the population. Such extreme concentration is an example of a more general phenomenon called the "catastrophe principle," which we discuss in detail in Chapter 3.

Curves like those in Figure 1.5 are referred to as Lorenz curves, after Max Lorenz, who developed them in 1905 as a way to represent the inequality of wealth distribution. The Gini coefficient, which is typically used to quantify wealth inequality today, is the ratio of the area between the line of perfect equality (the 45 degree line) and the Lorenz curve, and the area above the line of perfect equality. The greater the value of the Gini coefficient, the more pronounced the asymmetry in wealth distribution. Understanding properties of the Gini coefficient is still an area of active research, for example, [86] and the references therein.

Relationship to the Exponential Distribution

While heavy-tailed distributions often behave qualitatively differently than light-tailed distributions, there are still some connections between the two that can be useful. In particular, a heavy-tailed distribution can often be viewed as an exponential transformation of a light-tailed distribution. In the case of the Pareto, this connection is to the Exponential distribution. Specifically,

$$X \sim \text{Pareto}(x_m, \alpha) \iff \log(X/x_m) \sim \text{Exponential}(\alpha).$$

Or, equivalently,

$$Y \sim \text{Exponential}(\alpha) \iff x_m e^Y \sim \text{Pareto}(x_m, \alpha).$$

To see why this is true requires a simple change of variables. In particular, let $Y = \log(X/x_m)$ where $X \sim \text{Pareto}(x_m, \alpha)$. Then,

$$\Pr(Y > y) = \Pr(\log(X/x_m) > y) = \Pr(X > x_m e^y) = \left(\frac{x_m e^y}{x_m}\right)^{-\alpha} = e^{-\alpha y},$$

where the last expression is the c.c.d.f. of an Exponential distribution with rate α. This transformation turns out to be a powerful analytic tool, and we make use of it on multiple occasions in this book (e.g., Chapter 6, to study multiplicative processes, and Chapter 8, to derive properties of the maximum likelihood estimator for data from a Pareto distribution).

1.2.2 The Weibull Distribution

We just saw that the Pareto distribution has an intimate connection to the Exponential distribution – it is an *exponential* of the Exponential. The second heavy-tailed distribution we introduce has a similar connection to the Exponential distribution – it is a *polynomial* of the Exponential. Specifically, for $\alpha, \beta > 0$,

$$X \sim \text{Exponential}(1) \iff \frac{1}{\beta} X^{1/\alpha} \sim \text{Weibull}(\alpha, \beta).$$

From this relationship, one would expect that when $0 < \alpha < 1$, the Weibull distribution has a heavier tail than the Exponential (though lighter than the Pareto), making it heavy-tailed. On the other hand, when $\alpha > 1$, one would expect that the Weibull has a lighter tail than the Exponential.

It is straightforward to see what this transformation means for the c.c.d.f. of the Weibull distribution. In particular, a random variable follows a Weibull(α, β) distribution if

$$\bar{F}(x) = e^{-(\beta x)^\alpha}, \text{ for } x \geq 0. \tag{1.4}$$

Differentiating the c.c.d.f. gives us the p.d.f.

$$f(x) = \alpha\beta (\beta x)^{\alpha-1} e^{-(\beta x)^\alpha}, \text{ for } x \geq 0.$$

In these expressions, α is referred to as the *shape parameter* of the distribution, and β is the *scale parameter*. Note that when $\alpha = 1$ the Weibull is equivalent to an Exponential(β).

In fact, the Weibull distribution is an especially helpful distribution when seeking to contrast heavy tails with light tails because it can be either heavy-tailed or light-tailed depending

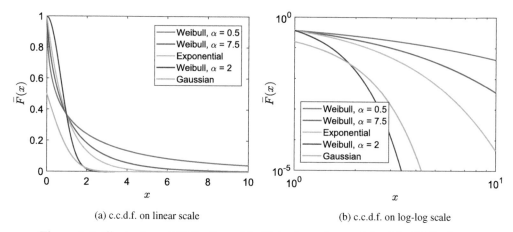

(a) c.c.d.f. on linear scale   (b) c.c.d.f. on log-log scale

Figure 1.6 Illustration of Weibull c.c.d.f.: Plots show the c.c.d.f. of the Weibull distribution with scale parameter $\beta = 1$ and different values of shape parameter α, alongside the Exponential distribution (with unit mean, which corresponds to $\alpha = \beta = 1$) and the standard Gaussian. Part (a) shows the c.c.d.f.s on a linear scale, while (b) plots them on a log-log scale.

on the shape parameter α. If $\alpha < 1$, then the Weibull is heavy-tailed, while if $\alpha \geq 1$, the Weibull is light-tailed. Mathematically, this can be verified by a quick calculation based on the definition of heavy-tailed distributions:

$$\limsup_{x \to \infty} \frac{\bar{F}(x)}{e^{-\mu x}} = \limsup_{x \to \infty} e^{\mu x - (\beta x)^\alpha}.$$

If $\alpha < 1$, this limit equals ∞ for any $\mu > 0$, while if $\alpha > 1$, it is 0 for any $\mu > 0$. (If $\alpha = 1$, the Weibull is equivalent to an Exponential distribution, which is, of course, light-tailed.)

Figure 1.6(a) illustrates the tail of the Weibull distribution for different values of α, contrasting the c.c.d.f. with those of the Gaussian and Exponential distributions. As was the case with the Pareto, the heaviness of the tail is clearly visible when we look at the log-log plot of the distribution; see Figure 1.6(b). While the Weibull looks nearly linear on a log-log plot when α is small (i.e., when the tail is heaviest) it is not perfectly linear like the Pareto distribution. To see why, we can take logarithms of both sides of (1.4) to obtain

$$\log \bar{F}(x) = -(\beta x)^\alpha.$$

While x^α gets close to $\log x$ as α shrinks to zero, it never entirely matches. However, if we move the negative sign to the other side and take logarithms again, we see that the Weibull looks linear according to a different scaling:

$$\underbrace{\log(-\log \bar{F}(x))}_{'y'} = \underbrace{\alpha \log \beta}_{y\text{-intercept}} + \underbrace{\alpha}_{\text{slope}} \underbrace{\log x}_{'x'}.$$

This tells us that the Weibull c.c.d.f. is linear on a $\log(-\log \bar{F}(x))$ versus $\log x$ plot. As with the Pareto distribution, this is a useful tool for exploratory analysis of data but one that must be used with care and should not be relied on for estimation.

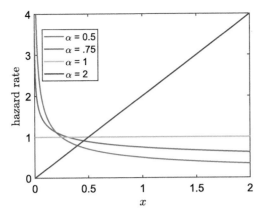

Figure 1.7 Weibull hazard rate for scale parameter $\beta = 1$ and different values of shape parameter α.

The Hazard Rate

The shape parameter α influences not just the tail behavior, but other properties of the Weibull distribution as well. One property that is of particular interest is the *hazard rate* (aka, failure rate) of the distribution. We study the hazard rate in detail in Chapter 4, and the Weibull is a particularly important distribution for that chapter because its hazard rate can have widely varying behaviors.

The hazard rate is defined as $q(t) = f(t)/\bar{F}(t)$ and has the following interpretation. Thinking of the distribution as capturing the *time to failure* (lifetime) of a component, $q(t)$ captures the instantaneous likelihood of a failure at time t of a component that entered into use at time 0, given that failure has not occurred until time t. Interestingly, when $\alpha > 1$, the hazard rate of the Weibull is increasing, meaning the likelihood of an impending failure increases with the age of the component; when $\alpha < 1$, the hazard rate is decreasing, meaning the likelihood of an impending failure actually decreases with the age of the component; and when $\alpha = 1$, the hazard rate is constant. We illustrate this in Figure 1.7.

The properties of the Weibull with respect to its hazard rate make it an extremely important distribution for survival analysis, reliability analysis, and failure analysis in a variety of areas. Additionally, the Weibull plays an important role in weather forecasting, specifically related to wind speed distributions and rainfall. As we discuss in Chapter 7, the Weibull (specifically, the mirror image of the Weibull distribution defined here) is an "extreme value distribution," which means that it is deeply connected to extreme events, such as the maximal rainfall in a day or year, the maximal overvoltage in an electrical system, or the maximal size of insurance claims. However, it was first used to describe the particle size distribution from milling and crushing operations in the 1930s [189]. Interestingly, though the distribution is named after Waloddi Weibull, who studied it in detail in the 1950s, it was introduced much earlier by Fréchet in 1927 in the context of extreme value theory [91].

Moments

An important difference between the Weibull distribution and the Pareto distribution is that all the moments of the Weibull distribution are finite. They can be large, especially when α is small, but they are not infinite.

To express the moments, we need to use the gamma function, Γ, which is a continuous extension of the factorial function. Specifically, $\Gamma(n) = (n-1)!$, for integer n. More generally,

$$\Gamma(z) = \int_0^\infty x^{z-1}e^{-x}dx, \text{ for } z > 0.$$

Using the gamma function, we can write the mean and variance of the Weibull as

$$\mathbb{E}[X] = \left(\frac{1}{\beta}\right)\Gamma(1+1/\alpha),$$

$$\text{Var}[X] = \left(\frac{1}{\beta}\right)^2 \left[\Gamma(1+2/\alpha) - (\Gamma(1+1/\alpha))^2\right].$$

Notice that the mean grows quickly as $\alpha \to 0$: it grows like the factorial of $1/\alpha$. More generally, the raw moments of the Weibull are given by

$$\mathbb{E}[X^n] = \left(\frac{1}{\beta}\right)^n \Gamma(1+n/\alpha).$$

1.2.3 The LogNormal distribution

While both the Pareto and the Weibull can be viewed as transformations of the Exponential distribution, as its name would suggest, the LogNormal distribution is a transformation of the Normal (aka Gaussian) distribution. In fact, the transformation of the Gaussian distribution that produces the LogNormal distribution is the same transformation that creates the Pareto from the Exponential – the LogNormal distribution is an *exponential* of the Gaussian distribution. Specifically,

$$X \sim \text{LogNormal}(\mu, \sigma^2) \iff \log(X) \sim \text{Gaussian}(\mu, \sigma^2).$$

Or, equivalently,

$$Z \sim \text{Gaussian}(\mu, \sigma^2) \iff e^Z \sim \text{LogNormal}(\mu, \sigma^2).$$

This means that the LogNormal distribution can be specified in terms of the Gaussian distribution via a logarithmic transformation. For example, the p.d.f. of a LogNormal(μ, σ^2) distribution is

$$f(x) = \frac{1}{\sqrt{2\pi}\sigma x}e^{-(\log x - \mu)^2/(2\sigma^2)}. \tag{1.5}$$

Note that the change of variables of $\log x$ introduces a $1/x$ term outside of the exponential in the p.d.f. as compared to the Gaussian distribution. This connection with the Gaussian distribution can also be used to show that the LogNormal distribution is heavy-tailed; this is left as an exercise for the reader (see Exercise 3).

While it may not be evident from the functional form of the p.d.f., the LogNormal distribution has a shape that is quite similar to that of the Pareto distribution. We illustrate the p.d.f. and c.c.d.f. in Figure 1.8. In fact, even when viewed on a log-log plot, the LogNormal and the Pareto can look similar. Specifically, when the variance parameter σ^2 is large, the

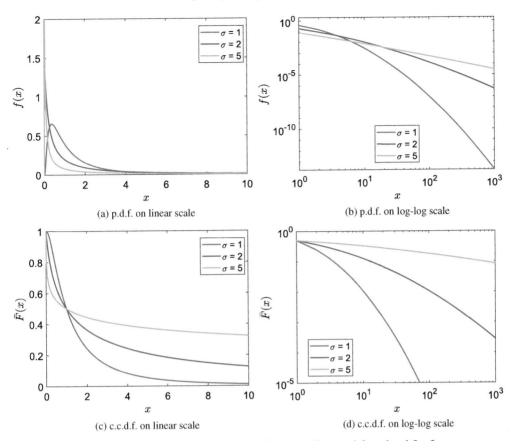

Figure 1.8 Illustration of LogNormal distribution: The c.c.d.f. and p.d.f. of LogNormal distributions with $\mu = 0$ and different values of σ are depicted. The p.d.f.s are plotted on a linear scale in (a), and on a log-log scale in (b). The corresponding c.c.d.f.s are plotted on a linear scale in (c), and on a log-log scale in (d). Note that the p.d.f. as well as the c.c.d.f. appears nearly linear on a log-log plot when σ is large.

LogNormal p.d.f. looks nearly linear on the log-log plot. To see why, let us take logarithms of both sides of (1.5):

$$\underbrace{\log f(x)}_{'y'} = -\log x - \log(\sigma\sqrt{2\pi}) - \frac{(\log x - \mu)^2}{2\sigma^2}$$

$$= -\frac{(\log x)^2}{2\sigma^2} + \left(\frac{\mu}{\sigma^2} - 1\right)\underbrace{\log x}_{'x'} - \log(\sigma\sqrt{2\pi}) - \frac{\mu^2}{2\sigma^2}.$$

This calculation shows that when σ is sufficiently large, the quadratic term above will be small for a large range of x and so the log-log plot will look nearly linear. Consequently, it is nearly impossible to distinguish the LogNormal from a Pareto using the log-log plot. Hence, one should be very careful when using the log-log plot as a statistical tool. We emphasize this point further in Chapter 8.

Properties

The LogNormal inherits many of the useful properties of the Gaussian distribution, with suitable adjustments owing to the exponential transformation between the distributions.

Perhaps the most important property of the Gaussian is that the sum of independent Gaussians is a Gaussian. Because of the exponential transformation, for the LogNormal, this property holds for the product rather than the sum. In particular, suppose $Y_i \sim$ LogNormal(μ_i, σ_i^2) are n independent random variables, Then

$$Y = \prod_{i=1}^{n} Y_i \quad \Longrightarrow \quad Y \sim \text{LogNormal}\left(\sum_{i=1}^{n} \mu_i, \sum_{i=1}^{n} \sigma_i^2\right).$$

This suggests that the LogNormal distribution is intimately tied to the growth of *multiplicative processes*. In particular, if a process grows multiplicatively, then it is additive on a logarithmic scale and, by the central limit theorem, it is likely to be Gaussian on the logarithmic scale. This, in turn, means that it is a LogNormal in the original scale. As a consequence, the LogNormal is a very common distribution in nature and human behavior. It has been used to model phenomena in finance, computer networks, hydrology, biology, medicine, and more. In fact, the LogNormal distribution was first studied by Robert Gibrat in the context of deriving a multiplicative version of the central limit theorem, which is sometimes termed "Gibrat's Law." Gibrat formulated this law during his study of the dynamics of firm sizes and industry structure [203]. We devote Chapter 6 to a discussion of multiplicative versions of the central limit theorem and their connections to heavy-tailed distributions.

Beyond products, LogNormal distributions also behave pleasantly with respect to other transformations. An important example is that

$$X \sim \text{LogNormal}(\mu, \sigma^2) \quad \Longrightarrow \quad X^a \sim \text{LogNormal}(a\mu, a^2\sigma^2) \text{ for } a \neq 0. \qquad (1.6)$$

Moments

Like the Weibull distribution, the moments of the LogNormal are always finite. They can be quite large but are never infinite. Perhaps the most counterintuitive thing about the moments of the LogNormal distribution is that, while we adopt the same parameter names as for the Gaussian, μ and σ^2 do not refer to the mean and variance of the LogNormal. Instead, they refer to the mean and variance of the Gaussian that is obtained by taking the $\log$ of the LogNormal. The mean and variance of the LogNormal are as follows:

$$\mathbb{E}\left[X\right] = e^{\mu+\sigma^2/2},$$
$$\text{Var}\left[X\right] = e^{2\mu+\sigma^2}\left(e^{\sigma^2} - 1\right).$$

The fact that mean and variance are exponentials of the distribution's parameters emphasizes that one should expect them to be large. More generally, the raw moments of the LogNormal distribution are given by

$$\mathbb{E}\left[X^n\right] = e^{n\mu+\frac{1}{2}n^2\sigma^2}.$$

Interestingly, though all the moments of the LogNormal distribution are finite, the distribution is not uniquely determined by its (integral) moments.

1.2.4 Other Heavy-Tailed Distributions

The Pareto, Weibull, and LogNormal are the most commonly used heavy-tailed distributions, but there are also other heavy-tailed distributions that appear frequently. We end this chapter by briefly introducing a few other distributions that come up later in the book as important examples of the concepts we discuss.

The Cauchy Distribution

The Cauchy distribution is an important distribution in statistics and is strongly connected to the central limit theorem, as we discuss in Chapter 5. However, it is most often used as a pathological example as a result of the fact that it does not have a well-defined mean (or variance). In fact, though it is named after Cauchy, the first explicit analysis of it was conducted by Poisson in 1824 in order to provide a counterexample showing that the variance condition in the central limit theorem cannot be dropped.

The c.d.f. and p.d.f. of a Cauchy(x_0, γ) distribution are given, for $x \in \mathbb{R}$, by

$$F(x) = \frac{1}{\pi} \arctan\left(\frac{x - x_0}{\gamma}\right) + \frac{1}{2},$$

$$f(x) = \frac{1}{\pi\gamma}\left(\frac{\gamma^2}{(x - x_0)^2 + \gamma^2}\right),$$

with location parameter $x_0 \in \mathbb{R}$ and scale parameter $\gamma > 0$. The distribution is plotted in Figure 1.9.

While the distribution function looks complicated, the Cauchy has a simple representation as the ratio of two Gaussian random variables. Specifically, if U and V are independent Gaussian random variables with mean 0 and variance 1, then $U/V \sim$ Cauchy$(0, 1)$ (see Exercise 12). A Cauchy$(0, 1)$ is referred to as the *standard Cauchy* and is important in its own right because it coincides with the *Student's t-distribution*, which is crucially important for estimating the mean and variance of a Gaussian distribution from data.

The Cauchy distribution's emergence in the context of the central limit theorem is a result of the fact that sums of Cauchy distributions have a property similar to sums of Gaussian distributions: if $X_1, \ldots, X_n$ are i.i.d. Cauchy$(0, 1)$ random variables, then the sum is also a Cauchy. Specifically, $\frac{1}{n}\sum_{i=1}^{n} X_i \sim$ Cauchy$(0, 1)$; we prove this property in Chapter 5 using characteristic functions.

Finally, a related distribution to the Cauchy is the LogCauchy, which has the same relationship to the Cauchy distribution that the LogNormal has to the Gaussian distribution, that is,

$$X \sim \text{Cauchy}(x_0, \gamma) \quad \iff \quad e^X \sim \text{LogCauchy}(x_0, \gamma),$$

or, equivalently,

$$Y \sim \text{LogCauchy}(x_0, \gamma) \quad \iff \quad \log(Y) \sim \text{Cauchy}(x_0, \gamma).$$

The LogCauchy is one of the few common distributions that has a heavier tail than the Pareto distribution – it has a logarithmically decaying tail. For this reason, it is sometimes referred to as a *super-heavy-tailed distribution*.

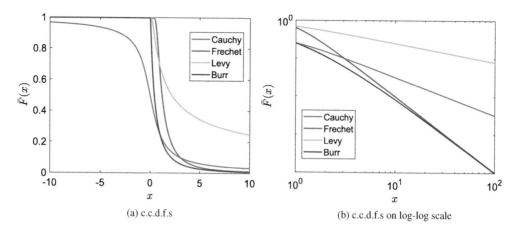

(a) c.c.d.f.s   (b) c.c.d.f.s on log-log scale

Figure 1.9 The plots show the c.c.d.f. of the standard Cauchy ($x_0 = 0$, $\gamma = 1$), the Fréchet (with $x_m = 0$, $\beta = 1$, $\alpha = 2$), the Lévy (with $\mu = 0$, $c = 1$), and the Burr distribution (with $c = \lambda = 1$, $k = 2$). Part (a) shows the plots on a linear scale, and (b) on a log-log scale. Note that all c.c.d.f.s look (asymptotically) linear on a log-log scale; we will formalize this property in Chapter 2.

The Fréchet Distribution

The Fréchet distribution plays a central role in extreme value theory, as we discuss in Chapter 7. It is commonly used in hydrology when studying the extremes of rainfall distributions.

The distribution is named after Maurice Fréchet, who introduced it in 1927; however, it is also referred to as the inverse Weibull distribution, which is a much more descriptive name since it is defined as exactly that. Specifically,

$$X \sim \text{Weibull}(\alpha, \beta) \quad \Longleftrightarrow \quad 1/X \sim \text{Fréchet}(\alpha, \beta, 0).$$

More generally, the c.d.f. and p.d.f. of the Fréchet(α, β, x_m) distribution are given, for $x > x_m$, by

$$F(x) = e^{-(\beta(x - x_m))^{-\alpha}},$$
$$f(x) = \alpha\beta\left(\beta(x - x_m)\right)^{-1-\alpha} e^{-(\beta(x - x_m))^{-\alpha}}.$$

Here, $\alpha > 0$ is the shape parameter, $\beta > 0$ is the scale parameter, and x_m is the minimum value taken by the distribution. The distribution is plotted in Figure 1.9.

The Lévy Distribution

The Lévy distribution is used most prominently in the study of financial models to explain stylized phenomena such as volatility clustering. Within mathematics and physics, it also plays an important role in the study of Brownian motion: the hitting time of a single point at a fixed distance from the starting point of a Brownian motion has a Lévy distribution. But perhaps its most prominent use is in the context of the generalized central limit theorem, which we discuss in Chapter 5.

Like the Cauchy distribution and the LogNormal distribution, the Lévy distribution is most conveniently defined as a transformation of the Gaussian distribution. In particular, a

Lévy distribution coincides with the square of the inverse of a Gaussian distribution (and is therefore sometimes also called the inverse Gaussian):

$$Z \sim \text{Gaussian}(\mu, \sigma^2) \quad \Longrightarrow \quad \frac{1}{(Z-\mu)^2} \sim \text{Lévy}(0, 1/\sigma^2).$$

More directly, the p.d.f. of the Lévy(μ, c) distribution is given, for $x \in \mathbb{R}$, by

$$f(x) = \sqrt{\frac{c}{2\pi(x-\mu)^3}} \, e^{-\frac{c}{2(x-\mu)}}. \tag{1.7}$$

From this equation, it is straightforward to see that the Lévy distribution is not just heavy-tailed, it is more specifically a "power law" distribution, since $f(x) \sim \sqrt{\frac{c}{2\pi}} x^{-3/2}$. A plot of the distribution is shown in Figure 1.9.

The Burr Distribution

The Burr distribution is a generalization of the Pareto distribution that often appears in statistics and econometrics. It is most frequently used in the study of household incomes and related wealth distributions. It was introduced by Irving Burr in 1942 as one of a family of 12 distributions, of which it is the *Burr Type XII distribution*. The c.c.d.f. and p.d.f. of a Burr(c, k, λ) distribution are given, for $x > 0$, by

$$\bar{F}(x) = (1 + \lambda x^c)^{-k}, \tag{1.8}$$

$$f(x) = \frac{ckx^{c-1}}{(1+x^c)^{k+1}}, \tag{1.9}$$

where $c, k, \lambda > 0$. The distribution is illustrated in Figure 1.9. When $c = 1$, the Burr corresponds to a so-called Type II Pareto distribution, and it is easy to see that it is a power law distribution, like the Cauchy, Fréchet, and Lévy distributions. In Chapter 7, we discuss an interesting connection between the Burr distribution and the residual life of the Pareto distribution. The hazard rate of the Burr distribution itself serves as an important counterexample in the same chapter.

The Zipf Distribution

We conclude by mentioning the Zipf distribution, which is a discrete version of the Pareto distribution. The Zipf distribution rose to prominence because of "Zipf's law," which states that, given a natural language corpus, the frequency of any word is inversely proportional to its rank in the frequency table of the corpus. That is, the most common word occurs twice as often as the second most common word, three times more than the third most common word, and so on. This law is named after George Zipf, who popularized it in 1935; however, the observation of the phenomenon predated his work by more than fifty years.

The Zipf(s, N) distribution is one example of a distribution that would explain Zipf's law, and is defined in terms of its probability mass function (p.m.f.):

$$p(n; s, N) = \frac{1/n^s}{\sum_{i=1}^{N} 1/i^s}, \tag{1.10}$$

where N can be thought of as the number of elements in the corpus, and s is the exponent characterizing the power law.

While the Zipf distribution is not heavy-tailed, given that it has a finite support, its generalization to the case $N = \infty$, which is called the Zeta distribution, is heavy-tailed (for $s > 1$).

1.3 What's Next

In this chapter we have introduced the definition of the class of heavy-tailed distributions, along with a few examples of common heavy-tailed distributions. Through these examples, you have already seen some illustrations of how heavy-tailed distributions behave differently from light-tailed distributions. But we have not yet sought to build intuition about these differences or to explain why heavy-tailed distributions are so common in the world around us. We have mentioned that controversy often surrounds heavy-tailed distributions because intuitive statistical approaches for identifying heavy tails in data are flawed, but we have not yet provided tools for correct identification and estimation of heavy-tailed phenomena.

The remainder of this book is organized to first provide intuition, both qualitative and mathematical, for the defining properties of heavy-tailed distributions (Part I: Properties), then explain why heavy-tailed distributions are so common in the world around us (Part II: Emergence), and finally develop the statistical tools for the estimation of heavy-tailed distributions (Part III: Estimation).

Given the mystique and excitement that surrounds the discovery of heavy-tailed phenomena, the detection and estimation of heavy tails in data is a task that is often (over)zealously pursued. While reading this book, you may be tempted to skip directly to Part III on estimation. However, the book is written so that the tools used in Part II are developed in Part I, and the tools used in Part III are developed in Parts I and II. Thus, we encourage readers to work through the book in order. That said, we have organized the material in each chapter so that there is a main body that presents the core ideas that are important for later chapters, followed by sections that present examples and/or variations of the main topic. These later sections can be viewed as enrichment opportunities that can be skipped as desired if the goal is to move quickly to Part III. However, if one is looking for the quickest path to understand the background needed before digging into Part III, then we recommend focusing on Chapter 2 from Part I, then Chapters 5 and 7 from Part II before moving to Part III.

Our goal for this book is that, through reading it, heavy-tailed distributions will be demystified for you. That their properties will be intuitive, not mysterious. That their emergence will be expected, not surprising. And that you will have the proper statistical tools for studying heavy-tailed phenomena and so will be able to resolve (or avoid) controversies rather than feed them. Happy reading!

1.4 Exercises

1. For a standard Gaussian random variable Z, show that for $x > 0$,

$$\Pr\left(Z > x\right) \leq \frac{e^{-x^2/2}}{\sqrt{2\pi}x}.$$

Note: In fact, the above bound can be shown to be asymptotically tight, that is, it can be shown that $\Pr\left(Z > x\right) \sim \frac{e^{-x^2/2}}{\sqrt{2\pi}x}$; *see [81, Chapter 7].*

2. Use the bound of Exercise 1 to prove that the Gaussian(μ, σ^2) distribution is light-tailed.

3. Prove that the LogNormal(μ, σ^2) distribution is heavy-tailed.

4. Consider a distribution F over $\mathbb{R}_+$ with finite mean μ. The *excess* distribution F_e corresponding to F is defined as

$$\bar{F}_e(x) = \frac{1}{\mu} \int_x^\infty \bar{F}(y)dy.$$

Prove that F is heavy-tailed if and only if F_e is heavy-tailed.

5. Let $X \sim$ Exponential(μ) and $Y = 1/X$. Prove that Y is heavy-tailed.

6. The random variable N takes values in $\mathbb{N}$. The distribution of N, conditioned on a uniformly distributed random variable U taking values in $(0, 1)$, is given by $\Pr(N > n \mid U) = U^n$. Assuming U is uniformly distributed, show that N is heavy-tailed.

   Note: Even though the conditional distribution of N given the value of U is light-tailed (in fact, Geometrically distributed), N itself is heavy-tailed!

7. In Exercise 6, you do not need U to be uniformly distributed for N to be heavy-tailed. Prove that, so long as $\Pr(U > x) > 0$ for all $x \in (0, 1)$, N is heavy-tailed.

8. Derive an expression for the Gini coefficient corresponding to the Pareto distribution. Show that the Gini coefficient converges to 1 as tail index $\alpha \downarrow 1$.

9. Compute the Lorenz curve corresponding to the Exponential distribution. Prove that the Gini coefficient in this case equals $1/2$.

10. Prove property (1.6) of the LogNormal distribution.

11. The goal of this exercise is to prove the following geometric interpretation of the standard Cauchy distribution. On the Cartesian plane, draw a random line passing through the origin, making an angle Θ with the x-axis as shown in the following figure, where Θ is uniformly distributed over $(-\pi/2, \pi/2)$. Let $(1, Y)$ denote the point where this random line intersects the vertical line $x = 1$. Prove that Y is a standard Cauchy random variable.

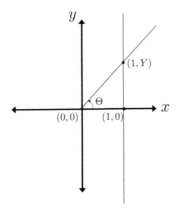

12. Prove that if U and V are independent, standard Gaussian random variables, then U/V is a standard Cauchy.

Hint: The geometric interpretation from Exercise 11 might help. Interpreting (U, V) as the Cartesian coordinates of a random point on the Cartesian plane, what is the joint distribution of the polar coordinates (R, Θ)?

13. The goal of this exercise is to compare the "heaviness" of the tails of the Pareto, Weibull, and LogNormal distributions. Let $X \sim \text{Pareto}(x_m, \alpha_1)$, $Y \sim \text{LogNormal}(\mu, \sigma^2)$, and $Z \sim \text{Weibull}(\alpha_2, \beta_2)$. Prove that

$$\lim_{x \to \infty} \frac{\Pr(Z > x)}{\Pr(Y > x)} = 0, \qquad \lim_{x \to \infty} \frac{\Pr(Y > x)}{\Pr(X > x)} = 0.$$

Note: This exercise shows that the Pareto has a heavier tail than the LogNormal, which has a heavier tail than the Weibull.

Part I

Properties

The mystique that surrounds heavy-tailed distributions is rooted, in part, in the ambiguous nature of the term "heavy-tailed." The precise mathematical definition is too opaque to provide insight and often too broad to correspond to the properties that are of interest in specific applications. Further, terms such as power-law, scale-free, self-similar, fat-tailed, long-tailed, stable, subexponential, and so on are often used synonymously with "heavy-tailed" when in fact they refer to particular properties of *some* heavy-tailed distributions. The result is a confusing and, at times, conflicting zoo of informal and formal terminology.

In Part I of this book we examine some of most mysterious and elegant properties of heavy-tailed distributions, connecting these properties to formal definitions of subclasses of heavy-tailed distributions. We focus on three illustrative properties: "scale-invariance" (aka scale-free), the "catastrophe principle," and "increasing residual life." We illustrate that these properties produce qualitatively different behaviors than what is seen under light-tailed distributions, and we describe how to formalize these properties mathematically as subclasses of heavy-tailed distributions. Combined, the chapters in Part I demystify some of the particularly exotic properties of heavy-tailed distributions and provide insight into the ways these properties interact with each other and with the broader class of heavy-tailed distributions.

Specifically, we introduce the classes of regularly varying distributions, subexponential distributions, and long-tailed distributions, which are perhaps the three most prominent classes of heavy-tailed distributions. These formalizations are particularly noteworthy because, while the general class of heavy-tailed distribution is difficult to work with, each of these subclasses has properties that make it appealing to work with analytically. These three classes also form the building blocks that allow us to study the emergence and estimation of heavy-tailed distributions in Parts II and III of this book.

2

Scale Invariance, Power Laws, and Regular Variation

In our daily lives, many things that we come across have a size, or scale, that we associate with them. For example, people's heights and weights differ, but they are not *that* different – they rarely differ by more than a factor of two or three and do not differ much from the population average. In contrast, the incomes of people we encounter in our daily lives do not have a typical size or scale – they may differ by a factor of *100 or more* and can be very far from the population average! This contrast is a consequence of the fact that many heavy-tailed phenomena, such as incomes, are *scale invariant*, aka, *scale-free*, while light-tailed phenomena, such as heights and weights, are not.

Scale invariance is a property that feels particularly magical the first time you observe it. An object is scale invariant if it looks the same regardless of what scale you look at it. Perhaps the easiest way to understand scale invariance is using fractals, like the one shown in Figure 2.1. If you zoom in or out, the picture will look the same. It turns out that Pareto distributions have the same property (see Figure 2.2). But Pareto distributions are even more special than fractals. With fractals, you have to zoom in or out in specific, discrete steps for the picture to look the same; with Pareto distributions, the invariance holds across a continuum of scale changes.

Scale invariance is a particularly mysterious aspect of heavy-tailed phenomena. It is natural to think of the average of a distribution as a good predictor of what samples will occur; but for scale invariant distributions the average is actually a very poor predictor. This fact leads to many of the counterintuitive properties of heavy-tailed distributions. For example, consider the old economics joke: "If Bill Gates walks into a bar . . . on average, everybody in the bar is a millionaire."

Though initially mysterious and counterintuitive, scale invariance is a beautiful and widely observed phenomenon that has attracted attention well beyond mathematics and statistics (e.g., in physics, computer science, economics, and even art). For example, scale invariance is an important concept in both classical and quantum field theory as well as statistical mechanics. In fact, it is closely tied to the notion of "universality" in physics, which relates to the fact that widely different systems can be described by the same underlying theory. Further, in the context of network science, scale invariance has received considerable attention. Widely varying networks have been found to have scale invariant degree distributions (and are thus termed "scale-free networks" [25, 41]) and this observation has had dramatic implications on our understanding of the structural properties of networks. For a discussion of scale invariance broadly, we recommend [220]. Here, we focus on scale invariance in the context of probability and statistics.

29

Figure 2.1 Illustration of the Sierpinski fractal [151].

In particular, in this chapter we explore the mathematics of the property of scale invariance and its connections with Pareto distributions and so-called *power law distributions*. Note that both "scale invariance" and "power law" are often (mis)used synonymously with "heavy-tailed," and thus it is important to point out that not all heavy-tailed distributions are scale invariant or have a power law (though all scale invariant distributions are heavy-tailed, as are all power law distributions). In this chapter, we formalize and generalize the notions of scale invariance and power laws as a subclass of heavy-tailed distributions termed "regularly varying distributions." In addition, we explain why the class of regularly varying distributions is particularly appealing from a mathematical perspective. The properties of this class shed light on many of the counterintuitive properties of heavy-tailed distributions, highlighting what properties of heavy-tailed distributions can be viewed as simple consequences of scale invariance. Further, to illustrate the usefulness of the class, we demonstrate how to apply properties of regular variation in order to analyze heavy-tailed phenomena more broadly. These examples show that it is not much more challenging to analyze the entire class of regularly varying distributions than it is to work with the specific case of the Pareto distribution.

2.1 Scale Invariance and Power Laws

To this point we have only introduced scale invariance informally as the property that something looks the "same" regardless of the scale at which it is observed. Given that our focus is on probability distributions, we can rephrase this idea as follows: if the scale (or units) with which the samples from the distribution are measured is changed, then the shape of the distribution is unchanged. This leads to the following formal definition.

Definition 2.1   A distribution function F is scale invariant if there exists an $x_0 > 0$ and a continuous positive function g such that

$$\bar{F}(\lambda x) = g(\lambda)\bar{F}(x),$$

for all x, λ satisfying $x, \lambda x \geq x_0$.

To interpret the definition of scale invariant, one can think of λ as the "change of scale" for the units being used. With this interpretation, the definition says that the shape of the

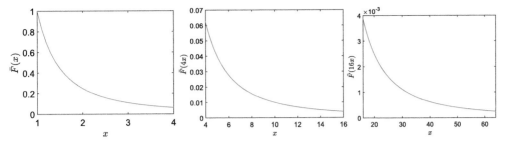

Figure 2.2 The complementary cumulative distribution function (c.c.d.f.) corresponding to the Pareto distribution ($\alpha = 2$, $x_m = 1$) plotted at different scales of the independent variable. Note that the shape of the curve is preserved up to a multiplicative scaling, consistent with scale invariance.

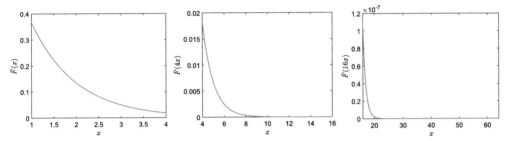

Figure 2.3 The complementary cumulative distribution function (c.c.d.f.) corresponding to the Exponential distribution (with mean 1) plotted at different scales of the independent variable. Note that the shape of the curve looks fundamentally altered at different scales.

distribution $\bar{F}$ remains unchanged up to a multiplicative factor $g(\lambda)$ if the measurements are scaled by λ. This is exactly what is shown in Figure 2.2 for the Pareto distribution.

More formally, to see that the Pareto is scale invariant, recall that a Pareto distribution has $\bar{F}(x) = (x/x_m)^{-\alpha}$ for $x > x_m$. Thus,

$$\bar{F}(\lambda x) = \left(\frac{\lambda x}{x_m}\right)^{-\alpha} = \bar{F}(x)\lambda^{-\alpha}, \text{ whenever } x, \lambda x > x_m.$$

Scale invariance is an elegant property, but it is also a fragile one. In particular, it does not hold for most probability distributions. For example, it is easy to see that the Exponential distribution is not scale invariant. Recall that an Exponential distribution has $\bar{F}(x) = e^{-\mu x}$ for $x \geq 0$. Therefore,

$$\bar{F}(\lambda x) = e^{-\mu \lambda x} = \bar{F}(x)e^{-\mu(\lambda-1)x}.$$

Thus, there is not a choice for g that is independent of x. This is also illustrated in Figure 2.3.

One may initially think that the lack of scale invariance of the Exponential distribution is a consequence of its being a light-tailed distribution. But that is not the case. For example, let us generalize the Exponential distribution to the Weibull distribution, $\bar{F}(x) = e^{-\beta x^\alpha}$ for $x \geq 0$, which is equivalent to the Exponential distribution when the shape parameter $\alpha = 1$ and is heavy-tailed when $\alpha < 1$. As the following calculation shows, the Weibull is also not scale invariant:

$$\bar{F}(\lambda x) = e^{-\beta(\lambda x)^\alpha} = \bar{F}(x)e^{-\beta x^\alpha(\lambda^\alpha - 1)}.$$

These examples start to give some intuition about scale invariance, but they leave open a fundamental, natural question:

Which distributions are scale invariant?

From the examples, we know that there is at least one scale invariant distribution (the Pareto distribution), but we also know that not all common distributions are scale invariant – not even all common heavy-tailed distributions. Perhaps surprisingly, it turns out that scale invariance is an extremely special property: distributions with "power law tails," (i.e., tails that match the Pareto distribution up to a multiplicative constant) are the *only* scale invariant distributions. That is, "scale invariance" can be thought of interchangeably with "power law." The following theorem states this formally.

Theorem 2.2   *A distribution function F is scale invariant if and only if F has a power law tail, that is, there exists $x_0 > 0$, $c \geq 0$, and $\alpha > 0$ such that $\bar{F}(x) = cx^{-\alpha}$ for $x \geq x_0$.*

Proof   Note that the case where $\bar{F}$ is identically zero over $[x_0, \infty)$ trivially satisfies the conditions of the lemma (this corresponds to the case $c = 0$).

Excluding the trivial case from consideration, it is easy to see that $\bar{F}(x)$ must be nonzero for all $x \geq x_0$. Indeed, if $\bar{F}(x') = 0$ for some $x' \geq x_0$, then for any $x \geq x_0$, $\bar{F}(x) = \bar{F}(x')g(x/x') = 0$.

Fix $x, y > 0$. We may then pick z large enough such that $z, zx, zxy \geq x_0$. From the scale invariant property of $\bar{F}$, $\bar{F}(xyz) = \bar{F}(z)g(xy)$. Of course, we may also write $\bar{F}(xyz) = \bar{F}(xz)g(y) = \bar{F}(z)g(x)g(y)$. Since $\bar{F}(z) \neq 0$, we can immediately see that the function g satisfies the following property:

$$g(xy) = g(x)g(y) \quad \text{for all } x, y > 0. \tag{2.1}$$

This is a very special property, and the only continuous positive functions satisfying the condition in (2.1) are $g(x) = x^{-\alpha}$ for some $\alpha \in \mathbb{R}$.[1] Noting that $\bar{F}(x) = \bar{F}(x_0)g(x/x_0)$ for all $x \geq x_0$, we conclude that $\alpha > 0$ (since $\bar{F}$ must be monotonically decreasing, with $\lim_{x\to\infty} \bar{F}(x) = 0$). Therefore, $\bar{F}(x) = cx^{-\alpha}$ for $x \geq x_0$ for some $c, \alpha > 0$. $\square$

2.2 Approximate Scale Invariance and Regular Variation

We have just seen that all scale invariant distributions are power law distributions, aka distributions with tails matching a Pareto distribution up to a multiplicative constant. This makes scale invariance a very special property, or a very fragile property depending on how you look at it. In fact, one interpretation of Theorem 2.2 is that we should not expect to see scale invariance in reality since it is so fragile.

---

[1] Defining $f(x) := \log g(e^x)$, the condition (2.1) is equivalent to $f(x + y) = f(x) + f(y)$ for $x, y \in \mathbb{R}$. This is known as *Cauchy's functional equation*. The stated claim now follows from the fact that the only continuous solutions of Cauchy's functional equation are of the form $f(x) = \alpha x$ for $\alpha \in \mathbb{R}$ (see, for example, [5]).

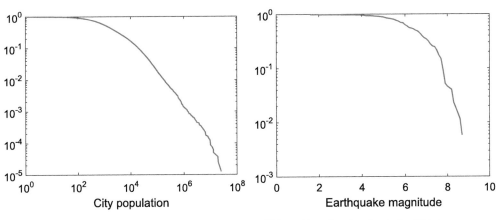

(a) Populations of US cities as per the 2010 census (data sourced from [2]).

(b) Intensities of earthquakes in the US between 1900 and 2017 (sourced from [1]). Earthquake intensity is measured on the Richter scale, which is inherently logarithmic. Thus, the values of the *x*-axis should be interpreted as being proportional to the logarithm of the intensity of the earthquake.

Figure 2.4 Empirical c.c.d.f. (aka rank plot) corresponding to two real-world datasets on a log-log scale.

In the strictest sense, that interpretation is correct. It is quite unusual for the distribution of an observed phenomenon to *exactly* match a power law distribution and thus be scale invariant. Instead, what tends to be observed in reality is that the *body* of the distribution is not an exact power law, but the *tail* of the distribution is *approximately* a power law. Consider the examples in Figure 2.4, which depicts the empirical c.c.d.f. corresponding to two real-world datasets on a log-log scale. Notice that the body of the empirical c.c.d.f. (on the log-log scale) does not look linear, which it would if the data were sampled from a power law distribution, but instead seems to approach a straight line *asymptotically*, which suggests that the c.c.d.f. behaves *asymptotically* like a power law.

Given that we should not expect to see precise scale invariance in real observations, it is natural to shift our focus from precise scale invariance to notions of approximate or asymptotic scale invariance; and it is natural not to focus on the whole distribution, but rather on just the tail of the distribution. In particular, the relevant formalism becomes *asymptotic scale invariance*, which we define here.

Definition 2.3 A distribution F is asymptotically scale invariant if there exists a continuous positive function g such that for any $\lambda > 0$,

$$\lim_{x \to \infty} \frac{\bar{F}(\lambda x)}{\bar{F}(x)} = g(\lambda).$$

The notion of asymptotic scale invariance almost exactly parallels the notion of scale invariance, except that it only requires the property to hold in the limit as $x \to \infty$, that is, it only requires the property to approximately hold for the tail: $\bar{F}(\lambda x) \sim g(\lambda)\bar{F}(x)$ as $x \to \infty$.[2]

As a result, it is immediately clear that Pareto distributions are asymptotically scale invariant:

$$\bar{F}(\lambda x)/\bar{F}(x) = \lambda^{-\alpha}.$$

Similarly, it is easy to see that asymptotic scale invariance is still quite a special property that is not satisfied by most distributions. For example, the Weibull and Exponential (Weibull with $\alpha = 1$) distributions are not asymptotically scale invariant since, as $x \to \infty$,

$$\frac{\bar{F}(\lambda x)}{\bar{F}(x)} = e^{-\beta(\lambda^\alpha - 1)x^\alpha} \to \begin{cases} \infty, & \lambda < 1, \\ 1, & \lambda = 1, \\ 0, & \lambda > 1. \end{cases}$$

However, asymptotic scale invariance is significantly broader than scale invariance, and it is easy to see that other distributions besides power law distributions are asymptotically scale invariant. For example, it follows from Exercise 8 that the convolution of a Pareto and an Exponential distribution is asymptotically scale invariant (though it is clearly not scale invariant). Similarly, consider the Fréchet distribution, which we introduced in Section 1.2.4 and will appear in our analysis of extremal processes in Chapter 7. This distribution, which is supported over the nonnegative reals, is defined by the distribution function $F(x) = e^{-x^{-\alpha}}$ for $x \geq 0$, where the parameter $\alpha > 0$. While this distribution is clearly not a power law, it is not hard to see that $\bar{F}(x) \sim x^{-\alpha}$ (see Exercise 1). In other words, the Fréchet distribution has an asymptotically power law tail, which in turn implies asymptotic scale invariance:

$$\lim_{x \to \infty} \frac{\bar{F}(\lambda x)}{\bar{F}(x)} = \lambda^{-\alpha}.$$

In general, since asymptotic scale invariance focuses only on the tail of the distribution, the body of such a distribution may behave in an arbitrary manner as long as the tail is approximately scale invariant.

As in the case of scale invariance, given the examples above, the natural question becomes:

Which distributions are asymptotically scale invariant?

It is clear that the class of asymptotically scale invariant distributions includes a variety of heavy-tailed distributions beyond the Pareto distribution, but it is also clear that it does not include all heavy-tailed distributions since it does not include the Weibull distribution. However, the fact that "scale invariant" can be thought of equivalently to "power law," leads to the suggestion that "asymptotically scale invariant" should correspond to some notion of "approximately power law," and this turns out to be true. In particular, it turns out that asymptotically scale invariant distributions have tails that are approximately power law in a rigorous sense that can be formalized via the class of *regularly varying distributions*.

---

[2] Throughout the book we use $a(x) \sim b(x)$ as $x \to \infty$ to mean $\lim_{x \to \infty} a(x)/b(x) = 1$.

Definition 2.4 A function $f\colon \mathbb{R}_+ \to \mathbb{R}_+$ is regularly varying of index $\rho \in \mathbb{R}$, denoted $f \in \mathcal{RV}(\rho)$, if for all $y > 0$,

$$\lim_{x \to \infty} \frac{f(xy)}{f(x)} = y^\rho.$$

Further, for $\rho \leq 0$, a distribution F is regularly varying of index ρ, denoted as $F \in \mathcal{RV}(\rho)$, if $\bar{F}(x) = 1 - F(x)$ is a regularly varying function of index ρ.[3]

The form of the definition immediately makes clear that regularly varying distributions are asymptotically scale invariant. Further, since $\lim_{x \to \infty} \bar{F}(xy)/\bar{F}(x) = y^\rho$, they seem to mimic the behavior of power law distributions, such as the Pareto distribution, asymptotically. This intuitively suggests that all asymptotically scale invariant distributions are regularly varying distributions – which turns out to be true.

Theorem 2.5 *A distribution F is asymptotically scale invariant if and only if it is regularly varying.*

Proof It is immediately clear that regularly varying distributions are asymptotically scale invariant, and so we need only prove the other direction. Fix $x, y > 0$. The asymptotic scale-free property implies that

$$\lim_{z \to \infty} \frac{\bar{F}(xyz)}{\bar{F}(z)} = g(xy).$$

We can also compute the same limit by writing $\frac{\bar{F}(xyz)}{\bar{F}(z)} = \frac{\bar{F}(xyz)}{\bar{F}(xz)} \frac{\bar{F}(xz)}{\bar{F}(z)}$. Note that $\frac{\bar{F}(xyz)}{\bar{F}(xz)} \to g(y)$ and $\frac{\bar{F}(xz)}{\bar{F}(z)} \to g(x)$ as $z \to \infty$, implying that

$$\lim_{z \to \infty} \frac{\bar{F}(xyz)}{\bar{F}(z)} = g(x)g(y).$$

From this, we can conclude that the function g satisfies

$$g(xy) = g(x)g(y) \quad \text{for all } x, y > 0.$$

This is the same relationship we used in the proof of Theorem 2.2 and, as in that case, it follows that there exists $\theta \in \mathbb{R}$ such that $g(x) = x^\theta$. Of course, by definition, this means that $\bar{F}$ is a regularly varying function, and F is a regularly varying distribution. □

Theorem 2.5 shows that the class of regularly varying distributions characterizes precisely those distributions that are asymptotically scale invariant, and from it we can immediately

---

[3] It is common practice to write the domain of regularly varying functions as $\mathbb{R}_+$. That said, it is important to understand that regular variation is an *asymptotic* property of a function as its argument tends to ∞. Thus, for a function f to be regularly varying, we only require that its domain includes $[x_0, \infty)$ for some $x_0 > 0$. Similarly, for a distribution to be regularly varying, we only require that its support include $[x_0, \infty)$ for some $x_0 > 0$. Specifically, regularly varying distributions need not be supported on the nonnegative reals. For example, the Cauchy distribution is regularly varying (see Exercise 1). Finally, why the index ρ must be nonpositive for regularly varying distributions will become apparent soon (see Lemma 2.7).

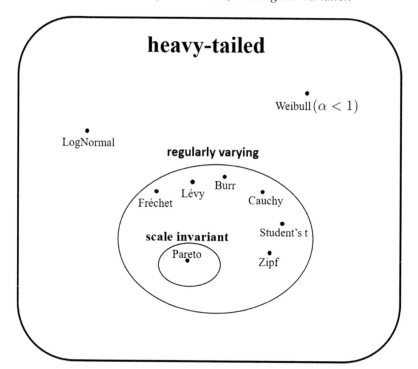

Figure 2.5 Scale invariant and regularly varying distributions.

see that a number of common heavy-tailed distributions are scale invariant. In particular, with a little effort, it is possible to verify that the Student's t-distribution, the Cauchy distribution, the Burr distribution, the Lévy, and also the Zipf distribution are all regularly varying and thus asymptotically scale invariant (see Exercise 1). This is summarized in Figure 2.5. We have not yet proven that regularly varying distributions are heavy-tailed; this follows from the analytic properties of regularly varying functions discussed next (see Lemma 2.9).

2.3 Analytic Properties of Regularly Varying Functions

The fact that regularly varying distributions are exactly those distributions that are asymptotically scale invariant suggests that they should be able to be analyzed (at least asymptotically) as if they are simply Pareto distributions. In fact, this intuition is correct and can be formalized explicitly, as we show in this section. Concretely, the properties we outline in this section provide the tools that enable regularly varying distributions to be analyzed "as if" they were polynomials, as far as the tail is concerned. This makes them remarkably easy to work with and shows that the added generality that comes from working with the class of regularly varying distributions, as opposed to working specifically with Pareto distributions, comes without too much added technical complexity.

To begin, it is important to formalize exactly what we mean when we say that regularly varying distributions have tails that are approximately power law. To do this, we need to first introduce the concept of a *slowly varying function*.

Definition 2.6 A function $L: \mathbb{R}_+ \to \mathbb{R}_+$ is said to be slowly varying if $\lim_{x\to\infty} \frac{L(xy)}{L(x)} = 1$ for all $y > 0$.

Slowly varying functions are simply regularly varying functions of index zero. So, intuitively, they can be thought of as functions that grow/decay asymptotically slower than any polynomial; for example, $\log x$, $\log \log x$, and so on. This can be formalized as follows.

Lemma 2.7 *If the function $L: \mathbb{R}_+ \to \mathbb{R}_+$ is slowly varying, then*

$$\lim_{x\to\infty} x^\rho L(x) = \begin{cases} 0 & \text{for } \rho < 0, \\ \infty & \text{for } \rho > 0. \end{cases}$$

We prove this lemma later in this section using properties of regularly varying functions. But we state it now in order to highlight an equivalent definition of regularly varying distributions as distributions that are "asymptotically power law." The following representation theorem for regularly varying functions makes this precise.

Theorem 2.8 *A function $f: \mathbb{R}_+ \to \mathbb{R}_+$ is regularly varying with index ρ if and only if $f(x) = x^\rho L(x)$, where $L(x)$ is a slowly varying function.*

Proof We start by proving the "if" direction. Suppose that $f \in \mathcal{RV}(\rho)$. Define $L(x) = f(x)/x^\rho$. To prove the result, it is enough to show that L is slowly varying, which can be argued as follows:

$$\lim_{x\to\infty} \frac{L(xy)}{L(x)} = \lim_{x\to\infty} \frac{f(xy)}{f(x)} \frac{x^\rho}{(xy)^\rho} = 1.$$

To prove the other direction, we need to show that, given $f(x) = x^\rho L(x)$, where $L(x)$ is a slowly varying function, $f \in \mathcal{RV}(\rho)$. For $y > 0$,

$$\lim_{x\to\infty} \frac{f(xy)}{f(x)} = \lim_{x\to\infty} \frac{(xy)^\rho}{x^\rho} \frac{L(xy)}{L(x)} = y^\rho,$$

which implies, by definition, that $f \in \mathcal{RV}(\rho)$. □

It is important to remember when applying this theorem that regularly varying *functions* can have arbitrary index ρ; however, regularly varying *distributions* must have index $\rho \le 0$.[4]

The key implication of Theorem 2.8 in the context of heavy-tailed distributions is that regularly varying distributions can be thought of as distributions with approximately power law tails in a rigorous sense. That is, they differ from a power law tail only by a slowly varying function $L(x)$, which can intuitively be treated as a constant when doing analysis. This intuition leads to many of the analytic properties that we discuss in the remainder of this section.

However, before we move to the analytic properties of regularly varying distributions, it is useful to illustrate how powerful the representation theorem is by itself. To illustrate

---

[4] One peculiarity of this notational convention is that a Pareto distribution with tail index α, where $\alpha > 0$, is regularly varying with index $-\alpha$.

this, we use it in Lemma 2.9 to argue that regularly varying distributions are heavy-tailed. Of course, this is not a surprising result, given the tie to Pareto distributions, but it is an important foundational result and it provides a simple illustration of how to work with the representation theorem.

Lemma 2.9 *All regularly varying distributions are heavy-tailed.*

Proof Suppose that the distribution F is regularly varying. We know then that $\bar{F}(x) = x^{-\alpha}L(x)$, where $\alpha \geq 0$, and $L(x)$ is a slowly varying function. Consider $\mu > 0$ and $\beta > \alpha$. Now,

$$\frac{\bar{F}(x)}{e^{-\mu x}} = (x^{-\beta}e^{\mu x})(x^{\beta-\alpha}L(x)).$$

Since $x^{-\beta}e^{\mu x} \to \infty$ as $x \to \infty$, and $x^{\beta-\alpha}L(x) \to \infty$ as $x \to \infty$ (via Lemma 2.7), we can conclude that $\lim_{x\to\infty} \frac{\bar{F}(x)}{e^{-\mu x}} = \infty$, which proves that F is heavy-tailed. □

The above provides an example of using the fact that regular varying distributions have tails that are approximately power law; however, this representation of regularly varying distributions has much broader implications as well. In particular, in the remainder of this section we illustrate a variety of analytic properties of regularly varying distributions that highlight how regularly varying distributions can be analyzed "as if" they were Pareto distributions, as far as the tail is concerned.

In the following, we focus on two crucial analytic properties: (i) integration/differentiation of regularly varying functions, which is formalized via Karamata's theorem, and (ii) inverting the integral transforms of regularly varying functions to understand properties of the tail of the distribution, which is formalized via Karamata's Tauberian theorem. In each case, we include a simple example application of the result. Then, in the following section (Section 2.4), we prove closure properties of regularly varying distributions with respect to various algebraic operations in order to provide further illustrations of how to apply these analytic properties.

2.3.1 Integration and Differentiation of Regularly Varying Distributions

Perhaps one of the most appealing aspects of working with power law and Pareto distributions is that when one needs to manipulate them to calculate moments, conditional probabilities, convolutions, and other such things, all that is required is the integration or differentiation of polynomials, which is quite straightforward. This is in stark contrast to distributions such as the Gaussian and LogNormal, which can be very difficult to work with in this way.

One of the nicest properties of regularly varying distributions is that, in a sense, you can treat them as if they were simply polynomials when integrating or differentiating them – as long as you only care about the tail – and so they are not much more difficult to work with than Pareto distributions. This is especially useful when calculating things like moments, conditional probabilities, convolutions, and so on, as we shall see repeatedly in the remainder of this chapter and throughout the book.

The foundational properties of regularly varying functions with respect to integration and differentiation are typically referred to as Karamata's theorem. This result provides the building block for working with regularly varying distributions.

Karamata's Theorem

Karamata's theorem is perhaps the most important result in the study of regular variation. We start our discussion of it by stating Karamata's theorem for integration of regularly varying functions, since its statement is a bit cleaner than that of differentiation.

It is useful to begin by anticipating what we should expect Karamata's theorem to say. To do this, begin by considering what would happen if $f(t) = t^\rho$. In that case,

$$\int_0^x f(t)dt = \frac{x^{\rho+1}}{\rho+1} = \frac{xf(x)}{\rho+1} \text{ if } \rho > -1, \text{ and}$$

$$\int_x^\infty f(t)dt = \frac{x^{\rho+1}}{-(\rho+1)} = \frac{xf(x)}{-(\rho+1)} \text{ if } \rho < -1.$$

Thus, we may expect that Karamata's theorem would say that, asymptotically, the integrals of regularly varying functions should behave as if the function were a polynomial as far as the tail is concerned (i.e., the $=$ above should be replaced by a $\sim$). In fact, this is exactly what Karamata's theorem states.

Theorem 2.10 (Karamata's Theorem)

(a) For $\rho > -1$, $f \in \mathcal{RV}(\rho)$ if and only if

$$\int_0^x f(t)dt \sim \frac{xf(x)}{\rho+1}.$$

(b) For $\rho < -1$, $f \in \mathcal{RV}(\rho)$ if and only if

$$\int_x^\infty f(t)dt \sim \frac{xf(x)}{-(\rho+1)}.$$

Not surprisingly, regularly varying distributions also asymptotically behave as if they were polynomials with respect to differentiation. In particular, if $f(x) = x^\alpha$, then $f'(x) = \alpha x^{\alpha-1}$, and so $\alpha f(x) = xf'(x)$. The following result, which is commonly referred to as the *monotone density theorem*, shows that exactly this relationship holds for regularly varying distributions with $=$ replaced by $\sim$, modulo some technical conditions.

Theorem 2.11 (Monotone Density Theorem) *Suppose that the function f is absolutely continuous with derivative f'. If $f \in \mathcal{RV}(\rho)$ and f' is eventually monotone, then $xf'(x) \sim \rho f(x)$. Moreover, if $\rho \neq 0$, then $|f'(x)| \in \mathcal{RV}(\rho-1)$.*[5]

---

[5] A function f is said to be *absolutely continuous* if it has a derivative f' almost everywhere that is integrable, such that

$$f(x) = f(0) + \int_0^x f'(t)dt \quad \forall x.$$

A function g is *eventually monotone* if there exists $x_0 > 0$ such that g is monotone over $[x_0, \infty)$.

In what follows, we give the proof of Theorem 2.11. The proof of Theorem 2.10 is a bit more cumbersome, and we refer the interested reader to [183, Section 2.3.2] for the proof.

Proof of Theorem 2.11   For simplicity, we assume that $f'(x)$ is nondecreasing over $x \geq x_0$ (the proof for the case of eventually nonincreasing f' is along similar lines). Fixing a, b such that $0 < a < b$, we may write

$$\int_{ax}^{bx} \frac{f'(t)}{f(x)} dt = \frac{f(bx) - f(ax)}{f(x)}.$$

For $x > x_0/a$, the monotonicity of f' implies that

$$\frac{f'(ax)x(b-a)}{f(x)} \leq \frac{f(bx) - f(ax)}{f(x)} \leq \frac{f'(bx)x(b-a)}{f(x)}. \tag{2.2}$$

Noting that $f \in \mathcal{RV}(\rho)$, the first inequality in (2.2) implies that

$$\limsup_{x \to \infty} \frac{f'(ax)x}{f(x)} \leq \frac{b^\rho - a^\rho}{b - a}.$$

Next, letting $b \downarrow a$ on the right side of the above inequality and noting that this corresponds to taking the derivative of the function x^ρ at $x = a$, we obtain

$$\limsup_{x \to \infty} \frac{f'(ax)x}{f(x)} \leq \rho a^{\rho-1}. \tag{2.3}$$

Similarly, using the second inequality in (2.2) and letting $a \uparrow b$, we obtain

$$\liminf_{x \to \infty} \frac{f'(bx)x}{f(x)} \geq \rho b^{\rho-1}. \tag{2.4}$$

Setting $a = 1$ in (2.3) and $b = 1$ in (2.4), we conclude that $x f'(x) \sim \rho f(x)$. Finally, when $\rho \neq 0$, it is easy to see that $f'(x) \sim \rho \left(\frac{f(x)}{x} \right)$ implies that $|f'(x)| \in \mathcal{RV}(\rho - 1)$. $\square$

Hopefully, it is already clear that Karamata's theorem is a particularly appealing and powerful property of regularly varying functions. But, it is worth considering a few examples in order to emphasize this further. Perhaps the most powerful example of the use of Karamata's theorem is in deriving the so-called "Karamata representation theorem" for regularly varying functions.

Karamata's Representation Theorem

We have already discussed one representation theorem for regularly varying functions (Theorem 2.8), which allows us to write any regularly varying function as $\bar{F}(x) = x^\rho L(x)$ for some slowly varying function $L(x)$. This is a particularly nice form since it highlights the view of regularly varying distributions as asymptotically power law; however, it is also a fairly implicit view of regularly varying functions since the form of $L(x)$ is not defined. Using Karamata's theorem, we can derive a much more precise representation theorem for regularly varying functions.

Theorem 2.12 (Karamata's Representation Theorem) $f \in \mathcal{RV}(\rho)$ *if and only if it can be represented as*

$$f(x) = c(x)\exp\left\{\int_1^x \frac{\beta(t)}{t}dt\right\} \tag{2.5}$$

for $x > 0$, where $\lim_{x\to\infty} c(x) = c \in (0, \infty)$ and $\lim_{x\to\infty} \beta(x) = \rho$.

The representation of regularly varying distributions in Karamata's representation theorem may initially seem surprising since it does not superficially look like a power law. However, note that if one treats $\beta(t)$ as if it is a constant ρ (which it converges to in the limit), then the exponent becomes $\rho \log x$ and so the power law form appears (since $e^{\rho \log x} = x^\rho$).

Proof To begin, let us first check that if a function f can be represented via (2.5), then $f \in \mathcal{RV}(\rho)$. Note that for $y > 0$,

$$\frac{f(xy)}{f(x)} = \frac{c(xy)}{c(x)}\exp\left\{\int_x^{xy} \frac{\beta(t)}{t}dt\right\}.$$

Now, since $\beta(x) \to \rho$ as $x \to \infty$, given $\epsilon > 0$, there exists $x_0 > 0$ such that $\rho - \epsilon < \beta(x) < \rho + \epsilon$ for all $x > x_0$. Therefore, for x large enough so that $x, xy > x_0$,

$$\frac{c(xy)}{c(x)}\exp\left\{\int_x^{xy} \frac{\rho - \epsilon}{t}dt\right\} < \frac{f(xy)}{f(x)} < \frac{c(xy)}{c(x)}\exp\left\{\int_x^{xy} \frac{\rho + \epsilon}{t}dt\right\},$$

which is equivalent to

$$\frac{c(xy)}{c(x)}y^{\rho-\epsilon} < \frac{f(xy)}{f(x)} < \frac{c(xy)}{c(x)}y^{\rho+\epsilon}.$$

Now, since $\frac{c(xy)}{c(x)} \to 1$ as $x \to \infty$, we obtain

$$y^{\rho-\epsilon} \le \liminf_{x\to\infty} \frac{f(xy)}{f(x)} \le \limsup_{x\to\infty} \frac{f(xy)}{f(x)} \le y^{\rho+\epsilon}.$$

Letting $\epsilon \downarrow 0$, we finally conclude that $\lim_{x\to\infty} \frac{f(xy)}{f(x)} = y^\rho$, which, of course, implies by definition that $f \in \mathcal{RV}(\rho)$.

Next, we prove that if $f \in \mathcal{RV}(\rho)$, then f has a representation of the form (2.5). We prove this first for the slowly varying case (i.e., $\rho = 0$), and then consider the case of general ρ.

Accordingly, suppose that $L \in \mathcal{RV}(0)$. Define $b(x) = \frac{xL(x)}{\int_0^x L(t)dt}$. Note that Karamata's theorem implies that $b(x) \to 1$ as $x \to \infty$. Let $\beta_L(x) = b(x) - 1$. Now,

$$\int_1^x \frac{\beta_L(t)}{t}dt = \int_1^x \frac{L(t)}{\int_0^t L(y)dy}dt - \log(x)$$

$$= \log\left(\int_0^x L(y)dy\right) - \log\left(\int_0^1 L(y)dy\right) - \log(x).$$

Now, using $\int_0^x L(y)dy = \frac{xL(x)}{b(x)}$, we obtain

$$\int_1^x \frac{\beta_L(t)}{t}dt = \log\left(\frac{L(x)}{b(x)\int_0^1 L(y)dy}\right),$$

which finally gives us

$$L(x) = c_L(x)\exp\left\{\int_1^x \frac{\beta_L(t)}{t}dt\right\}, \tag{2.6}$$

where $c_L(x) = b(x)\int_0^1 L(y)dy$. Noting that $c_L(x) \to \int_0^1 L(y)dy$ and $\beta_L(x) \to 0$ as $x \to \infty$, we have proved that L has the postulated representation.

Finally, moving to the case of general ρ, suppose that $f \in \mathcal{RV}(\rho)$. We know then that $f(x) = x^\rho L(x)$, where $L(x)$ is a slowly varying function. We have already established that $L(x)$ can be represented as (2.6), with $\beta_L(x) \to 0$ and $c_L(x) \to c \in (0, \infty)$ as $x \to \infty$. It then follows immediately that

$$f(x) = c_L(x)\exp\left\{\int_1^x \frac{(\beta_L(t) + \rho)}{t}dt\right\},$$

which gives us the desired representation. □

Karamata's representation theorem is an extremely powerful tool for working with regularly varying distributions. To see this, note that it is straightforward to prove Lemma 2.7 using Karamata's representation theorem (see Exercise 4). Additionally, Karamata's representation theorem can be used to show a number of other properties of regularly varying distributions that connect them to power law and Pareto distributions. We illustrate two of these here: (i) the observation that regularly varying distributions appear approximately linear on a log-log plot, and (ii) properties of the moments of regularly varying distributions.

Let us start by considering the behavior of regularly varying distributions in logarithmic scale. We have seen earlier that one distinguishing property of Pareto distributions is that they are exactly linear when viewed on a log-log scale. Specifically, recall that for Pareto distributions $\bar{F}(x) = (x/x_m)^{-\alpha}$ for $x > x_m$, and so

$$\log \bar{F}(x) = -\alpha \log(x) + \alpha \log(x_m).$$

Thus, $\log \bar{F}(x)$ is exactly linear in terms of $\log x$, with slope α. This is a property that allows for easy preliminary identification of them in data, as we have seen in Chapter 1 and explore in detail in Chapter 8. Note that one must be very cautious using this approach for estimation, as we illustrate in Chapter 8.

Using Karamata's representation theorem, we can easily obtain the corresponding property for the tail of regularly varying distributions. In particular, we have the following result, which shows that the tail of regularly varying distributions with index α is asymptotically linear with slope α when viewed on a log-log plot.

Lemma 2.13 *If f is a regularly varying function with index ρ, then*

$$\lim_{x\to\infty} \frac{\log f(x)}{\log(x)} = \rho.$$

Proof From Karamata's representation theorem, we know that

$$f(x) = c(x)\exp\left\{\int_1^x \frac{\beta(t)}{t}dt\right\},$$

where $\lim_{x\to\infty} c(x) = c \in (0, \infty)$ and $\lim_{x\to\infty} \beta(x) = \rho$.

Given $\epsilon > 0$, there exists $x_0 > 1$ such that for all $x \geq x_0$, $\rho - \epsilon < \beta(x) < \rho + \epsilon$. Therefore, for $x > x_0$,

$$\log f(x) \leq \log c(x) + \int_1^{x_0} \frac{\beta(t)}{t} dt + \int_{x_0}^{x} \frac{\rho + \epsilon}{t} dt$$

$$= \log c(x) + \int_1^{x_0} \frac{\beta(t)}{t} dt + (\rho + \epsilon)(\log(x) - \log(x_0)).$$

From the preceding inequality, it follows that

$$\limsup_{x \to \infty} \frac{\log f(x)}{\log(x)} \leq \rho + \epsilon.$$

Using similar arguments, it can be shown that

$$\liminf_{x \to \infty} \frac{\log f(x)}{\log(x)} \geq \rho - \epsilon.$$

Letting ϵ approach zero completes the proof. $\square$

Next, let us move to studying the moments of regularly varying distributions. Recall that the moments of Pareto distributions are a bit peculiar: for Pareto(x_m, ρ) distributions, the ith moment is finite if $i < \rho$ and infinite if $i > \rho$. The fact that moments can be infinite is, as we have seen in Chapter 1, not just of theoretical interest. Data from a variety of situations has been shown to exhibit power law tails with ρ around 1.2–2.1, and so is often well approximated by distributions with infinite variance.

Using the above result, it is not hard to show that regularly varying distributions have moments that parallel those of Pareto distributions. In particular, we have the following result.

Theorem 2.14 *Suppose that a nonnegative random variable X is regularly varying of index $-\rho$. Then*

$$\mathbb{E}\left[X^i\right] < \infty \text{ for } 0 \leq i < \rho,$$
$$\mathbb{E}\left[X^i\right] = \infty \text{ for } i > \rho.$$

The moment conditions in the theorem above should not be particularly surprising at this point since computing moments has to do with integration and the finiteness of moments has to do mainly with the tail, which means that Karamata's theorem should ensure that regularly varying distributions behave like power laws. This intuition serves as a good guide for the proof, which makes use of Lemma 2.13, which was a consequence of Karamata's theorem. We leave the proof of the result as an exercise for the reader (see Exercise 5).

2.3.2 Integral Transforms of Regularly Varying Distributions

Integral transforms like the moment generating function, the characteristic function, and the Laplace–Stieltjes transform are of fundamental importance in probability, as well as in many physical problems in applied mathematics. It is often easier to study probabilistic and sto-chastic models using transforms than it is to study them directly as a result of the ease of computing convolutions, moments, time scalings, performing integration of the distribution, and so on. Thus, one can typically complete the analysis in "transform space" and then invert the transform to understand the distribution itself, taking advantage of the uniqueness of the representation.

In the context of this book, we have already seen the importance of transforms in the def-inition of heavy-tailed distributions. Recall that the definition of heavy-tailed distributions explicitly uses the moment generating function (m.g.f.) and defines heavy-tailed distribu-tions as those distributions for which $M_X(t) := \mathbb{E}\left[e^{tX}\right] = \infty$ for all $t > 0$. This means that, while moment generating functions are often a powerful analytic tool, working with the m.g.f. of heavy-tailed distributions is problematic. Thus, one needs to consider other inte-gral transforms in the case of heavy-tailed distributions. This section provides the tools for working with transforms in the heavy-tailed setting.

Though the m.g.f. is not appropriate for heavy-tailed distributions, one can instead use other transforms. When the distribution corresponds to a nonnegative random variable, the Laplace–Stieltjes transform (LST) is appropriate and, more generally, the characteristic func-tion is the appropriate tool. The *Laplace–Stieltjes transform* of a function f is defined as

$$\psi_f(s) := \int_{-\infty}^{\infty} e^{-sx} df(x).$$

Specializing to probability distributions, given a random variable X following distribution F, the *Laplace–Stieltjes transform* of F (or X) is defined as

$$\psi_X(s) := \int_{-\infty}^{\infty} e^{-sx} dF(x) = \mathbb{E}\left[e^{-sX}\right].$$

Notice that the LST of a distribution is related to the m.g.f. via a change of variable: we replace the argument t in the definition of $M(t)$ by $-s$. Similarly, the characteristic function can be obtained by replacing t with it, where i is the imaginary unit. So, given a random variable X following distribution F, the *characteristic function* of F (or X) is defined as

$$\phi_X(t) := \mathbb{E}\left[e^{itX}\right] = \mathbb{E}\left[\cos(tX) + i\sin(tX)\right].$$

Note that if X is nonnegative, the LST $\psi_X(s)$ is well defined and finite for $s \geq 0$. On the other hand, the characteristic function $\phi(t)$ associated with any distribution is well defined and finite for all t.

To get a feel for the behavior of transforms in the case of heavy-tailed distributions it is useful to look at the specific case of a power-law function. To keep things simple, we consider power laws of the form

$$f(x) = \begin{cases} x^\rho & x \geq 0, \\ 0 & x < 0, \end{cases}$$

where $\rho > 0$. Note that with $\rho > 0$, f cannot capture a probability distribution, but limiting our attention to positive indices makes things simpler, so we do that for now and then extend the analysis to probability distributions later in the section. We do this because the case of $\rho > 0$ turns out to be quite instructive. In this case, the LST of f can be written as follows:

$$
\begin{aligned}
\psi_f(s) &= \int_0^\infty e^{-sx} df(x) \\
&= \rho \int_0^\infty e^{-sx} x^{\rho-1} dx \\
&= \rho s^{-\rho} \int_0^\infty e^{-sx} (sx)^{\rho-1} d(sx) \\
&= \rho s^{-\rho} \Gamma(\rho) = s^{-\rho} \Gamma(\rho+1),
\end{aligned}
$$

where the last line uses the Gamma function Γ, which is a continuous extension of the factorial function to the real numbers defined as $\Gamma(z) := \int_0^\infty e^{-t} t^{z-1} dt$ and satisfying $\Gamma(n) = (n-1)!$ for $n \in \mathbb{Z}$ and $z\Gamma(z) = \Gamma(z+1)$.

The calculation in this example reveals something exciting. The LST of a function f with a power law (as $x \to \infty$) also behaves like a power law (as $s \downarrow 0$). This is exciting because it suggests that one can potentially understand properties of the tail of a distribution by studying properties of the LST near zero. Of course, one could potentially obtain this information by inverting the LST, but that is typically very involved and cannot be done in closed form apart from a few special cases. In contrast, in this example, the tail behavior of f can be obtained with a simple observation about ψ.

However, before getting too excited, it is important to remember that, so far, we have only seen this behavior in the case of specific f following a power law with a positive index. Thus, the question becomes:

Do regularly varying distributions have regularly varying transforms?

If so, it would be quite powerful since it is often the case that one can derive the LST in situations where it is not tractable to work directly with the distribution.

Fortunately. the answer is "yes." Results of this form are called *Abelian* and *Tauberian* theorems, and there is a wide variety of these theorems for the LST and the characteristic function. As a first example, we present Karamata's Tauberian theorem, which is the direct extension of the power-law example above for the case of increasing functions.

Theorem 2.15 (Karamata's Tauberian Theorem) *Let f be a nonnegative right-continuous increasing function such that $f(x) = 0$ for $x < 0$, and let $\rho \geq 0$. Then, for slowly varying $L(x)$, the following are equivalent:*

$$
\begin{aligned}
f(x) &\sim L(x) x^\rho & (x \to \infty), & \qquad (2.7) \\
\psi_f(s) &\sim \Gamma(\rho+1) L(1/s) s^{-\rho} & (s \downarrow 0). & \qquad (2.8)
\end{aligned}
$$

This result says something very powerful. Informally, it says that if the behavior of the LST as $s \downarrow 0$ is approximately a power law, then the corresponding function is also a power law *with the same index*. Further, given the representation of regularly varying functions in Theorem 2.8, this can be interpreted in a different light as well. In particular, since (2.7) characterizes regularly varying functions, the theorem states that regularly varying functions are exactly those that have LSTs that are regularly varying around zero.

Though Theorem 2.15 is commonly called a Tauberian theorem, it actually includes both a Tauberian theorem and an Abelian theorem. In particular, the direction showing that (2.7) implies (2.8) is called an Abelian theorem, and the reverse direction is a Tauberian theorem. The Tauberian direction is typically harder to prove, which is why such theorems are typically referred to as Tauberian theorems. We omit the proof of Theorem 2.15 here; the interested reader is referred to Theorem 1.7.1 in [31].

Theorem 2.15 is powerful but does not yet give us exactly what we would like since it still assumes that the index ρ is positive. Thus, it does not apply to regularly varying *distributions* directly. However, it is possible to remedy this. In particular, the following is a more general version of Karamata's Tauberian theorem that uses a Taylor expansion of the LST in terms of the moments of the distribution.

Theorem 2.16   *Consider a nonnegative random variable X with distribution F. For $n \in \mathbb{Z}_+$, suppose that $\mathbb{E}[X^n] < \infty$. Then for slowly varying $L(x)$ and $\alpha = n + \beta$ where $\beta \in (0, 1)$, the following are equivalent:*

$$\bar{F}(x) \sim \frac{(-1)^n}{\Gamma(1-\alpha)} x^{-\alpha} L(x) \qquad (x \to \infty), \qquad (2.9)$$

$$(-1)^{n+1} \left[\psi_X(s) - \sum_{i=0}^{n} \frac{\mathbb{E}[X^i](-s)^i}{i!} \right] \sim s^\alpha L(1/s) \qquad (s \downarrow 0). \qquad (2.10)$$

To interpret the statement of Theorem 2.16, note that if a nonnegative random variable X satisfies $\mathbb{E}[X^n] < \infty$, then its LST can be expressed via a Taylor expansion as follows:

$$\psi_X(s) = \sum_{i=0}^{n} \frac{\mathbb{E}[X^i](-s)^i}{i!} + o(s^n). \qquad (2.11)$$

The Abelian component of Theorem 2.16 states that if X is regularly varying with index $-\alpha$, where $\alpha \in (n, n+1)$, then the $o(s^n)$ correction term in (2.11) is of the order of s^α. The Tauberian component makes the converse implication.

To understand this connection better, let us first consider the case $n = 0$. For this case, Theorem 2.16 states that for $\alpha \in (0, 1)$,

$$\bar{F}(x) \sim \frac{1}{\Gamma(1-\alpha)} x^{-\alpha} L(x) \iff \psi_X(s) - 1 \sim -s^\alpha L(1/s).$$

Clearly, this statement applies to distributions with infinite mean. Consider, for example, the Lévy distribution, which is parameterized by $c > 0$ and has LST $\psi(s) = e^{-\sqrt{2sc}}$ for $s \geq 0$ [120]. Noting that as $s \downarrow 0$, $1 - \psi(s) \sim -\sqrt{2cs}$, Theorem 2.16 (specifically, the Tauberian part) implies that the Lévy tail satisfies $\bar{F}(x) \sim \frac{\sqrt{2c}}{\Gamma(1/2)} x^{-1/2}$, that is, $\bar{F}(x) \sim \sqrt{\frac{2c}{\pi}} x^{-1/2}$. The same conclusion can be arrived at by applying Karamata's theorem to the Lévy density function; see Exercise 1.

Next, consider the case $n = 1$. In this case, Theorem 2.16 states that for $\alpha \in (1, 2)$,

$$\bar{F}(x) \sim \frac{-1}{\Gamma(1 - \alpha)} x^{-\alpha} L(x) \iff \psi_X(s) - 1 + \mathbb{E}[X] s \sim s^{\alpha} L(1/s).$$

This statement in turn is applicable to distributions with a finite first moment, but an infinite second moment.

Proof sketch of Theorem 2.16   We present here the proof of Theorem 2.16 for the case $n = 0$ to illustrate how Theorem 2.16 actually follows from Theorem 2.15. The case $n \geq 1$ is slightly more cumbersome, but follows along similar lines (see Exercise 10).

Let us first consider the Abelian direction. Accordingly, suppose that $\bar{F}(x) \sim \frac{1}{\Gamma(1-\alpha)} x^{-\alpha} L(x)$ for $\alpha \in (0, 1)$. Consider now the function $g(x) = \int_0^x \bar{F}(y) dy$, which has LST $\psi_g(s) = \frac{1 - \psi_X(s)}{s}$ (checking this claim is left as an exercise for the reader). From Karamata's theorem (Theorem 2.10), note that

$$g(x) \sim \frac{1}{(1 - \alpha)\Gamma(1 - \alpha)} x^{1-\alpha} L(x).$$

Since $1 - \alpha > 0$, we can now invoke Theorem 2.15 (specifically, the Abelian part) to conclude that

$$\psi_g(s) = \frac{1 - \psi_X(s)}{s} \sim \frac{\Gamma(2 - \alpha)}{(1 - \alpha)\Gamma(1 - \alpha)} s^{\alpha-1} L(1/s),$$

which in turn implies that $\psi_X(s) - 1 \sim -s^{\alpha} L(1/s)$ (note that $\Gamma(2-\alpha) = (1-\alpha)\Gamma(1-\alpha)$).

Next, consider the Tauberian direction, that is, suppose that $\psi_X(s) - 1 \sim -s^{\alpha} L(1/s)$, which is equivalent to $\psi_g(s) \sim s^{\alpha-1} L(1/s)$. Invoking Theorem 2.15 (specifically, the Tauberian part), we conclude that $g(x) \sim \frac{1}{\Gamma(2-\alpha)} x^{1-\alpha} L(x)$. Finally, the monotone density theorem (Theorem 2.11) implies that

$$\bar{F}(x) \sim \frac{1 - \alpha}{\Gamma(2 - \alpha)} x^{-\alpha} L(x) = \frac{1}{\Gamma(1 - \alpha)} x^{-\alpha} L(x).$$

This completes the proof.

The proof for general n follows along similar lines; the Abelian direction involves applying Karamata's theorem $n + 1$ times, while the Tauberian direction involves applying the monotone density theorem $n + 1$ times (see Exercise 10). □

Karamata's Tauberian theorem is only one of many Tauberian theorems that are useful when studying heavy-tailed distributions. In particular, given that it relies on the LST, the versions we have stated above are only relevant for distributions of nonnegative random variables. For other distributions, one needs Tauberian theorems for the characteristic function. An example of such a Tauberian theorem is the following, which is due to Pitman [176] (see also page 336 of [31]). Note that this Tauberian theorem uses only the real component of the characteristic function, $U_X(t)$, that is,

$$U_X(t) := \mathrm{Re}(\phi_X(t)) = \int_{-\infty}^{\infty} \cos(tx) dF(x).$$

Theorem 2.17 (Pitman's Tauberian Theorem)   *For slowly varying $L(x)$, and $\alpha \in (0, 2)$, the following are equivalent:*

$$\Pr\left(|X| > x\right) \sim x^{-\alpha} L(x) \text{ as } x \to \infty,$$

$$1 - U_X(t) \sim \frac{\pi}{2\Gamma(\alpha)\sin(\pi\alpha/2)} t^{\alpha} L(1/t) \text{ as } t \downarrow 0.$$

While there are many versions of Abelian and Tauberian theorems for the characteristic function, we choose to highlight this one because we make use of it later in the book in Chapter 5 when introducing and proving the generalized central limit theorem. Like Karamata's Tauberian theorem, this result connects the tail of the distribution to the behavior of a "transform" around zero, only in this case the "transform" considered is the characteristic function. Note that, because this Tauberian theorem applies to the tail of $|X|$ rather than X, it cannot be used to distinguish the behavior of the right and left tails of the distribution. Rather, it provides information only about the sum of the two tails. However, because of this fact, it deals only with the real part of the characteristic function, which makes it much simpler to work with analytically. The interested reader can find more general Abelian and Tauberian theorems in [31].

We have not focused on examples in this section; however, there are a number of illustrative examples of how to apply the theorems in this section scattered throughout the book. Two particularly important examples are in Chapter 5: we apply the Abelian part of Theorem 2.17 to prove the generalized central limit theorem, and the Tauberian part of Theorem 2.16 to study the return time of a one-dimensional random walk.

2.4 An Example: Closure Properties of Regularly Varying Distributions

Regularly varying distributions play a central role in this book, showing up in nearly every chapter. So, as you work through the book you will encounter a variety of applications of the properties and theorems discussed in the previous sections of this chapter. For example, regularly varying distributions play a foundational role in the generalized central limit theorem discussed in Chapter 5, the analysis of a multiplicative processes in Chapter 6, and the discussion of the extremal central limit theorem in Chapter 7.

Here, as a "warm-up" to those applications we provide some simple illustrations of the properties we have studied so far in order to prove some important *closure* properties about the set of regularly varying distributions. These closure properties, while intuitive, should not be taken for granted. In fact, these closure properties do not always hold for the more general classes of heavy-tailed distributions we study in the next two chapters.

Lemma 2.18   *Suppose that the random variables X and Y are independent, and regularly varying of index $-\alpha_X$ and $-\alpha_Y$ respectively.*

(i) $\min(X, Y)$ is regularly varying with index $-(\alpha_X + \alpha_Y)$.
(ii) $\max(X, Y)$ is regularly varying with index $-\min\{\alpha_X, \alpha_Y\}$.
(iii) $X + Y$ is regularly varying with index $-\min\{\alpha_X, \alpha_Y\}$. Moreover, $\Pr\left(X + Y > t\right) \sim \Pr\left(\max(X, Y) > t\right)$.

Lemma 2.18 shows that the class of regularly varying distributions is closed with respect to min, max, and convolution. These properties should be exactly what you should expect, given the intuition that regularly varying distributions are generalizations of Pareto distributions. For example, the convolution of two Pareto distributions does not yield another Pareto distribution, of course, but when one considers only the tail, the resulting convolution will certainly continue to have a tail that decays like a polynomial, and thus be regularly varying. A similar intuition holds for both the min and max of two Pareto random variables. Though the resulting distributions are certainly not Pareto distributions, they still have a tail that decays like a polynomial, and thus are regularly varying.

While simple and intuitive, these closure properties often turn out to be powerful. For example, the third property in Lemma 2.18 can be extended to the case of n i.i.d. regularly varying random variables $Y_i, i \geq 1$ easily, that is, $\Pr(Y_1 + Y_2 + \cdots + Y_n > t) \sim n \Pr(Y_1 > t)$ (see Exercise 7). This fact is used crucially in our analysis of random walks in Chapter 7, specifically in the proof of Theorem 7.6.

Proof of Lemma 2.18   We begin by using the representation of regularly varying distributions given by Theorem 2.8. Since X and Y are regularly varying, there exist slowly varying functions L_X and L_Y such that $\Pr(X > t) = t^{-\alpha_X} L_X(t)$ and $\Pr(Y > t) = t^{-\alpha_Y} L_Y(t)$. Now, using these representations, we can prove each closure property in turn.

(i)   Note that $\Pr(\min(X, Y) > t) = \Pr(X > t) \Pr(Y > t) = t^{-(\alpha_X + \alpha_Y)} L_X(t) L_Y(t)$. Since the product of slowly varying functions is also slowly varying, Claim *(i)* of the lemma follows.

(ii)   Since $\{\max(X, Y) > t\} = \{X > t\} \cup \{Y > t\}$, we have

$$\Pr(\max(X, Y) > t) = \Pr(X > t) + \Pr(Y > t) - \Pr(X > t) \Pr(Y > t). \quad (2.12)$$

Without loss of generality, we can consider the following cases separately: $\alpha_X < \alpha_Y$, and $\alpha_X = \alpha_Y$.

If $\alpha_X < \alpha_Y$, it follows from (2.12) that $\Pr(\max(X, Y) > t) \sim \Pr(X > t)$, which then implies that $\max(X, Y)$ is regularly varying with index $-\alpha_X$.

If $\alpha_X = \alpha_Y$, then it follows from (2.12) that $\Pr(\max(X, Y) > t) \sim \Pr(X > t) + \Pr(Y > t)$, that is, $\Pr(\max(X, Y) > t) \sim t^{-\alpha_X} (L_X(t) + L_Y(t))$. Since the sum of slowly varying functions is also slowly varying, it follows that $\max(X, Y)$ is regularly varying with index $-\alpha_X$.

This completes the proof of Claim *(ii)*.

(iii)   This final claim is the most involved. The first step in our proof is to establish an upper bound and a lower bound on the probability of the event $\{X + Y > t\}$. Then we analyze those bounds to obtain the result.

To begin, note that the event $\{X > t\} \cup \{Y > t\}$ implies $\{X + Y > t\}$. This gives us the following lower bound.

$$\Pr(X + Y > t) \geq \Pr(X > t) + \Pr(Y > t) - \Pr(X > t) \Pr(Y > t). \quad (2.13)$$

Next, let us fix $\delta \in (0, 1/2)$. It is easy to see that the event $\{X + Y > t\}$ implies the event $\{X > (1 - \delta)t\} \cup \{Y > (1 - \delta)t\} \cup \{X > \delta t, Y > \delta t\}$. This implication, along with the union bound, leads to the following upper bound.

$$\Pr\left(X+Y>t\right) \leq \Pr\left(X>(1-\delta)t\right) + \Pr\left(Y>(1-\delta)t\right) + \Pr\left(X>\delta t\right)\Pr\left(Y>\delta t\right).$$
$$(2.14)$$

Now, to complete the proof we consider the following two cases separately: $\alpha_X < \alpha_Y$, and $\alpha_X = \alpha_Y$.

Let us first consider the case $\alpha_X < \alpha_Y$. It follows from (2.13) that

$$\liminf_{t\to\infty} \frac{\Pr\left(X+Y>t\right)}{\Pr\left(X>t\right)} \geq 1.$$

Similarly, it follows from (2.14) that

$$\limsup_{t\to\infty} \frac{\Pr\left(X+Y>t\right)}{\Pr\left(X>t\right)} \leq \lim_{t\to\infty} \frac{\Pr\left(X>(1-\delta)t\right)}{\Pr\left(X>t\right)} = (1-\delta)^{-\alpha_X}.$$

Letting δ approach zero, we conclude that $\Pr\left(X+Y>t\right) \sim \Pr\left(X>t\right)$. This implies that $X+Y$ is regularly varying with index $-\alpha_X$, and also that $\Pr\left(X+Y>t\right) \sim \Pr\left(\max(X,Y)>t\right)$ (since we have established in the proof of Claim *(ii)* that $\Pr\left(\max(X,Y)>t\right) \sim \Pr\left(X>t\right)$).

Finally, we consider the case $\alpha_X = \alpha_Y$. In this case, using the same steps as above, it can be shown that

$$\Pr\left(X+Y>t\right) \sim \Pr\left(X>t\right) + \Pr\left(Y>t\right).$$

This, of course, implies that $X+Y$ is regularly varying with index $-\alpha_X$, and also that $\Pr\left(X+Y>t\right) \sim \Pr\left(\max(X,Y)>t\right)$ (we have established in the proof of Claim *(ii)* that $\Pr\left(\max(X,Y)>t\right) \sim \Pr\left(X>t\right) + \Pr\left(Y>t\right)$).

This completes the proof of Claim *(iii)*.

□

2.5 An Example: Branching Processes

Branching processes are a fundamental and widely applicable area of stochastic processes. While they were born from the study of surnames in genealogy, at this point they have found applications broadly in the study of reproduction, epidemiology, queueing theory, statistics, and many other areas. Here we use one of the first, and most famous branching process models – the Galton–Watson process – as an illustrative example of the power of the properties of regular variation that we have explored in this chapter.

Not only is the Galton–Watson model one of the most prominent examples of a branching process, it has an interesting story behind it. As the story goes, Victorian aristocrats were concerned about keeping their surnames from going extinct and wanted to understand how many children they needed to have to ensure the survival of their name. This prompted Sir Francis Galton to pose the following question in the *Educational Times* in 1873 [95]:

> *How many children (on average) must each generation of a family have in order for the family name to continue in perpetuity?*

Just a year later, Reverend Henry William Watson came up with the answer, and the two wrote a paper [215]. By now, the model named after them has become the canonical model of branching processes and has been used in wide-reaching areas from biology (see [11]), to the analysis of algorithms (see [65]), to the spread of epidemics (see, for example, [35, 164]).

The modern version of this model is defined formally as follows. In particular, a Galton–Watson process $\{X_n\}_{n \geq 0}$ is defined by

$$X_0 = 1,$$

$$X_{n+1} = \sum_{j=1}^{X_n} N_j^{(n+1)} \quad (n \geq 0),$$

where $N_j^{(n+1)}$ are i.i.d. random variables taking nonnegative integer values, distributed as N. In the Victorian context, N was interpreted as the number of male children (since the woman took the man's surname at marriage) in a family, and X_n as the number of men in the $n + 1$st generation. Given this model, the question asked by Victorian aristocrats can be studied by asking, given the distribution of N, will the process go on forever (i.e., $X_n > 0$ for all n) or will it go extinct (i.e., for some n_o, $X_n = 0$ for all $n \geq n_0$)? And, if it goes extinct, how many total distinct descendants (across all generations) would exist?

It turns out that the answers to these questions depend on the expected number of male children each person has, that is, $\mu := \mathbb{E}[N]$. It is not hard to see that the probability of extinction, η, is given by $\eta = \lim_{n \to \infty} \Pr(X_n = 0)$. The foundational theorem for Galton–Watson processes illustrates that there are three cases, depending on whether μ is greater than, less than, or equal to 1. Basically, to have a positive probability of avoiding extinction, the expected number of children of each generation needs to be strictly greater than one.

Theorem 2.19 *The probability of extinction, η, in a Galton–Watson branching process satisfies the following:*

(i)   Subcritical case: If $\mu < 1$ then $\eta = 1$.
(ii) Critical case: If $\mu = 1$ and N has positive variance, then $\eta = 1$.
(iii) Supercritical case: If $\mu > 1$, then $\eta \in (0, 1)$.

Note that in the subcritical case and the critical case, extinction is guaranteed (except in the trivial case where N equals 1 with probability 1). On the other hand, in the supercritical case, the lineage has a positive probability of surviving in perpetuity. Theorem 2.19 is classically proven using an approach based on probability generation functions; and we refer the interested reader to [104, Section 5.4] for the proof.

While the subcritical and critical cases are identical from the standpoint of extinction probability, they differ in terms of the distribution of the total number of distinct male descendants $Z := \sum_{n \geq 0} X_n$ as well as the time to extinction $\tau := \min\{n : X_n = 0\}$ (see [20, 103]). Here, our goal in studying branching processes is to illustrate the power of the Tauberian theorems we have introduced in this chapter; thus, we study the tail of Z in the critical case. Specifically, we prove the following result.

Theorem 2.20 *Suppose that $\mu = 1$ and that N has finite, positive variance. Then, the total number of distinct male descendants, Z, is regularly varying with*

$$\Pr(Z > t) \sim \frac{1}{\sqrt{\pi(\mathbb{E}[N^2] - \mathbb{E}[N])}} t^{-1/2}.$$

Before moving to the proof, it is important to notice that the total number of distinct male descendants has the following recursive structure:

$$Z \overset{d}{=} 1 + \sum_{i=1}^{N} Z_i, \qquad (2.15)$$

where $\{Z_i\}$ are i.i.d. random variables with the same distribution as Z and independent of N. To see this, think of N as the number of descendants of the "first" individual, and Z_i as the number of descendants of his ith child. That each Z_i has the same distribution as Z is referred to as the *branching property*. In the proof that follows, we exploit this branching property to characterize the tail of Z.

Proof of Theorem 2.20   Since we are going to apply a Tauberian theorem, the transform of Z is important. Here, we use the LST of Z, denoted by ψ_Z. Let $G_N(t) := \mathbb{E}\left[t^N\right] = \sum_{i=0}^{\infty} t^i \mathrm{Pr}\,(N = i)$ denote the *probability generating function* of N. It is now not hard to show, using (2.15), that ψ_Z satisfies the following functional equation:

$$\psi_Z(s) = e^{-s} G_N(\psi_Z(s)); \qquad (2.16)$$

see Exercise 13.

Analogously to the LST, the probability generating function admits the following Taylor expansion around $t = 1$:

$$G_N(t) = 1 + m_1(t-1) + m_2(t-1)^2(1+o(1)) \quad \text{as } t \uparrow 1,$$

where $m_1 = \mathbb{E}\,[N]$, $m_2 = \mathbb{E}\,[N^2] - \mathbb{E}\,[N]$. Given that N has finite, positive variance, $m_2 > 0$ (see Exercise 14) and using the fact that $m_1 = \mu = 1$ in the critical case that we are studying, the above expansion simplifies to

$$G_N(t) = t + m_2(t-1)^2(1+o(1)).$$

Now, we combine this with the functional equation for $\psi_Z(s)$ to obtain

$$\psi_Z(s) = e^{-s}[\psi_Z(s) + m_2(1 - \psi_Z(s))^2(1+o(1)) \quad \text{as } s \downarrow 0. \qquad (2.17)$$

Our goal is to apply Theorem 2.16, but we must first simplify the above expression. To do so, we use a few Taylor expansions. First, note that $e^{-s} = 1 - s(1+o(1))$. Similarly, we also know that $\psi_Z(s) = 1 - o(1)$, since the total size of Z is finite with probability 1 by Theorem 2.19. Now, if we first move the term $e^{-s}\psi_Z(s)$ to the left-hand side of (2.16) and then use the two expansions, we get

$$s = m_2(1 - \psi_Z(s))^2(1+o(1)) \quad \text{as } s \downarrow 0.$$

It follows that

$$1 - \psi_Z(s) \sim \sqrt{s/m_2} \quad \text{as } s \downarrow 0.$$

Finally, we are ready to apply Theorem 2.16 with $n = 0$, $\alpha = 1/2$, and $L(x) = 1/\sqrt{m_2}$. Using the identity $\Gamma(1/2) = \sqrt{\pi}$, we get

$$\mathrm{Pr}\,(Z > x) \sim \frac{1}{\sqrt{\pi m_2}} x^{-1/2} \quad \text{as } x \to \infty.$$

$\square$

2.6 Additional Notes

In the chapter we gave an overview of several properties of regularly varying *distributions*. While we did not focus much on regularly varying *functions*, the properties we described also apply more broadly. However, the interested reader can find much more on regularly varying *functions* in [31]. That book also contains results for sums that are discrete analogues of Section 2.3.1. That section is the only section in this chapter that assumes continuity.

While we described many analytic and closure properties of regularly varying distributions within this chapter, there are many other useful properties we did not have space to cover. For example, with respect to closure properties, an additional important property is the closure of products of random variables, for which we refer to one of the exercises at the end of the chapter; see also [63]. In addition, it can be shown that certain generalized inverses of regularly varying distributions are still regularly varying. On the analytical side, an important property that should be mentioned is the *uniform convergence theorem*, stating that the convergence of $L(at)/L(t)$ in the definition of slowly varying functions is necessarily uniform on any interval $a \in [g, d]$ for $0 < g < d < \infty$. For an overview of these and many other properties, we refer to the landmark monograph on regular variation [31].

We have focused on the most classical version of regularly varying distributions in this chapter, but it is important to be aware that there are several important extensions of regular variation, some of which will appear in later chapters. Two particularly useful extensions are (i) *intermediate regular variation* and (ii) *dominated variation*. A function f is of intermediate regular variation if

$$\lim_{\epsilon \downarrow 0} \limsup_{x \to \infty} \frac{f(x(1+\epsilon))}{f(x)} = \lim_{\epsilon \downarrow 0} \liminf_{x \to \infty} \frac{f(x(1+\epsilon))}{f(x)} = 1; \qquad (2.18)$$

and a function f is of dominated variation if

$$\limsup_{x \to \infty} \frac{f(xa)}{f(x)} < \infty, \liminf_{x \to \infty} \frac{f(xa)}{f(x)} > 0, \qquad (2.19)$$

for every $a > 0$.

It is straightforward to see that regularly varying functions of index $\neq 0$ satisfy both intermediate regular variation and dominated variation. In some particular situations, for example in queueing theory, the assumption of intermediate regular variation is the most general possible assumption for particular approximations of the tail to hold, or for particular proof methods to work; see [33, 165] for examples.

In statistical applications such as establishing asymptotic normality of estimators, it is convenient to consider a subclass of slowly varying functions, which is the class of second-order slowly varying functions. In particular, a function L is second-order slowly varying of index γ if there exists a function g that is regularly varying with index γ such that

$$\lim_{t \to \infty} \frac{\frac{L(tx)}{L(t)} - 1}{g(t)} = K \frac{x^\gamma - 1}{\gamma}. \qquad (2.20)$$

See, for example, Chapter 2 in [31].

As we have mentioned already, regular variation plays an important in queuing theory. In addition, the concept of regular variation is also paramount in financial and insurance mathematics [75], [17]. Another application area where the concept of scale-freeness and regular variation is important is that of complex networks, where the definition of power laws versus the more flexible class of regularly varying distributions sometimes seems to cause some confusion; see the discussion in [212]. For an introduction to the field of complex networks, see [23]. In Chapter 6 of this book, we come back to this application, when we discuss the mechanism of preferential attachment as a classical example of the emergence of heavy-tailed phenomena.

Regular variation will reappear at many other places in this book. A non-exhaustive list of examples is

- Chapter 3, where we use regularly varying distributions to investigate the behavior of random sums.
- Chapter 4, where we apply regularly varying distributions to study residual life and illustrate a connection between slowly varying functions and long-tailed distributions.
- Chapter 5, where we use Tauberian theorems to derive the generalized Central Limit Theorem.
- Chapter 6, where we use regular variation to understand variations of the multiplicative Central Limit Theorem.
- Chapter 7, where characterizing the classes of distributions that admit a limit law for their maxima rely on analytic tools of regular variation, as does the analysis of the all-time maximum of a random walk with negative drift.
- Chapters 8 and 9, where regular variation plays a key role in the development of statistic tools for estimating heavy-tailed distributions from data.

2.7 Exercises

1. Show that the following distributions are asymptotically scale invariant (i.e., regularly varying).
    (a) The Cauchy distribution.
    (b) The Burr distribution.
    (c) The Lévy distribution.
    The definitions of these distributions can be found in Section 1.2.4.
2. Consider a distribution F over $\mathbb{R}_+$ with finite mean μ. Recall that the *excess* distribution corresponding to F, denoted by F_e, is defined as

$$\bar{F}_e(x) = \frac{1}{\mu} \int_x^\infty \bar{F}(y)dy.$$

If $F \in \mathcal{RV}(-\alpha)$ for $\alpha > 1$, show that $F_e \in \mathcal{RV}(-(\alpha - 1))$. Specifically, show that

$$\bar{F}_e(x) \sim \frac{x}{\alpha - 1}\bar{F}(x).$$

3. Prove that the LogNormal distribution is *not* regularly varying.
4. Prove Lemma 2.7.
5. Prove Theorem 2.14.
6. Prove that if the function f satisfies $xf'(x) \sim \rho f(x)$, then $f \in \mathcal{RV}(\rho)$.
    Hint: Use the Karamata representation theorem (Theorem 2.12).

7. Suppose that $X_1, X_2, \ldots, X_n$ are i.i.d. regularly varying random variables with index $-\alpha$, where $n \geq 2$. Prove that

$$\Pr\left(X_1 + X_2 + \cdots + X_n > x\right) \sim n\Pr\left(X_1 > x\right).$$

8. Suppose that the random variables X and Y are independent, with $X \in \mathcal{RV}(-\alpha_X)$ and

$$\Pr\left(Y > t\right) = o(\Pr\left(X > t\right)) \quad (t \to \infty).$$

Prove that $X + Y \in \mathcal{RV}(-\alpha_X)$.

9. Suppose that the random variable $X \in \mathcal{RV}(-\alpha)$. Show that the same property holds for its integer part $\lfloor X \rfloor$.

10. Prove Theorem 2.16 for the case $n = 1$. Specifically, for a nonnegative random variable with finite mean, prove that for $\alpha \in (1, 2)$,

$$\bar{F}(x) \sim \frac{-1}{\Gamma(1-\alpha)} x^{-\alpha} L(x) \iff \psi_X(s) - 1 + \mathbb{E}\left[X\right] s \sim s^{\alpha} L(1/s).$$

Note: The above exercise should give the reader an idea of how to prove Theorem 2.16 for general n.

11. Let X be a nonnegative random variable of which the distribution function F is regularly varying with index $-\alpha$, and let Y be a random variable independent of X for which $E[Y^{\alpha+\epsilon}] < \infty$ for some $\epsilon > 0$. Show that

$$\frac{\Pr\left(XY > t\right)}{\Pr\left(X > t\right)} \to \mathbb{E}\left[Y^{\alpha}\right] \tag{2.21}$$

as $t \to \infty$. [Hint: condition on the value of Y and use the condition $E[Y^{\alpha+\epsilon}] < \infty$ and a useful property of slowly varying functions to justify the interchange of limit and integral.]

12. Suppose that $f(t)$ is regularly varying of index $\alpha > 0$. Define $f^{\leftarrow}(x) = \inf\{t : f(t) = x\}$. Prove that $f^{\leftarrow}(x)$ is regularly varying of index $1/\alpha$.

13. Suppose that $\{X_i\}_{i \geq 1}$ is a sequence of i.i.d. random variables. The random variable N takes nonnegative integer values, and is independent of $\{X_i\}_{i \geq 1}$. Let $G_N(\cdot)$ denote the probability generating function corresponding to N. Define $S_N = \sum_{i=1}^{N} X_i$. Prove that

$$\psi_{S_N}(s) = G_N(\psi_X(s)).$$

Here, $\psi_Y(\cdot)$ denotes the LST corresponding to random variable Y.

14. Suppose that the random variable N takes nonnegative integer values. If the variance of N is positive and finite, show that $\mathbb{E}\left[N^2\right] > \mathbb{E}\left[N\right]$.

3

Catastrophes, Conspiracies, and Subexponential Distributions

Suppose you are in a class with fifty students and the professor does an experiment. She records the height and the number of Twitter followers of every student in the class. It turns out that both the sum of the heights and the sum of the numbers of followers are unexpectedly large. The sum of the student heights comes to more than 310 feet, and the sum of the number of Twitter followers comes to just over half a million! The average height of a male in the US is 5 feet 9 inches, and the average number of Twitter followers is 700, so these two totals are quite surprising. Since the sums are significantly larger than what the averages would suggest, the professor asks the class:

What do you think led to the unexpectedly large sums?

Of course, there are many possible explanations, but they fall into two general categories: either a lot of students in the class have slightly larger values than average, or a few students in the class have extremely large values and everyone else is nearly average.

Let us first think about heights. In this case, the more intuitive of these explanations is clearly the first: that the sum is large because a lot of students are slightly taller than average. For example, maybe the basketball and volleyball teams are taking the class. It is certainly not very likely that the sum of the heights is large because a few students are 20-foot-tall giants!

However, in the case of Twitter followers, it is the other way around – the most likely explanation for the large sum is that one person in the class is a Twitter celebrity, and that person alone has nearly half a million followers.

The contrast between these two explanations gives us the title of this chapter: the most likely explanation for a large sum of heights is a "conspiracy" where many people are slightly taller than average and the combination leads to a large sum, while the most likely explanation for a large sum of Twitter followers is a "catastrophe" where one person has an extremely large number of followers. The difference between these explanations is quite striking and, of course, it is not limited to the comparison of heights and number of Twitter followers. The fundamental reason for this difference is that the distribution of heights is light-tailed, and the distribution of Twitter followers is heavy-tailed. Light-tailed distributions tend to follow a "conspiracy principle," while heavy-tailed distributions tend to follow a "catastrophe principle."

To drive the point home, consider a classical example of a heavy-tailed phenomenon – earthquakes. A priori, one might expect that if there are a lot of deaths due to earthquakes in a given year, it is most likely due to a large number of earthquakes. However, the catastrophe principle for heavy-tailed distributions tells us that we should expect something different. As in the case of Twitter followers, we should expect that the most likely reason that a year has an unexpectedly large number of deaths due to earthquakes is that one earthquake was extremely deadly. In fact, we should expect that one earthquake led to nearly all of the deaths during the year (i.e., an extreme catastrophe). To make this concrete, let us look back at deaths caused by earthquakes in recent years: 2010 and 2011 were particularly bad years for earthquakes. It is estimated that there were 320,627 deaths worldwide from earthquakes in 2010 and 22,053 deaths worldwide from earthquakes in 2011. Both of these death tolls are considerably larger than in any other year in the recent past [218, 219]. In both cases, just as the catastrophe principle predicts, the unusually large number of deaths is due to one catastrophic event that led to a large fraction of the deaths for the year. In 2010, a 7.0 magnitude earthquake in Lèogâne, Haiti led to an estimated 316,000 deaths (out of 320,627 deaths for the year worldwide), and in 2011, a 9.1 magnitude earthquake in Tōhoku, Japan led to an estimated 20,896 deaths (out of 22,053 deaths for the year worldwide).

This example highlights an interpretation of the catastrophe principle as a form of *Occam's razor* – the simplest explanation for a large sum is that one large event happened, not that a collection of many slightly larger than expected events conspired to make the sum large. Another simplistic but useful view of heavy-tailed distributions is that they are made of "many mice and a few elephants." That is, heavy-tailed distributions yield a lot of small samples (many mice), and when there are large samples, they are *very* large (a few elephants).[1] This view of heavy-tailed distributions provides a very tangible explanation of the catastrophe principle – if the sum is unexpectedly large, it is most likely because an elephant arrived. Further, since there are only a few elephants, it is unlikely that two elephants arrived, so the sum is probably dominated by a single elephant. This is sometimes referred to as "the principle of a single big jump," which we formalize in Section 3.4.

However, the catastrophe principle is completely contrary to what happens under the "typical" light-tailed distributions we learn about in introductory probability courses (e.g., the Gaussian and Exponential distributions). To understand a bit more about why the conspiracy principle often feels more intuitive, it is useful to think a bit more mathematically. In particular, an overly simplistic but still useful characterization of light-tailed distributions is that the samples never differ too much from the mean of the distribution (i.e., samples are concentrated around the mean). This means that all the samples from a light-tailed distribution are of a similar size, that is, on the same scale. The conspiracy principle is quite natural given this view of light-tailed distributions – since all samples are similar, any large sum is most likely to be a combination of many slightly larger than average samples rather than the result of one big outlier, which would be extremely unlikely.

While we have been casual in our description of conspiracies and catastrophes so far, these concepts are precise, formal properties. They are not only useful for intuitively reasoning

---

[1] The view of heavy-tailed distributions in terms of mice and elephants came out of the computer networks community, where it provided useful intuition for guiding protocol design, e.g., [76, 108, 170]. The perspective highlights that one often needs to handle small events in a different way than large events, or at least one needs to separate the mice from the elephants when designing systems. Given that computing workloads are typically heavy-tailed, this insight has played a key role in the design of routing and scheduling policies within computer networks and distributed systems, e.g., [32, 112, 191].

about heavy-tailed and light-tailed distributions but are also important analytic tools, as we show in this chapter. In particular, in this chapter we formally present the conspiracy and catastrophe principles and connect the catastrophe principle to a general class of heavy-tailed distributions termed *subexponential distributions*.

3.1 Conspiracies and Catastrophes

To this point we have only described the conspiracy and catastrophe principles informally. However, these principles can be made rigorous and can serve as powerful analytic tools when studying heavy-tailed and light-tailed distributions. It is important to note that there is not really one catastrophe principle and one conspiracy principle. Instead, there are many variations of these principles that can be defined and used, each with varying strengths and generality. In this section we introduce the simplest statements of each in order to highlight how these properties can be formalized. Later in the chapter, in Section 3.4, we give examples of variations of both catastrophe and conspiracy principles.

3.1.1 A Catastrophe Principle

As we have already described, the idea behind the catastrophe principle is that an unexpectedly large sum of random variables is likely the result of a catastrophe, that is, a result of one unexpectedly large event (sample). We can formalize this idea in terms of the tail of a sum of random variables. In particular, if a large sum is most likely the result of a single "catastrophic event," that means that the tail of the sum is *on the same order as* the tail of the maximum element in the sum. Formally, we state this property as follows.

Definition 3.1   A distribution F over the nonnegative reals is said to satisfy the catastrophe principle if, for $n \geq 2$ independent random variables $X_1, X_2, \ldots, X_n$ with distribution F,

$$\Pr\left(\max(X_1, X_2, \ldots, X_n) > t\right) \sim \Pr\left(X_1 + X_2 + \cdots + X_n > t\right) \quad \text{as } t \to \infty.$$

The catastrophe principle is a particularly powerful property because, a priori, there are many things that could have led to the sum being large, but the catastrophe principle specifies precisely how it happened. That is, a priori the sum could have been large because every sample was slightly bigger than average, but the catastrophe principle specifies that the sum is large because of exactly one very large sample.

This catastrophe principle can be rephrased in many different ways, and another appealing one is the following. Assuming that the random variables are nonnegative (i.e., $X_i \geq 0$), the catastrophe principle can be reformulated in terms of a conditional probability as follows:

$$\Pr\left(\max(X_1, \ldots, X_n) > t \mid X_1 + \cdots + X_n > t\right)$$
$$= \frac{\Pr\left(\max(X_1, \ldots, X_n) > t, \ X_1 + \cdots + X_n > t\right)}{\Pr\left(X_1 + \cdots + X_n > t\right)}$$
$$= \frac{\Pr\left(\max(X_1, \ldots, X_n) > t\right)}{\Pr\left(X_1 + \cdots + X_n > t\right)} \to 1 \text{ as } t \to \infty.$$

This shows that if the catastrophe principle holds, then the maximum of n samples is very likely to be bigger than t, given that the sum of n samples is bigger than t. So, the event that the sum is unusually large is, with very high probability, the result of precisely one large "catastrophic" event occurring.

The definition of this catastrophe principle is simple and general enough that it is satisfied by almost all common heavy-tailed distributions. For example, the class of regularly varying distributions satisfies this catastrophe principle. Indeed, Lemma 2.18 states that for i.i.d. regularly varying random variables X_1 and X_2, $\Pr\left(\max(X_1, X_2) > t\right) \sim \Pr\left(X_1 + X_2 > t\right)$. It is easy to see that the same argument extends to the case of n i.i.d. regularly varying random variables (see Exercise 1). Additionally, many other common heavy-tailed distributions satisfy this catastrophe principle; for example, the Weibull (with $\alpha < 1$) and the LogNormal distribution, although it is more difficult to verify these.[2]

The generality of this catastrophe principle has further led to the definition of a formal class of heavy-tailed distributions – the class of subexponential distributions, which contains the class of regularly varying distributions we studied in the previous chapter. We discuss the class of subexponential distributions in depth in Section 3.2.

Though most common heavy-tailed distributions satisfy this catastrophe principle, it is important to note that not all heavy-tailed distributions do. Thus, as is the case with scale-invariance, one needs to be careful to separate the catastrophe principle from the notion of heavy-tailed distributions as a whole. However, it is worth noting that distributions that are heavy-tailed but do not satisfy the catastrophe principle are rather pathological; we comment on constructing such examples in Chapter 4.

3.1.2 A Conspiracy Principle

In contrast to the catastrophe principle, the conspiracy principle aligns with our intuition about how unexpectedly large events happen – as the combination of a large number of factors. This makes the conspiracy principle more intuitive than the catastrophe principle for most people. However, it is also less powerful in many cases because it does not provide as "simple" an explanation for the rare event.

As in the case of the catastrophe principle, formalizing the notion of the conspiracy principle is most naturally done in terms of the tail of sums of random variables. In particular, the conspiracy principle implies that the tail of a sum of random variables *dominates* the tail of the maximum element in the sum.

Definition 3.2 A distribution F over the nonnegative reals is said to satisfy the conspiracy principle if, for $n \geq 2$ independent random variables $X_1, X_2, \ldots, X_n$ with distribution F,

$$\Pr\left(\max(X_1, X_2, \ldots, X_n) > t\right) = o(\Pr\left(X_1 + X_2 + \cdots + X_n > t\right)) \quad \text{as } t \to \infty.[3]$$

As in the case of the catastrophe principle, this definition can also be interpreted in terms of the conditional probability when the random variables are nonnegative (i.e., $X_i \geq 0$).

---

[2] We prove that the Weibull distribution satisfies the catastrophe principle in Definition 3.1 using properties of the class of subexponential distributions in Section 3.2. We leave the case of the LogNormal as Exercise 2.
[3] Recall that $f(t) = o(g(t))$ as $t \to \infty$ if $\lim_{t\to\infty} \frac{f(t)}{g(t)} = 0$.

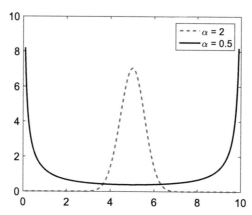

Figure 3.1 A visualization of the conspiracy and catastrophe principles under the Weibull distribution. The conditional density function of X_1 given that $X_1 + X_2 = 10$ is shown, where X_1 and X_2 are i.i.d. Weibull random variables that are either light-tailed ($\alpha = 2$) or heavy-tailed ($\alpha = 0.5$).

$$\Pr\left(\max(X_1, \ldots, X_n) > t | X_1 + \cdots + X_n > t\right)$$
$$= \frac{\Pr\left(\max(X_1, \ldots, X_n) > t \cap X_1 + \cdots + X_n > t\right)}{\Pr\left(X_1 + \cdots + X_n > t\right)}$$
$$= \frac{\Pr\left(\max(X_1, \ldots, X_n) > t\right)}{\Pr\left(X_1 + \cdots + X_n > t\right)} \to 0 \text{ as } t \to \infty.$$

The above definition of a conspiracy principle is simple and broad enough that it is easy to show that it is satisfied by all common light-tailed distributions. For example, consider the exponential distribution. For simplicity, we just consider the case of two independent Exponential(μ) random variables X_1 and X_2. In this case we have $\Pr\left(X_1 + X_2 > t\right) = e^{-\mu t}\left(1 + \mu t\right)$. On the other hand, $\Pr\left(\max(X_1, X_2) > t\right) \sim 2e^{-\mu t}$. Therefore, $\Pr\left(\max(X_1, X_2) > t\right) = o(\Pr\left(X_1 + X_2 > t\right))$, which shows that the conspiracy principle holds. It is easy to extend this to the sum of n i.i.d. exponentials (see Exercise 4). Similarly, the Gaussian distribution and the Weibull distribution ($\alpha > 1$) can easily be shown to satisfy the conspiracy principle too (see Exercises 5 and 6).

In fact, the Weibull distribution is a particularly nice example to use to contrast the conspiracy principle with the catastrophe principle since, depending on α, it can satisfy either. Figure 3.1 illustrates the conspiracy and catastrophe principles under different Weibull distributions. For i.i.d. Weibull random variables X_1 and X_2, Figure 3.1 plots the conditional density function of X_1, given that $X_1 + X_2 = 10$, for different values of α. This sum is unexpectedly large since the $\mathbb{E}\left[X_1\right] = \mathbb{E}\left[X_2\right] = 1$. For $\alpha = 0.5$ (X_i are heavy-tailed), we see the catastrophe principle in action. Indeed, given that the sum of X_1 and X_2 is large, then with high probability, X_1 either contributes almost the entire sum or almost nothing. On the other hand, for $\alpha = 2$ (X_1 are light-tailed), we see that the density peaks at the center of its support, implying that with high probability, X_1 and X_2 contribute comparably to the sum.

An important insight from Figure 3.1 is that the definition of a conspiracy principle we have stated is, in fact, fairly weak – there is a much stronger notion of a "conspiracy" that is happening under the Weibull distribution with $\alpha > 1$. In particular, not only is $\Pr\left(\max(X_1, X_2) > t\right) = o(\Pr\left(X_1 + X_2 > t\right))$, but neither X_1 or X_2 is contributing much more than the other with respect to the sum. That is, both X_1 and X_2 are equal partners in the conspiracy. It turns out that this can be proven formally in the case of the Weibull as well as other light-tailed distributions, for example, the Gaussian (see Exercise 7). To highlight this point, the following result shows that, if a sum of two light-tailed Weibull random variables is unexpectedly large, then it is most likely that each of the random variables contributes equally to the sum (i.e., they have "conspired" to make the sum large). This is in stark contrast to the case of heavy-tailed Weibull random variables, where it is most likely that one of the random variables contributes most of the sum (i.e., a "catastrophe" occurs).

Proposition 3.3 *Suppose X_1 and X_2 are independent and identically distributed light-tailed Weibull random variables with shape parameter $\alpha > 1$. Then, for any $\delta \in (1/2, 1)$,*

$$\Pr\left(X_1 + X_2 > t,\ X_1 > \delta t\right) = o(\Pr\left(X_1 + X_2 > t\right)) \quad \text{as } t \to \infty.[4]$$

Proof Recall that the c.c.d.f. of X_i is given by $\Pr\left(X_i > t\right) = e^{-\lambda t^\alpha}$.

To begin, we establish a lower bound on $\Pr\left(X_1 + X_2 > t\right)$, since we need to compare this quantity in order to establish the lemma. To bound this quantity we consider one specific way in which the sum could be large – both X_1 and X_2 could have been bigger than $t/2$. Since there are many other ways the sum could have been large, this is a lower bound. This yields

$$\begin{aligned}
\Pr\left(X_1 + X_2 > t\right) &\geq \Pr\left(X_1 > t/2,\ X_2 > t/2\right) \\
&= \Pr\left(X_1 > t/2\right)\Pr\left(X_2 > t/2\right) \\
&= e^{-\lambda 2^{1-\alpha} t^\alpha},
\end{aligned} \tag{3.1}$$

where the second step follows from independence of X_1 and X_2.

Now, let us move to bounding $\Pr\left(X_1 + X_2 > t,\ X_1 > \delta t\right)$. We can rewrite this probability as

$$\Pr\left(X_1 + X_2 > t,\ X_1 > \delta t\right) = \Pr\left(X_1 > t\right) + \Pr\left(X_1 + X_2 > t,\ \delta t < X_1 \leq t\right).$$

Thus, to prove the lemma, it suffices to show that

$$\Pr\left(X_1 > t\right) = o(\Pr\left(X_1 + X_2 > t\right)), \tag{3.2}$$

and

$$\Pr\left(X_1 + X_2 > t,\ \delta t < X_1 \leq t\right) = o(\Pr\left(X_1 + X_2 > t\right)). \tag{3.3}$$

Our bound of $\Pr\left(X_1 + X_2 > t\right)$ in (3.1) is useful for both of these steps. In particular (3.2) follows easily from (3.1):

$$\frac{\Pr\left(X_1 > t\right)}{\Pr\left(X_1 + X_2 > t\right)} \leq \exp\{-\lambda(1 - 2^{1-\alpha})t^\alpha\},$$

which implies (3.2) since $\alpha > 1$.

---

[4] Interestingly, the exponential distribution ($\alpha = 1$), which satisfies the conspiracy principle of Definition 3.2, does not satisfy this stronger conspiracy principle; see Exercise 8.

All that remains is to prove (3.3). In this case we can directly compute the probability of interest:

$$
\Pr\left(X_1 + X_2 > t,\ \delta t < X_1 \leq t\right) = \int_{\delta t}^{t} f_{X_1}(x)\bar{F}_{X_2}(t - x)dx
$$

$$
= \int_{\delta t}^{t} \lambda\alpha x^{\alpha-1}e^{-\lambda x^{\alpha}}e^{-\lambda(t-x)^{\alpha}}dx
$$

$$
= \lambda\alpha \int_{\delta t}^{t} x^{\alpha-1}\exp\left\{-\lambda t^{\alpha}\left[\left(\frac{x}{t}\right)^{\alpha} + \left(1 - \frac{x}{t}\right)^{\alpha}\right]\right\}dx.
$$

The form in the last step is chosen in order to facilitate rewriting the equation using the change of variables $g(t) = t^{\alpha} + (1 - t)^{\alpha}$, where $g(t)$ is defined over the interval $[1/2, 1]$. It is easy to show that the function g is strictly increasing over $[1/2, 1]$, with $g(1/2) = 2^{1-\alpha}$. With this definition, the calculation yields

$$
\Pr\left(X_1 + X_2 > t,\ \delta t < X_1 \leq t\right) = \lambda\alpha \int_{\delta t}^{t} x^{\alpha-1}\exp\left\{-\lambda t^{\alpha}g(x/t)\right\}dx
$$

$$
\leq \lambda\alpha \exp\left\{-\lambda t^{\alpha}g(\delta)\right\} \int_{\delta t}^{t} x^{\alpha-1}dx
$$

$$
\leq \lambda \exp\left\{-\lambda t^{\alpha}g(\delta)\right\} t^{\alpha}.
$$

Finally, combining the above bound with (3.1), we obtain

$$
\frac{\Pr\left(X_1 + X_2 > t,\ \delta t < X_1 \leq t\right)}{\Pr\left(X_1 + X_2 > t\right)} \leq \lambda t^{\alpha} \exp\left\{-\lambda t^{\alpha}(g(\delta) - g(1/2))\right\},
$$

which implies (3.3) and completes the proof.   □

The above result provides both an example of how to apply the formal statement of the conspiracy principle in a proof *and* an example of how to formalize a stronger version of the conspiracy principle in the specific case of the Weibull distribution. Note that even stronger versions of the conspiracy principle exist for many light-tailed distributions. We discuss an example of one such variation in detail in Section 3.4.

A final remark about this first conspiracy principle is that, although all common light-tailed distributions satisfy the conspiracy principle, not all light-tailed distributions do (as was the case for the catastrophe principle and heavy-tailed distributions). To illustrate this, consider the following light-tailed distribution, created by mixing common heavy-tailed and light-tailed distributions. Let $X = \min(Y, Z)$, where Y and Z are independent, $Y \sim \text{Exponential}(\mu)$, and $Z \sim \text{Pareto}(x_m, \alpha)$ with $\alpha > 1$. It is not hard to show that X is light-tailed but does not satisfy the conspiracy principle (see Exercise 10).

3.2 Subexponential Distributions

Since our focus in this book is on heavy-tailed distributions, we will dwell a little longer on the catastrophe principle. The importance (and usefulness) of the catastrophe principle has led to the definition of a formal subclass of heavy-tailed distributions called "subexponential distributions" that correspond to those distributions for which a catastrophe principle holds, though the connection between the definition of subexponential distributions and the catastrophe principle will not be obvious initially. The most classical definition of the class of subexponential distributions is the following.

Definition 3.4 A distribution F with support $\mathbb{R}_+$ is subexponential ($F \in \mathcal{S}$) if, for all $n \geq 2$ independent random variables $X_1, X_2, \ldots, X_n$ with distribution F,

$$\Pr(X_1 + X_2 + \cdots + X_n > t) \sim n\Pr(X_1 > t),$$

which we can state more compactly as $\bar{F}^{n*}(t) \sim n\bar{F}(t)$.[5]

This classical definition does not immediately make it clear how subexponential distributions are related to the catastrophe principle. However, with a simple calculation it is easy to see that they are intimately related. In particular, with a little effort, it is possible to see that $n\Pr(X_1 > t)$ is asymptotically equivalent to the $\Pr(\max(X_1, \ldots, X_n) > t)$. To see this, we simply need to expand $\Pr(\max(X_1, \ldots, X_n) > t)$ as follows:

$$\lim_{t \to \infty} \frac{\Pr(\max(X_1, \ldots, X_n) > t)}{\Pr(X_1 > t)} = \lim_{t \to \infty} \frac{1 - (1 - \bar{F}(t))^n}{\bar{F}(t)}$$

$$= \lim_{t \to \infty} \frac{1 - (1 - n\bar{F}(t) + \binom{n}{2}\bar{F}(t)^2 - \cdots)}{\bar{F}(t)}$$

$$= \lim_{t \to \infty} \frac{n\bar{F}(t) + o(\bar{F}(t))}{\bar{F}(t)} = n.$$

The above calculation shows that the tail of the max of n random variables is proportional to n times the tail of a single random variable, that is,

$$\Pr(\max(X_1, \ldots, X_n) > t) \sim n\Pr(X_1 > t), \tag{3.4}$$

and so the definition of subexponential distributions *exactly* matches the catastrophe principle. We state this formally in the following lemma.

Lemma 3.5 *Consider $X_1, X_2, \ldots$ independent random variables with distribution F having support $\mathbb{R}_+$. The following statements are equivalent.*

(i) *F is subexponential, i.e., $\Pr(X_1 + X_2 + \cdots + X_n > t) \sim n\Pr(X_1 > t)$ for all $n \geq 2$.*

(ii) *F satisfies the catastrophe principle, i.e., $\Pr(X_1 + X_2 + \cdots + X_n > t) \sim \Pr(\max(X_1, X_2, \ldots, X_n) > t)$ for all $n \geq 2$.*

We have already pointed out that most common heavy-tailed distributions satisfy the catastrophe principle, and so a consequence of the above is that they are subexponential distributions. That is, the class of subexponential distributions includes all regularly varying distributions (including the Pareto), the LogNormal, the Weibull (with $\alpha < 1$), and many others; see Figure 3.2.

---

[5] Recall that $\bar{F}^{n*}(t) := \Pr(X_1 + X_2 + \cdots + X_n > t)$ denotes the c.c.d.f. of the sum of n i.i.d. random variables having distribution F.

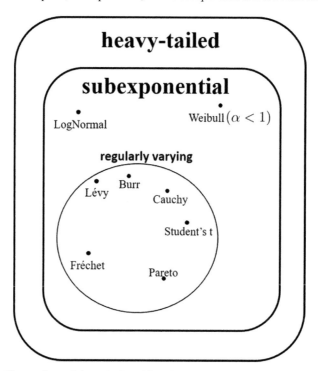

Figure 3.2 Illustration of the relationship of the class of subexponential distributions to other heavy-tailed distributions.

Though we have already proven that regularly varying distributions satisfy the catastrophe principle (see Lemma 2.18), we have not actually proven that the same is true for the LogNormal, the Weibull, or any other distribution. This is not an accident, it is not straightforward to prove inclusion of these distributions directly from the definitions of the class we have seen so far. In particular, though the definitions we have given so far have intuitive forms and provide useful structure for the analysis of sums of subexponential distributions, they do not provide an easy approach for proving that distributions are subexponential in the first place.

However, there are a number of more easily verifiable conditions that can be used to show that distributions are subexponential. The first such condition is yet another equivalent definition of the class of subexponential distributions. Specifically, it turns out that it is not required for the definition of subexponentiality to hold for *all* $n \geq 2$; if it holds for $n = 2$, then it necessarily holds for all $n \geq 2$. Further, if it can be shown that if it holds for *some* $n \geq 2$, it necessarily holds for $n = 2$ and, consequently, for all $n \geq 2$. This is all summarized in the following lemma.[6]

Lemma 3.6 *Consider* $X_1, X_2, \ldots$ *independent random variables with distribution* F *having support* $\mathbb{R}_+$. *The following statements are equivalent.*

---

[6] We state Lemma 3.6 only for the classical notion of subexponentiality $\Pr(X_1 + X_2 + \cdots + X_n > t)$ $\sim n\Pr(X_1 > t)$, but the parallel equivalence holds for $\Pr(X_1 + X_2 + \cdots + X_n > t) \sim$ $\Pr(\max(X_1, X_2, \ldots, X_n) > t)$ as well.

(i) F *is subexponential, i.e.,* $\Pr\left(X_1 + X_2 + \cdots + X_n > t\right) \sim n\Pr\left(X_1 > t\right)$ *for all* $n \geq 2$.

(ii) $\Pr\left(X_1 + X_2 > t\right) \sim 2\Pr\left(X_1 > t\right)$.

(iii) $\Pr\left(X_1 + X_2 + \cdots + X_n > t\right) \sim n\Pr\left(X_1 > t\right)$ *for some* $n \geq 2$.

We do not give the proof of this lemma here. That (i) and (ii) are equivalent was originally proved in [46] and is the goal of Exercise 13. A proof of the equivalence of (ii) and (iii) can be found in [75, Appendix A].

Clearly, Lemma 3.6 makes the task of verifying subexponentiality easier; however, it can still be difficult to work with. Often, the most natural approach for verifying subexponentiality comes through the use of the hazard rate. Recall that the hazard rate, aka the failure rate, of a distribution F with density f is defined as $q(x) = f(x)/\bar{F}(x)$. Heavy-tailed distributions often have a decreasing hazard rate. In fact, this is such an important property of heavy-tailed distributions that it is the focus of Chapter 4. However, for now, the role of the hazard rate is simply as a tool for checking subexponentiality – as long as the hazard rate decays to zero quickly enough, the distribution is subexponential.

Lemma 3.7 *Suppose that* $q(x)$ *is eventually decreasing, with* $q(x) \to 0$ *as* $x \to \infty$. *If* $\int_0^\infty e^{xq(x)} f(x)dx < \infty$, *then* X *is subexponential.*

This result is due to Pitman [177] and allows one to exploit a representation of the convolution in terms of hazard rates. The condition in the lemma makes a certain crucial interchange of limit and integration permissible. We do not go into the details, and instead refer to a proof which can be found in [177] or [75, Proposition A3.16].

The form of this condition is not particularly intuitive; however, it provides a clear approach for verifying that a distribution is subexponential. In particular, it is quite effective for showing that the Weibull (with $\alpha < 1$) and the LogNormal distributions are subexponential. For example, recall that the c.c.d.f. of the Weibull distribution is given by $\bar{F}(x) = e^{-\beta x^\alpha}$, and so its corresponding hazard rate is given by

$$q(x) = \frac{\alpha\beta x^{\alpha-1}e^{-\beta x^\alpha}}{e^{-\beta x^\alpha}} = \alpha\beta x^{\alpha-1}.$$

Clearly, this hazard rate is decreasing to zero monotonically. So, plugging this hazard rate into the condition of Lemma 3.7, we can simply calculate the integral to show that the Weibull with $\alpha < 1$ is subexponential.

$$\int_0^\infty e^{xq(x)} f(x)dx = \int_0^\infty e^{x\alpha\beta x^{\alpha-1}} \alpha\beta x^{\alpha-1} e^{-\beta x^\alpha}\, dx = \frac{1}{1-\alpha} = \int_0^\infty \beta e^{\alpha\beta y} e^{-\beta y}dy < \infty.$$

Note that if $\alpha \geq 1$, then the integral is infinite.

As one would expect from the name "subexponential," all subexponential distributions have a tail that is "sub" exponential, that is, their tail decays more slowly than an Exponential and thus the tail is heavier than an exponential. However, to this point, we have not actually proven that this is true. In fact, it is not obvious from any of the definitions we have given of subexponential distributions that they are *necessarily* heavy-tailed.

Of course, all subexponential distributions are indeed heavy-tailed, as the name suggests. However, proving this statement is not as easy as one might expect. The difficulty of the proof highlights some of the challenges of working with subexponential distributions. They are well-suited for analysis of random sums and extrema, where the definition provides useful analytic structure; however, they can be difficult to work with in general. To end this section, we prove the seemingly obvious statement that subexponential distributions are heavy-tailed and, in the process, demonstrate some common techniques for working with subexponential distributions.

Lemma 3.8   *Subexponential distributions are heavy-tailed.*

Proof   Our proof has two steps. First, we show that if a distribution $F \in \mathcal{S}$, then it satisfies the following property

$$\lim_{t \to \infty} \frac{\bar{F}(t - y)}{\bar{F}(t)} = 1 \qquad \forall \, y > 0. \tag{3.5}$$

While this property may seem mysterious now, it is actually the defining property of the class of long-tailed distributions, which are the focus of Chapter 4. However, that will not be important for us here. At this point, Condition (3.5) is important for us because it implies that F is heavy-tailed, and the second step in the proof is showing that implication.

To begin, suppose that the distribution $F \in \mathcal{S}$, and let X_1, X_2 be independent random variables with distribution F. Recall that $\bar{F}^{2*}(t) := \Pr(X_1 + X_2 > t)$. From the definition of subexponentiality we have $\lim_{t \to \infty} \frac{\bar{F}^{2*}(t)}{\bar{F}(t)} = 2$. Combining this with the fact that we can write

$$\frac{\bar{F}^{2*}(t)}{\bar{F}(t)} = \frac{\bar{F}(t) + \int_0^t \bar{F}(t - u) dF(u)}{\bar{F}(t)},$$

it follows that

$$\lim_{t \to \infty} \int_0^t \frac{\bar{F}(t - u)}{\bar{F}(t)} dF(u) = 1.$$

Let us denote the function in the above limit by $I(t)$. Fixing $y > 0$, we may bound $I(t)$ as follows:

$$I(t) = \int_0^y \frac{\bar{F}(t - u)}{\bar{F}(t)} dF(u) + \int_y^t \frac{\bar{F}(t - u)}{\bar{F}(t)} dF(u)$$

$$\geq \int_0^y dF(u) + \frac{\bar{F}(t - y)}{\bar{F}(t)} \int_y^t dF(u)$$

$$= F(y) + \frac{\bar{F}(t - y)}{\bar{F}(t)} (F(t) - F(y)).$$

This bound implies that

$$\frac{\bar{F}(t - y)}{\bar{F}(t)} \leq \frac{I(t) - F(y)}{F(t) - F(y)}.$$

Now, since $\lim_{t\to\infty} I(t) = 1$, we obtain

$$\limsup_{t\to\infty} \frac{\bar{F}(t-y)}{\bar{F}(t)} \leq 1.$$

Of course, since $\bar{F}(t-y) \geq \bar{F}(t)$, we also have

$$\liminf_{t\to\infty} \frac{\bar{F}(t-y)}{\bar{F}(t)} \geq 1.$$

This proves that F satisfies Condition (3.5), completing the first step of the proof.

We now move to the second step of the proof: proving that Condition (3.5) implies that F is heavy-tailed. To show this we will prove that F is heavy-tailed using Lemma 1.2, which says that it is enough to prove that $\liminf_{x\to\infty} -\frac{\log \Pr(X>x)}{x} = 0$.

To this end, define $\Psi(t) = -\log \Pr(X > t)$. It is easy to see that Condition (3.5) implies that $\lim_{t\to\infty} (\Psi(t) - \Psi(t-1)) = 0$. Therefore, for $\epsilon > 0$, there exists $t_0 > 0$ such that $\Psi(t) - \Psi(t-1) < \epsilon$ for all $t \geq t_0$. Since Ψ is a nondecreasing function, it follows that for $t \geq t_0$,

$$\Psi(t) \leq \Psi(t_0) + \lceil t - t_0 \rceil \epsilon,$$

which implies that $\limsup_{t\to\infty} \frac{\Psi(t)}{t} \leq \epsilon$. Letting $\epsilon \downarrow 0$, we get

$$\limsup_{t\to\infty} \frac{\Psi(t)}{t} \leq 0.$$

Moreover, since $\Psi(t) \geq 0$, we can conclude that

$$\lim_{t\to\infty} \frac{\Psi(t)}{t} = 0.$$

From Lemma 1.2, it follows now that F is heavy-tailed, which completes the proof. $\square$

3.3 An Example: Random sums

Given the overview of subexponential distributions provided in the previous section, the goal of this section is to provide an example of how properties of subexponential distributions can be powerful analytic tools, enabling analysis in a wide array of applications. Since the class of subexponential distributions serves as a formalization of the catastrophe principle, it is natural that it finds application most readily in settings that are fundamentally related to some form of random sum. Of course, such applications are common in domains such as finance, insurance, scheduling, and queueing (among many others).

In many such settings the core of the analysis relies on understanding a very simple process – a sum of a random number of independent and identically distributed random variables. For example, if one considers the number of deaths from earthquakes in a given year, which we discussed at the beginning of this chapter, it is reasonable to consider that there are a random number of earthquakes in a year and that the number of deaths from each one is independent and identically distributed. Of course, both the distribution of the number of earthquakes and the number of deaths from an earthquake are heavy-tailed [129, 175]. A similar model could capture the amount of money paid out in insurance claims in a given year, the amount of load arriving to a cloud service in a given day, and many other situations.

In this section, we illustrate the power of the class of subexponential distributions by considering a simple and classical model of random sums that underlies such situations. Specifically, suppose $\{X_i\}_{i \geq 1}$ is a sequence of independent and identically distributed random variables with mean $\mathbb{E}[X]$ and the random variable N takes values in $\mathbb{N}$ and is independent of $\{X_i\}_{i \geq 1}$. Our goal will be to characterize

$$S_N = \sum_{i=1}^{N} X_i.$$

You have likely studied the expectation of this random sum in an introductory probability course. In particular, you have likely heard of Wald's equation, which gives a simple and pleasing formula for $\mathbb{E}[S_N]$:

$$\mathbb{E}[S_N] = \mathbb{E}\left[\sum_{i=1}^{N} X_i\right] = \mathbb{E}[N]\,\mathbb{E}[X].$$

Wald's equation is particularly pleasing because it tells us that, with respect to the expectation, we can basically ignore the fact that N is random. That is, if we had just considered $E[S_n]$ for some fixed constant n, then $E[S_n] = nE[X]$, and Wald's equation simply replaces n with $E[N]$.

While Wald's equation is a very useful result, it is not always enough to have a characterization of the expectation S_N. We often want to understand the variance of S_N, or even the distribution of S_N. Luckily, it is not hard to generalize Wald's equation. For example, the variance of the random sum S_N still has a pleasing form:

$$\mathrm{Var}\left[\sum_{i=1}^{N} X_i\right] = \mathbb{E}[N]\,\mathrm{Var}[X] + (\mathbb{E}[X])^2\mathrm{Var}[N].$$

It is even possible to go much further than just the variance and to derive Wald-like inequalities for the tail of random sums. However, results about the tail of S_N are not as general as Wald's equation and rely on using particular properties of distributions. In fact, as we show below, the tail of random sums can behave very differently depending on whether X_i and/or N are heavy-tailed or light-tailed. It is in the derivation of these results that the class of subexponential distributions shows its value (as will the class of regularly varying distributions).

The Tail of Random Sums

Before talking about formal results characterizing the tail of random sums, it is useful to think about how we should expect that tail to behave. One natural suggestion is that we should expect something like we saw with Wald's equation – it should be "as if" the random variable N was simply a constant n. If this were the case, and if the X_i are subexponential, we would have a very simple equation for the tail of the sum:

$$\mathrm{Pr}\left(\sum_{i=1}^{n} X_i > t\right) \sim n\mathrm{Pr}\left(X_1 > t\right).$$

Thus, a natural guess is that this formula should hold for the tail of the random sum as well, by simply replacing n with $\mathbb{E}[N]$.

The above guess is indeed correct when N is light-tailed. In this case, since the X_i are heavy-tailed they "dominate" the behavior of the tail of the random sum, and so only the expectation of N plays a role. We formalize this in the following theorem.

Theorem 3.9 *Consider an infinite i.i.d. sequence of subexponential random variables $X_1, X_2, \ldots$, and a light-tailed random variable $N \in \mathbb{N}$ that is independent of $\{X_i\}_{i \geq 1}$. Then,*

$$\Pr\left(\sum_{i=1}^{N} X_i > t\right) \sim \mathbb{E}[N]\Pr(X_1 > t).$$

Proof The idea behind this proof is simple. We rewrite the tail of our random sum taking advantage of the definition of subexponential distributions. Denoting the distribution of X by F, conditioning on N gives

$$\Pr\left(\sum_{i=1}^{N} X_i > t\right) = \sum_{i \in \mathbb{N}} \Pr(N = i)\,\bar{F}^{i*}(t).$$

Looking at the limit we can rewrite this in a form where we can apply the definition of subexponential distributions as follows:

$$\lim_{t \to \infty} \frac{\Pr\left(\sum_{i=1}^{N} X_i > t\right)}{\Pr(X_1 > t)} = \lim_{t \to \infty} \sum_{i \in \mathbb{N}} \frac{\bar{F}^{i*}(t)}{\bar{F}_1(t)}\Pr(N = i).$$

Applying the definition of subexponential distributions, specifically that $\lim_{t \to \infty} \frac{\bar{F}^{i*}(t)}{\bar{F}_1(t)} = i$, gives the following, where we assume for the moment that we can interchange the order of the limit and the summation:

$$\lim_{t \to \infty} \sum_{i \in \mathbb{N}} \frac{\bar{F}^{i*}(t)}{\bar{F}_1(t)}\Pr(N = i) = \sum_{i \in \mathbb{N}} i\Pr(N = i) = \mathbb{E}[N].$$

All that remains is to justify the interchange of the limit and the summation. This can be done via an application of the dominated convergence theorem, which states that if we can find a t-independent upper bound β_i of $\frac{\bar{F}^{i*}(t)}{\bar{F}_1(t)}\Pr(N = i)$ such that $\sum_{i \in \mathbb{N}} \beta_i < \infty$, then the limit and the summation can be interchanged. The upper bound β_i can be obtained as follows. It can be proved that for any $\epsilon > 0$, there exists $K > 0$ such that $\frac{\bar{F}^{i*}(t)}{\bar{F}_1(t)} \leq K(1 + \epsilon)^i$ (see Exercise 11). Taking $\beta_i = K(1 + \epsilon)^i \Pr(N = i)$, we get $\sum_{i \in \mathbb{N}} \beta_i = \mathbb{E}[(1 + \epsilon)^N]$, which is finite for small enough ϵ since N is light-tailed.

$\square$

The characterization of random sums in Theorem 3.9 relies on the fact that the distribution of X_i is dominant, that is, X_i are heavy-tailed and N is light-tailed. In this case, we obtained a simple form for the tail of the random sum that follows the intuition we derived from Wald's equation. What remains now is to understand what happens when things are reversed and the distribution of N is dominant, that is, when N is heavy-tailed and X_i is light-tailed.

Intuitively, in this case, we should expect the tail of the random sum to be determined by the tail of N. To get intuition, let us consider what would happen if the X_i were deterministically x. In such a case, the behavior of the sum would be simple:

$$\Pr\left(\sum_{i=1}^{N} x > t\right) = \Pr\left(Nx > t\right) = \Pr\left(N > t/x\right).$$

Thus, one might guess that this formula should continue to hold for the tail of the random sum as well, simply by replacing x with $\mathbb{E}\left[X\right]$.

This guess is again correct. In particular, when X_i is light-tailed and N is heavy-tailed, specifically when N is regularly varying, we can characterize the random sum as follows. The proof of this result serves as a good example of the usefulness of the properties of regularly varying distributions that we discussed in Chapter 2.

Theorem 3.10   *If the X_i are light-tailed and N is regularly varying with index $-\alpha$ ($\alpha > 0$), then*

$$\Pr\left(\sum_{i=1}^{N} X_i > t\right) \sim \Pr\left(N > \frac{t}{\mathbb{E}\left[X\right]}\right).$$

Before we prove this result, it is important to point out that this characterization does not hold in general for heavy-tailed N. To get intuition as to why not, consider the case when N is deterministically n. In this case, the central limit theorem implies that for large n, $S_n \approx n\mathbb{E}\left[X_1\right] + \sqrt{n}Z$ for a Gaussian random variable Z. Thus, intuitively one could guess that

$$\Pr\left(S_N > t\right) \approx \Pr\left(N\mathbb{E}\left[X_1\right] + \sqrt{N}Z > t\right) \approx \Pr\left(N > t/\mathbb{E}\left[X_1\right] - O(\sqrt{t})\right). \quad (3.6)$$

This heuristic turns out to be correct, and it can be shown that, if the X_i are not deterministic, a necessary condition for $\Pr\left(\sum_{i=1}^{N} X_i > t\right) \sim \Pr\left(N > \frac{t}{\mathbb{E}[X]}\right)$ to hold is that

$$\Pr\left(N > x\right) \sim \Pr\left(N > x - \sqrt{x}\right). \quad (3.7)$$

This property is referred to as *square-root insensitivity* and plays a prominent role in the analysis of heavy-tailed distributions. Many common heavy-tailed distributions are square-root insensitive, including the Pareto, the LogNormal, and the Weibull (if $\alpha < 0.5$) (see Exercise 9). For a more in-depth discussion of square-root insensitivity and its applications, see [62, 155].

Proof of Theorem 3.10   To prove the theorem, we establish asymptotically matching upper and lower bounds on $\Pr\left(\sum_{i=1}^{N} X_i > t\right)$. In both cases we rely on properties of regularly varying distributions introduced in Chapter 2.

We start with the lower bound. To obtain a lower bound we characterize the likelihood that the sum is large as a result of N being large. The intuition behind this is that, since N is heavy-tailed, this is likely to be the dominant event. Specifically, let $\epsilon > 0$, and let $n(t) = \frac{t(1+\epsilon)}{\mathbb{E}[X_1]}$. We now condition on $N > n(t)$ as follows:

$$\Pr\left(\sum_{i=1}^{N} X_i > t\right) \geq \Pr\left(\sum_{i=1}^{N} X_i > t, \ N > n(t)\right)$$

$$\geq \Pr\left(\sum_{i=1}^{\lceil n(t)\rceil} X_i > t, \ N > n(t)\right).$$

In the second step we used the fact that the event $\{\sum_{i=1}^{\lceil n(t)\rceil} X_i > t, \ N > n(t)\}$ implies the event $\{\sum_{i=1}^{N} X_i > t, \ N > n(t)\}$ in order to eliminate N from the sum. This gives independence between the two events in the probability and allows us to simplify the expression as follows:

$$\Pr\left(\sum_{i=1}^{N} X_i > t\right) \geq \Pr\left(\sum_{i=1}^{\lceil n(t)\rceil} X_i > t\right) \Pr\left(N > n(t)\right)$$

$$\geq \Pr\left(\sum_{i=1}^{\lceil n(t)\rceil} X_i > \frac{\mathbb{E}\left[X_1\right]}{1+\epsilon} \lceil n(t)\rceil\right) \Pr\left(N > n(t)\right),$$

where the second step follows from the definition of $n(t)$. Now, the weak law of large numbers allows us to conclude that the first probability approaches 1 as $t \to \infty$. Therefore, we have

$$\liminf_{t\to\infty} \frac{\Pr\left(\sum_{i=1}^{N} X_i > t\right)}{\Pr\left(N > \frac{t}{\mathbb{E}[X_1]}\right)} \geq \lim_{t\to\infty} \frac{\Pr\left(N > \frac{t(1+\epsilon)}{\mathbb{E}[X_1]}\right)}{\Pr\left(N > \frac{t}{\mathbb{E}[X_1]}\right)} = (1+\epsilon)^{-\alpha},$$

where the last step follows from the fact that N is regularly varying with index $-\alpha$. Finally, letting $\epsilon \downarrow 0$ gives the desired lower bound

$$\liminf_{t\to\infty} \frac{\Pr\left(\sum_{i=1}^{N} X_i > t\right)}{\Pr\left(N > \frac{t}{\mathbb{E}[X_1]}\right)} \geq 1.$$

We now turn to the upper bound. We partition the event of interest based on the size of N. In particular, we pick $\epsilon \in (0,1)$ and let $m(t) = \frac{t(1-\epsilon)}{\mathbb{E}[X_1]}$. Then, we partition the event that the random sum exceeds t based on whether or not N exceeds $m(t)$:

$$\Pr\left(\sum_{i=1}^{N} X_i > t\right) = \Pr\left(\sum_{i=1}^{N} X_i > t, \ N > m(t)\right) + \Pr\left(\sum_{i=1}^{N} X_i > t, \ N \leq m(t)\right)$$

$$\leq \Pr\left(N > m(t)\right) + \Pr\left(\sum_{i=1}^{\lfloor m(t)\rfloor} X_i > t\right).$$

In the last step we have simplified the expression by using the fact that the event $\{\sum_{i=1}^{N} X_i > t, \ N \leq m(t)\}$ implies the event $\{\sum_{i=1}^{\lfloor m(t)\rfloor} X_i > t\}$. Next, plugging in the definition of $m(t)$ yields

$$\Pr\left(\sum_{i=1}^{N} X_i > t\right) \leq \Pr\left(N > m(t)\right) + \Pr\left(\sum_{i=1}^{\lfloor m(t)\rfloor} X_i > \frac{\mathbb{E}\left[X_1\right]}{1-\epsilon} \lfloor m(t)\rfloor\right).$$

Now, to bound the second term we apply a Chernoff bound. Specifically, we use the following result: given light-tailed i.i.d. random variables $\{Y_i\}_{i\geq 1}$ having positive mean, for $\alpha > \mathbb{E}\left[Y_1\right]$, there exists $\phi > 0$ such that $\Pr\left(Y_1 + Y_2 + \cdots + Y_n > n\alpha\right) \leq e^{-n\phi}$ for all $n \geq 1$. Applying this result to $\{X_i\}_{i\geq 1}$, we conclude that there exists a positive constant ϕ such that

$$\Pr\left(\sum_{i=1}^{\lfloor m(t)\rfloor} X_i > \frac{\mathbb{E}\left[X_1\right]}{1-\epsilon}\lfloor m(t)\rfloor\right) \leq e^{-\phi\lfloor m(t)\rfloor} \leq e^{-\phi(m(t)-1)} = ce^{-\mu t},$$

where $c = e^{\phi}$ and $\mu = \frac{\phi(1-\epsilon)}{\mathbb{E}[X_1]}$. Therefore, we have

$$\frac{\Pr\left(\sum_{i=1}^{N} X_i > t\right)}{\Pr\left(N > \frac{t}{\mathbb{E}[X_1]}\right)} \leq \frac{\Pr\left(N > \frac{t(1-\epsilon)}{\mathbb{E}[X_1]}\right)}{\Pr\left(N > \frac{t}{\mathbb{E}[X_1]}\right)} + \frac{ce^{-\mu t}}{\Pr\left(N > \frac{t}{\mathbb{E}[X_1]}\right)}.$$

Since N is regularly varying, its tail decays approximately like a polynomial, and thus it is easy to see that

$$\lim_{t\to\infty} \frac{ce^{-\mu t}}{\Pr\left(N > \frac{t}{\mathbb{E}[X_1]}\right)} = 0,$$

which implies that

$$\limsup_{t\to\infty} \frac{\Pr\left(\sum_{i=1}^{N} X_i > t\right)}{\Pr\left(N > \frac{t}{\mathbb{E}[X_1]}\right)} \leq \lim_{t\to\infty} \frac{\Pr\left(N > \frac{t(1-\epsilon)}{\mathbb{E}[X_1]}\right)}{\Pr\left(N > \frac{t}{\mathbb{E}[X_1]}\right)}$$
$$= (1-\epsilon)^{-\alpha},$$

where the last step follows from the fact that N is regularly varying with index $-\alpha$. Letting $\epsilon \downarrow 0$, we get our desired upper bound

$$\limsup_{t\to\infty} \frac{\Pr\left(\sum_{i=1}^{N} X_i > t\right)}{\Pr\left(N > \frac{t}{\mathbb{E}[X_1]}\right)} \leq 1.$$

$\square$

3.4 An Example: Conspiracies and Catastrophes in Random Walks

In order to introduce the catastrophe principle and the conspiracy principle as fundamental properties of many heavy-tailed and light-tailed distributions, we have focused on simple, intuitive forms of these properties up until this point. The goal of this section is to show that more precise versions of these principles can be obtained, at the cost of technical complexity. To illustrate this we highlight variations of both the catastrophe principle and the conspiracy principle in the context of *random walks*.

Random walks are one of the most classical examples of a stochastic process and have found application in remarkably diverse settings, including finance, computer science, physics, biology, and more. In this section, we study a generic random walk of the following form:

$$S_n = X_1 + X_2 + \cdots + X_n,$$

for i.i.d. X_i with mean μ. Here, S_n is the position of a walker who is taking steps of size X_i at each stage.

Most typically, analyses of random walks focus on the expected behavior of the random walk; however, that is not our goal here. The power of the conspiracy and catastrophe principle comes from providing explanations for rare events, situations where an unexpectedly large deviation from the "normal" behavior happens. Thus, the goal of the results here is to understand the large deviations of random walks, that is, $\Pr(S_n > t)$ for large t. But, it is not enough to consider a fixed t since what qualifies as a large deviation should depend on n. Thus, more precisely, in this section we study

$$\Pr(S_n > an) \text{ as } n \to \infty \text{ for } a > \mu.$$

Events where $S_n > an$ for $a > \mu$ truly are "large" in the sense that they cannot be characterized by the central limit theorem, which focuses on "small" deviations, on the order of $\sqrt{n}$, around the mean μn. This means that understanding $\Pr(S_n > an)$ requires understanding *rare events*. To this end, the results we present in this section both characterize how likely such rare events are *and* what leads to such rare events.

In particular, in the setting where X_i are heavy-tailed (specifically regularly varying), we strengthen the catastrophe principle to provide the "principle of a single big jump," which has been applied to scheduling, queueing, insurance, and finance. In the setting where X_i are light-tailed, we strengthen the conspiracy principle to obtain Cramér's theorem, which is a fundamental concentration inequality in the theory of large deviations whose applications cannot be overstated. Cramér's theorem can be thought of as a third fundamental limit theorem, alongside the law of large numbers and the central limit theorem.

Of course, these are only two such examples of stronger catastrophe and conspiracy principles, and there are many others available, depending on the assumptions and requirements of the desired setting. We discuss some other extensions in the additional notes at the end of the chapter.

3.4.1 The Principle of a Single Big Jump

The principle of a single big jump is a variation of the catastrophe principle that adapts the ideas we have discussed so far to the setting of random walks. To begin, let us consider what insight we can already obtain for random walks, given the catastrophe principle in Definition 3.1. If the steps of the walk, X_i, are subexponential, the catastrophe principle already gives us a powerful insight into the behavior of the random walk. In particular, if n is fixed, then

$$\frac{\Pr(S_n > t)}{n\Pr(X_1 > t)} \to 1, \text{ as } t \to \infty. \tag{3.8}$$

The intuition behind this statement should be very familiar by this point in the chapter: it is highly likely that, when random walk has a large deviation from the mean, a single large X_i is responsible.

However, in the context of random walks, the statement in (3.8) is not completely satisfying since it considers only the case when n is fixed. When studying random walks, the goal is typically to understand the behavior of the random walk in the long run as n grows large. Thus, what we really want is a more powerful statement that bounds $\Pr(S_n > an)$ as $n \to \infty$ for $a > \mu$.

If you have internalized the ideas in this chapter, you, of course, expect the intuition from the catastrophe principle to still hold in this case. This intuition allows us to heuristically work out what we should expect a bound on $\Pr(S_n > an)$ to look like. In particular, if we expect that exactly one X_i will be large and the others will be approximately the expected value μ, we obtain

$$\Pr(S_n > an) \approx n\Pr(X_1 > (an - \mu(n-1)), X_2 + \cdots + X_n \approx \mu(n-1))$$
$$\approx n\Pr(X_1 > (a - \mu)n), \text{ for large } n.$$

The form suggested by this intuition turns out to hold, and is termed the "principle of a single big jump." Not only does the intuition bound the likelihood of such a rare event, it also suggests that a large deviation of a random walk is most likely the result of one, and not more than one, big jump in the walk. For example, the likelihood of two jumps of the order n would be of the order $(n\Pr(X_1 > n))^2$, which is of a smaller order of magnitude than $n\Pr(X_1 > (a - \mu)n)$.

In the rest of this section we make this rigorous and state and prove the principle of a single big jump formally. Note that the principle does not hold generally for subexponential distributions. We prove it here in the case of regularly varying distributions, though some generalizations can be found in [62].

Theorem 3.11 (The principle of a single big jump) *Suppose X_i are i.i.d. with mean μ and follow a regularly varying distribution with index $-\alpha$, where $\alpha > 1$. Then for $a > \mu$, the random walk $S_n = X_1 + \cdots + X_n$ satisfies*

$$\lim_{n\to\infty} \frac{\Pr(S_n > an)}{n\Pr(X_1 > (a - \mu)n)} = 1. \tag{3.9}$$

Proof   We prove this result by constructing matching upper and lower bounds. Each of the bounds provides important intuition for the result. In particular, we prove the lower bound by constructing a simple event that is sufficient to cause the event $\{S_n > an\}$ to happen, and at the same time has a probability large enough to match the desired asymptotic behavior. This event is exactly the one provided by the intuition behind the catastrophe principle – that a single big jump occurred. In contrast, the upper bound shows that, without a big jump, the random walk could not have been so large with such a high probability.

Throughout, without loss of generality we let $\mu = 0$ and $a > 0$. Otherwise, replace X_i with $X_i - \mu, i \geq 1$ and a with $a - \mu$.

Lower bound: Our first goal is to show that an asymptotic lower bound holds, that is, to show that

$$\liminf_{n\to\infty} \frac{\Pr(S_n > an)}{n\Pr(X_1 > an)} \geq 1. \tag{3.10}$$

To do this, we formalize the event that the random walk is large because of a single big jump. This is one way a large deviation could have occurred, and so it provides a lower bound on the probability. Formally, pick an auxiliary constant $\delta > a$ and observe that

$$\Pr\left(S_n > an\right) \geq \Pr\left(\cup_{i=1}^{n} B_i^n\right), \tag{3.11}$$

with $B_i^n = \{X_i > \delta n, \sum_{j=1, j \neq i}^{n} X_j > (a - \delta)n\}$. Observe that $\Pr\left(B_i^n\right)$ is constant in i for fixed n. Thus,

$$
\begin{aligned}
\Pr\left(S_n > an\right) &\geq \Pr\left(\cup_{i=1}^{n} B_i^n\right) \\
&\geq \sum_{i=1}^{n} \Pr\left(B_i^n\right) - \sum_{i,j:i\neq j} \Pr\left(B_i^n \cap B_j^n\right) \\
&= n\Pr\left(B_1^n\right) - \frac{n(n-1)}{2}\Pr\left(B_1^n \cap B_2^n\right).
\end{aligned}
$$

Since

$$\Pr\left(B_1^n \cap B_2^n\right) \leq P(X_1 > \delta n; X_2 > \delta n) = P(X_1 > \delta n)^2,$$

we see that, by decomposing $\Pr\left(B_1\right)$,

$$\Pr\left(S_n > an\right) \geq n\Pr\left(X_1 > \delta n\right)\Pr\left(C_1^n\right) - \frac{n(n-1)}{2}\Pr\left(X_1 > \delta n\right)^2, \tag{3.12}$$

with $C_1^n = \{\sum_{j=2}^{n} X_j > (a - \delta)n\}$. Since $a < \delta$,

$$\Pr\left(C_1^n\right) \to 1$$

by the weak law of large numbers. In addition, we see that

$$\lim_{n\to\infty} \frac{n(n-1)\Pr\left(X_1 > \delta n\right)^2}{n\Pr\left(X_1 > an\right)} = 0, \tag{3.13}$$

since the numerator is regularly varying with index $2 - 2\alpha$ and the denominator is regularly varying with index $1 - \alpha$. Combining the last three displays we obtain

$$\liminf_{n\to\infty} \frac{\Pr\left(S_n > an\right)}{n\Pr\left(X_1 > an\right)} \geq \liminf_{n\to\infty} \frac{n\Pr\left(X_1 > \delta n\right)}{n\Pr\left(X_1 > an\right)} = (\delta/a)^{-\alpha}.$$

As this conclusion holds for any $\delta > a$, (3.10) follows by having $\delta \downarrow a$.

Upper bound: We now turn to the corresponding asymptotic upper bound, that is, we aim to show that

$$\limsup_{n\to\infty} \frac{\Pr\left(S_n > an\right)}{n\Pr\left(X_1 > an\right)} \leq 1. \tag{3.14}$$

Let us denote $A_n = \{S_n > an\}$. To prove (3.14), we have to show that the event A_n is most likely caused by a single big jump. Our strategy will be to partition A_n into two events, one in which a single jump contributes significantly to the sum S_n, and another in which no single jump contributes significantly to S_n. The crux of the proof is then to show that the latter event has an asymptotically negligible probability relative to the former.

Let $\tau \in (0, a)$. We partition A_n as follows:

$$\Pr\left(A_n\right) = \Pr\left(A_n, \max_i X_i > \tau n\right) + \Pr\left(A_n, \max_i X_i \leq \tau n\right)$$
$$=: \Pr\left(A_{1,n}\right) + \Pr\left(A_{2,n}\right).$$

We now deal with both terms separately.

The first term is easy to handle. Take $\delta \in (\tau, a)$ and observe that, since $\delta > \tau$,

$$\Pr\left(A_{1,n}\right) = \Pr\left(A_n, \max_i X_i > \delta n\right) + \Pr\left(A_n, \tau n < \max_i X_i \leq \delta n\right)$$

$$\leq \Pr\left(\cup_{i=1}^n \{X_i > \delta n\}\right) + \Pr\left(\cup_{i=1}^n \{X_i > \tau n, \sum_{j=1, j\neq i}^n X_j > (a-\delta)n\}\right)$$

$$\leq n\Pr\left(X_1 > \delta n\right) + n\Pr\left(X_1 > \tau n\right)\Pr\left(\sum_{j=2}^n X_j > n(a-\delta)\right).$$

By the weak law of large numbers we see that $\Pr\left(\sum_{j=2}^n X_j > n(a-\delta)\right) \to 0$. Since $\Pr\left(X_1 > \tau n\right)/\Pr\left(X_1 > an\right)$ stays bounded in n for any $\tau > 0$, we have

$$\limsup_{n\to\infty} \frac{\Pr\left(A_{1n}\right)}{n\Pr\left(X_1 > an\right)} \leq (\delta/a)^{-\alpha}. \tag{3.15}$$

This can be made arbitrarily close to 1 by having δ approach a.

Thus, the proof of (3.14) is complete once we show there exists a $\tau \in (0, a)$ such that

$$\lim_{n\to\infty} \frac{\Pr\left(A_{2,n}\right)}{n\Pr\left(X_1 > an\right)} = 0. \tag{3.16}$$

Note that

$$\Pr\left(A_{2n}\right) \leq \Pr\left(S_n > an \mid X_i \leq \tau n, \, i \leq n\right).$$

Thus, it suffices to have an upper bound on the probability that the sum of n i.i.d. random variables exceeds a threshold, given that each random variable is bounded from above by a different threshold. Such a bound is provided by the following *concentration inequality*, which dates back to Prokhorov [180]. Let $Y_i, i \geq 1$ be an i.i.d. sequence such that $E[Y_i] = 0$ and $\Pr\left(Y_i \leq c\right) = 1$ for some $c \in (0, \infty)$. For any $t > 0$ and $n \geq 1$,

$$\Pr\left(Y_1 + \cdots + Y_n > t\right) \leq \left(\frac{ct}{n\text{Var}(Y_1)}\right)^{-t/2c}. \tag{3.17}$$

Note that the random variables Y_i in the above inequality are bounded from above by c. We apply the concentration inequality to $\Pr\left(A_{2n}\right)$ as follows. Let $X_i^*, i \geq 1$ be an i.i.d. sequence such that

$$\Pr\left(X_1^* \leq y\right) = \Pr\left(X_i \leq y \mid X_i \leq \tau n\right).$$

Observe that the random variables X_i^* are bounded from above by τn, and $\mu^* = E[X_1^*] \leq E[X_1] = 0$. Therefore,

$$\Pr\left(A_{2n}\right) \leq \Pr\left(S_n > an \mid X_i \leq \tau n, i \leq n\right) = \Pr\left(\sum_{i=1}^n X_i^* > an\right).$$

We can now apply (3.17) with $Y_i = X_i - \mu^*$, $c = \tau n$, $t = an$ to obtain

$$\Pr\left(\sum_{i=1}^{n} X_i^* > an\right) \leq \Pr\left(\sum_{i=1}^{n} Y_i > an\right) \leq \left(\frac{\tau an}{\mathrm{Var}(Y_1)}\right)^{-a/2\tau}. \qquad (3.18)$$

To simplify bounding $Var(Y_1)$, we assume that $\alpha > 2$. In this case, $Var(Y_1) \leq \mathbb{E}\left[(X_1^*)^2\right] \leq \mathbb{E}\left[X_1^2\right] < \infty$. Thus, if we choose τ small enough to satisfy $a/2\tau > \alpha - 1$, we arrive at the desired conclusion (3.16), completing the proof.

The case $\alpha \in (1,2]$ requires a more delicate analysis, since $Var(Y_1)$ can grow with n in this case; see [230] for the details. $\qquad \square$

3.4.2 Cramér's Theorem

Cramér's theorem strengthens and refines the conspiracy principle for random walks in a parallel way to how the principle of a single big jump strengthens and refines the catastrophe principle. As we have already highlighted, the conspiracy principle in Definition 3.2 is too general to provide a precise characterization. Cramér's theorem provides a much more powerful characterization.

Like the principle of a single big jump, Cramér's theorem provides a bound on $\Pr(S_n > an)$ though, of course, the bound has a very different form since it applies to light-tailed distributions instead of heavy-tailed distributions. Cramér's theorem is tightly connected to concentration inequalities, specifically the Chernoff bound. In fact, a crisp statement of Cramér's theorem is that it proves that the Chernoff bound is tight. Thus, to motivate the form of Cramér's theorem we must start with the Chernoff bound.

The Chernoff bound itself is best viewed in the context of another fundamental result – Markov's inequality. Markov's inequality is the simplest and most fundamental concentration inequality. It provides a bound on the tail of a probability distribution in terms of the mean of the distribution. Specifically, Markov's inequality states that

$$\Pr(X \geq a) \leq \frac{\mathbb{E}[X]}{a},$$

for any nonnegative random variable X and constant $a > 0$. Though simple, Markov's inequality is extremely powerful and is the building block for many other more sophisticated concentration inequalities, such as the Chernoff bound. In fact, the Chernoff bound is a simple extension of Markov's inequality that focuses on obtaining a tighter bound by applying Markov's inequality to e^{sX} instead of X. In the case of the random walk S_n this gives

$$\Pr(S_n \geq an) = \Pr\left(e^{sS_n} \geq e^{ans}\right) \leq \frac{\mathbb{E}\left[e^{sS_n}\right]}{e^{ans}}.$$

Since $S_n = X_1 + \cdots + X_n$, this can be expanded further to yield

$$\Pr(S_n \geq an) \leq e^{-san}\mathbb{E}\left[e^{sX_1}\cdots e^{sX_n}\right] = e^{-ans}\left(\mathbb{E}\left[e^{sX_i}\right]\right)^n.$$

Moving the expectation into the exponent then yields

$$\Pr(S_n \geq an) \leq e^{-n(as-\log\mathbb{E}[e^{sX_i}])}.$$

Finally, since the bound holds for all $s \geq 0$, we can optimize over s. This gives a Chernoff bound for S_n:

$$\Pr\left(S_n \geq an\right) \leq e^{-n \sup_{s \geq 0} \left(as - \log \mathbb{E}\left[e^{sX_i}\right]\right)}. \tag{3.19}$$

The key to understanding why this Chernoff bound is so much more powerful than Markov's inequality is to look at what information about the distribution is considered by each. Markov's inequality uses only the mean to determine a bound, but the transformation to the Chernoff bound brings the whole of the moment generating function into the bound. This means that the sup in (3.19) chooses an optimal bound that depends on the whole distribution, thus yielding a much tighter bound on the tail.

Chernoff's bound has proven to be a powerful tool across a wide variety of applications. However, it only provides insight into the likelihood of rare events; it provides no information about what the rare events "look" like. This is because Chernoff's bound provides only an *upper bound* on $\Pr\left(S_n \geq an\right)$. This is enough if all that you want to do is show that rare events are unlikely; but often it is also important to understand the cause of rare events. To get insight into what leads to the rare events requires understanding the events that lead to a tight *lower bound*, as in the case of the principle of a single big jump.

Cramér's theorem provides this insight or, more specifically, the proof of Cramér's theorem does. Specifically, Cramér's theorem shows that the Chernoff bound is tight, and the proof provides insight into what events led to the rare event. The intuition is crisp and serves as a refinement of the conspiracy principle in Definition 3.2. In particular, the proof highlights that during rare events, it is as if each X_i is sampled i.i.d. from a slightly different distribution – a "twisted" distribution – which has mean slightly larger than that of the original distribution, just large enough to make the previously rare event become likely. Thus, the X_i truly conspired together to create the rare event.

Theorem 3.12 (Cramér's theorem)   *Consider* $S_n = X_1 + \cdots + X_n$, *where* X_i *are i.i.d. and light-tailed. Then for* $a > \mathbb{E}\left[X_1\right]$,

$$\lim_{n \to \infty} \frac{-\log \Pr\left(S_n > an\right)}{n} = \sup_{s > 0}[as - \log \mathbb{E}\left[e^{sX_i}\right]]. \tag{3.20}$$

Proof   Chernoff's bound already implies the upper bound on the tail in the theorem, that is, that

$$\liminf_{n \to \infty} \frac{-\log \Pr\left(S_n > an\right)}{n} \geq \sup_{s > 0}[as - \log \mathbb{E}\left[e^{sX_1}\right]].$$

In fact, Chernoff's bound implies that this inequality holds for every finite n.

Thus, we can focus on the lower bound on the tail. We assume, for convenience, that X_i has a density f. Additionally, to eliminate some technical complexities, we make the simplifying assumption that the moment generating function $\mathbb{E}\left[e^{sX_1}\right]$ is "steep." Specifically, setting $\bar{s} = \sup\{s : \mathbb{E}\left[e^{sX_1}\right] < \infty\}$, we assume that $\lim_{s \uparrow \bar{s}} \mathbb{E}\left[e^{sX_1}\right] = \infty$. Under this assumption the optimization problem $\sup_{s>0}[as - \log \mathbb{E}\left[e^{sX_1}\right]]$ has a unique solution s_a for which $s_a < \bar{s}$. This means that s_a solves the first-order optimality condition

$$a = \frac{\mathbb{E}\left[X_1 e^{s_a X_1}\right]}{\mathbb{E}\left[e^{s_a X_1}\right]}.$$

The key to the proof is the "twisted" distribution with density function $\tilde{f}(x_i) = \frac{e^{s_a x_i} f(x_i)}{\mathbb{E}\left[e^{s_a X_1}\right]}$, which characterizes the distribution from which X_i are sampled in the rare event. This twisted distribution can be shown to have mean a. To see this, let $\tilde{X}$ denote a random variable with the twisted density $\tilde{f}(\cdot)$. Note that $\mathbb{E}\left[e^{s\tilde{X}}\right] = \mathbb{E}\left[e^{(s_a+s)X_1}\right]/\mathbb{E}\left[e^{s_a X_1}\right]$. This implies that

$$\mathbb{E}\left[\tilde{X}\right] = \frac{d}{ds}\mathbb{E}\left[e^{s\tilde{X}_i}\right]\big|_{s=0} = \frac{\mathbb{E}\left[X_1 e^{s_a X_1}\right]}{\mathbb{E}\left[e^{s_a X_1}\right]} = a.$$

Let $\tilde{X}_i$ denote an i.i.d. sequence of random variables following the twisted distribution. We bound the desired probability as follows. To begin, note that

$$\Pr\left(S_n > an\right) = \int_{(x_i):x_1+\cdots+x_n>an} \prod_{i=1}^{n}[f(x_i)dx_i].$$

The main step is to rewrite the above integral in terms of the twisted densities as follows:

$$\Pr\left(S_n > an\right) = \int_{(x_i):\, x_1+\cdots+x_n>an} e^{-s_a \sum_i x_i} \prod_{i=1}^{n}[e^{s_a x_i} f(x_i)dx_i]$$

$$= \mathbb{E}\left[e^{s_a X_1}\right]^n \int_{(x_i):\, x_1+\cdots+x_n>an} e^{-s_a \sum_i x_i} \prod_{i=1}^{n}\left[\tilde{f}(x_i)dx_i\right]$$

$$\geq \mathbb{E}\left[e^{s_a X_1}\right]^n \int_{(x_i):\, an+\sqrt{n}>x_1+\cdots+x_n>an} e^{-s_a \sum_i x_i} \prod_{i=1}^{n}\left[\tilde{f}(x_i)dx_i\right]$$

$$\geq e^{-s_a(an+\sqrt{n})}\mathbb{E}\left[e^{s_a X_1}\right]^n \int_{(x_i):\, an+\sqrt{n}>x_1+\cdots+x_n>an} \prod_{i=1}^{n}\left[\tilde{f}(x_i)dx_i\right]$$

$$= e^{-s_a(an+\sqrt{n})}\mathbb{E}\left[e^{s_a X_1}\right]^n \Pr\left(an < \tilde{X}_1 + \tilde{X}_2 + \cdots + \tilde{X}_n < an + \sqrt{n}\right).$$

$$(3.21)$$

Next, we take the log and look at the limit to complete the proof. In doing so, the first two terms remain, but the third term disappears. To show this, we apply the central limit theorem to obtain

$$\Pr\left(an < \tilde{X}_1 + \tilde{X}_2 + \cdots + \tilde{X}_n < an + \sqrt{n}\right) \to \Pr\left(0 < N < 1\right), \qquad (3.22)$$

where N is a Gaussian random variable with mean zero and variance equal to $\mathrm{Var}(\tilde{X}_i)$. Note that this application requires that $\mathrm{Var}(\tilde{X}_i)$ is finite. To verify this, note that since $s_a < \bar{s}$, the moment generating function of $\tilde{X}_i$ is finite in a neighborhood of the origin, which implies the finiteness of the variance of $\tilde{X}_i$.

Now, taking the log and looking at the limit of (3.21) gives

$$\limsup_{n\to\infty} \frac{-\log \Pr\left(S_n > an\right)}{n} \le s_a a - \log \mathbb{E}\left[e^{s_a X_1}\right]$$

$$= \sup_{s>0}[sa - \log \mathbb{E}\left[e^{s X_1}\right]],$$

which is the desired lower bound. □

3.5 Additional Notes

This chapter has introduced the basics of the class of subexponential distributions, and their connection to the catastrophe and conspiracy principles. Our treatment only scratches the surface of these topics, and so we point the interested reader to more in-depth treatments of each below.

Subexponential distributions: The class of subexponential distributions was introduced in Chistyakov [46] in 1964 and originally found applications to problems in branching processes [21]. Its significance for applications in insurance and queueing theory was recognized in [167, 207, 210]. Readers in search of more technical details about the class of properties and examples of modern applications can find excellent surveys in [75, 89, 101].

Subclasses of subexponential distributions have also proven interesting and useful. In particular, technical issues with the full class of subexponential distributions have led to many variations. For example, the fact that the class of subexponential distributions is not closed under addition (i.e., if X and Y are independent nonnegative subexponential random variables, then $X + Y$ may not be subexponential, see [138]) led to the important subclass termed S^*, introduced by [127]. A distribution function F of a nonnegative random variable X belongs to S^* if it has finite mean μ and satisfies

$$\lim_{x\to\infty} \frac{\int_0^x \bar{F}(x-y)\bar{F}(y)dy}{\mu\bar{F}(x)} = 2. \tag{3.23}$$

Essentially, this entails that the density $\bar{F}/\mu$ can be seen as a subexponential distribution. This property implies that F as well as $\int_0^x F(y)dy/\mu$ are subexponential and that the class is closed under addition. Other subclasses of subexponential distributions, for example, strong subexponential distributions and local subexponential distributions, have received attention too and are discussed in detail in [89].

Another important, related class of distributions is the class of *subexponentially concave* distributions, which are distributions for which $-\log \bar{F}$ is concave. The motivation for this class is the existence of powerful concentration inequalities (e.g., see [121] and the survey of Sergey Nagaev [159]). Note that this class of random variables is particularly useful when extending Theorems 3.9 and 3.10 from the regularly varying case toward more general subexponential distributions. The most general properties for Theorem 3.10 can be found in [62, 155]. These references also have discussions on extensions of square-root insensitivity and its connection with extreme value theory, which is connected to the material we discuss in Chapter 7.

Stronger conspiracy and catastrophe principles: The bulk of the chapter focused on simple, general versions of conspiracy and catastrophe principles. We gave examples of stronger versions of these principles in Section 3.4 (Cramér's Theorem and the principle of a single big jump); however, there are many other such generalizations.

In 1938, Cramer [51] developed the result that eventually was named after him. The initial work was motivated by a problem in insurance; however, after that, versions of his theorem found applications in a wide variety of fields, such as statistics, information theory, and communication networks. For an overview of these applications, as well as extensions of Cramér's theorem, we refer interested readers to the literature on large deviations. A concise introduction to both theory and applications of large deviations is [61]. Monographs that are focused on applications in communication and computer networks are [38, 96, 195]. Other valuable sources with more technical presentations are [60, 64, 72, 83].

Though the phrase itself has become popular relatively recently, the principle of a single big jump appeared first in [46, 158]. Variations of the principle are many and varied and can be found in [19, 22, 87, 90, 226, 229]. These are often motivated by various applications in insurance mathematics and queueing theory as well as applications to communication networks [88, 113, 232].

One interesting variation that we use later in this book is the following. In the version we presented here (Theorem 3.11) it was shown that the limit

$$\frac{\Pr(S_n > t)}{n\Pr(X_1 > t)} \to 1, \text{ as } t \to \infty \qquad (3.24)$$

holds if n is growing proportional with t. The result can actually be extended to cases where n is growing faster than t. To be precise, it can be shown (see [155]) for $\alpha > 2$ that there exists a constant $C > 0$ such that

$$\lim_{n \to \infty} \sup_{t:t>C\sqrt{n \log n}} \left|\frac{\Pr(S_n > t)}{n\Pr(X_1 > t)} - 1\right| = 0. \qquad (3.25)$$

Another important set of variations are extensions to cases where there is more than one big jump. In particular, in some cases it is likely that there are multiple big jumps, and recent research has focused on developing a generic mathematical framework to handle such extensions (e.g., [142, 187]).

3.6 Exercises

(The exercises marked *** are particularly challenging.)

1. Prove that regularly varying random variables are subexponential. Specifically, in Chapter 2, we proved that for i.i.d. regularly varying random variables X_1 and X_2, $\Pr(X_1 + X_2 > t) \sim \Pr(\max(X_1, X_2) > t)$ (see Lemma 2.18). Your task is to extend the same argument to n random variables, that is, prove that

$$\Pr(X_1 + X_2 + \cdots + X_n > t) \sim \Pr(\max(X_1, X_2, \ldots, X_n) > t),$$

where $n \geq 2$, and $X_1, X_2, \ldots, X_n$ are i.i.d., regularly varying random variables.

2. Prove that the LogNormal distribution is subexponential.
   Hint: One approach is to use Lemma 3.7 in conjunction with the statement of Exercise 12 below. You may also find the following asymptotic approximation of the tail of the standard Gaussian useful: For a standard Gaussian random variable N,

$$\Pr\left(N > t\right) \sim \frac{1}{\sqrt{2\pi}} \frac{e^{-t^2/2}}{t} \quad \textit{as } t \to \infty. \tag{3.26}$$

3. Consider a distribution F with support $[0, \infty)$ satisfying

$$\bar{F}(x) \sim e^{-cx/\log(x)} \quad \text{as } x \to \infty.$$

   Prove that F is subexponential.
4. Prove that the exponential distribution satisfies the conspiracy principle in Definition 3.2.
5. Prove that the light-tailed Weibull distribution (with shape parameter $\alpha > 1$) satisfies the conspiracy principle in Definition 3.2.
   Hint: One approach is to develop a simple lower bound on $\Pr\left(X_1 + X_2 + \cdots + X_n > t\right)$, *where $X_1, X_2, \ldots, X_n$ are i.i.d. Weibull with $\alpha > 1$.*
6. Prove that the Gaussian distribution satisfies the conspiracy principle in Definition 3.2.
   Hint: You may find the asymptotic approximation (3.26) of the tail of the standard Gaussian useful.
7. Prove the following stronger conspiracy principle for the standard Gaussian (similar to the result of Proposition 3.3 for the light-tailed Weibull). Given i.i.d. standard Gaussian random variables X_1 and X_2, and $\delta \in (1/2, 1)$, prove that

$$\Pr\left(X_1 + X_2 > t, X_1 > \delta t\right) = o(\Pr\left(X_1 + X_2 > t\right)) \quad \text{as } t \to \infty.$$

8. Prove that the exponential distribution does not satisfy the stronger conspiracy principle in the statement of Proposition 3.3. Specifically, prove that, if X_1, X_2 i.i.d. exponential random variables and $\delta \in (0, 1)$,

$$\lim_{t \to \infty} \Pr\left(X_1 > \delta t \mid X_1 + X_2 > t\right) = 1 - \delta.$$

9. A distribution with support over $[0, \infty)$ is said to be *square-root insensitive* if

$$\bar{F}(x) \sim \bar{F}(x - \sqrt{x}) \quad \text{as } x \to \infty.$$

   Prove that the following distributions are square-root insensitive:
   a. regularly varying distributions;
   b. the LogNormal distribution;
   c. the Weibull distribution with shape parameter $\alpha < 1/2$.
10. Let $X = \min(Y, Z)$, where Y and Z are independent, $Y \sim \text{Exponential}(\mu)$, and $Z \sim \text{Pareto}(x_m, \alpha)$ with $\alpha > 1$.
    a. Prove that X is light-tailed.
    b. Prove that X does not satisfy the conspiracy principle in Definition 3.2. Specifically, prove that

$$\limsup_{x \to \infty} \frac{\bar{F}^{2*}(x)}{\bar{F}(x)} < \infty,$$

    where F denotes the distribution of X.

11. *** Your task in this exercise is to prove a non-asymptotic bound on the ratio of $\bar{F}^{n*}(x)$ to $\bar{F}(x)$ for subexponential F called *Kesten's bound*. This bound is a technical result that is often useful in proofs; for example, it is used in the proof of Theorem 3.9.
    Suppose that F is the distribution function corresponding to a nonnegative subexponential random variable. Prove that for any $\epsilon > 0$, there exists $K > 0$ such that for any $n \geq 2$ and $x \geq 0$,

$$\frac{\bar{F}^{*n}(x)}{\bar{F}(x)} \leq K(1+\epsilon)^n.$$

Hint: Let $\alpha_n := \sup_{x \geq 0} \frac{\bar{F}^{*n}(x)}{\bar{F}(x)}$. Prove that for large enough T,

$$\alpha_{n+1} \leq 1 + \frac{1}{\bar{F}(T)} + \alpha_n(1+\epsilon).$$

From this, it follows inductively that

$$\alpha_n \leq \frac{1}{\epsilon}\left(1 + \frac{1}{\bar{F}(T)}\right)(1+\epsilon)^n,$$

which completes the proof taking $K = \frac{1}{\epsilon}\left(1 + \frac{1}{\bar{F}(T)}\right)$.

12. *** Prove that the class of subexponential distributions is closed under tail equivalence. Specifically, prove that if the distribution F is subexponential, and the distribution G satisfies

$$\bar{G}(x) \sim C\bar{F}(x)$$

for some $C > 0$ (i.e., G is tail-equivalent to F), then G is also subexponential.
    Hint: Let Y_1 and Y_2 denote two i.i.d. random variables with distribution G. For a constant but large enough v, partition the event $\{Y_1 + Y_2 > x\}$ as

$$\{Y_1 + Y_2 > x\} = \{Y_1 \leq v, \ Y_1 + Y_2 > x\} \cup \{Y_2 \leq v, \ Y_1 + Y_2 > x\}$$
$$\cup \ \{v < Y_2 \leq x - v, \ Y_2 + Y_2 > x\} \cup \{Y_1 > v, \ Y_2 > x - v\}$$

so that

$$\frac{\bar{G}^{*2}(x)}{\bar{G}(x)} = \frac{2}{\bar{G}(x)}\int_0^v \bar{G}(x-t)dG(t) + \frac{1}{\bar{G}(x)}\int_v^{x-v} \bar{G}(x-t)dG(t) + \frac{\bar{G}(v)\bar{G}(x-v)}{\bar{G}(x)}$$
$$=: T_1 + T_2 + T_3.$$

Use tail equivalence and the fact that $\bar{F}(x) \sim \bar{F}(x-y)$ for constant y (see the proof of Lemma 3.8) to argue that

$$\lim_{x \to \infty} T_1 = 2\bar{G}(v), \quad \lim_{x \to \infty} T_3 = \bar{G}(v).$$

Next, and this is the tricky part of the proof, use tail equivalence and the subexponentiality of F to show that $\limsup_{x \to \infty} T_2$ can be made arbitrarily small for large enough v.

13. *** Prove the equivalence of statements (i) and (ii) of Lemma 3.6. Specifically, let $\{X_i\}_{i \geq 1}$ denote a sequence of i.i.d. subexponential random variables. For $n \geq 2$, recall that the defining property of subexponential distributions is

$$\Pr\left(X_1 + X_2 + \cdots + X_n > t\right) \sim n\Pr\left(X_1 > t\right) \quad \text{as } t \to \infty.$$

Prove that, if the above property holds for $n = 2$, then it holds for all $n \geq 2$.

Hint: Of course, this result is proved inductively. Note that it is easy to see that

$$\liminf_{t \to \infty} \frac{\Pr\left(X_1 + X_2 + \cdots + X_n > t\right)}{\Pr\left(X_1 > t\right)}$$

$$\geq \liminf_{t \to \infty} \frac{\Pr\left(\max(X_1, X_2, \ldots, X_n) > t\right)}{\Pr\left(X_1 > t\right)} = n.$$

The challenge in the proof is therefore to show the upper bound, that is,

$$\limsup_{t \to \infty} \frac{\Pr\left(X_1 + X_2 + \cdots + X_n > t\right)}{\Pr\left(X_1 > t\right)} \leq n.$$

4

Residual Lives, Hazard Rates, and Long Tails

Over the course of our days we spend a lot of time waiting for things – we wait for tables at restaurants, we wait for a subway train or a bus to show up, we wait for people to respond to our emails. In these instances, we hold on to the belief that, as we wait, the likely length of the remaining wait is getting smaller. For example, we believe that, if we have waited ten minutes for a table at a restaurant, the expected time we have left to wait should be smaller than it was when we arrived and that, if we have waited five minutes for the subway, then our expected remaining wait time should be less than it was when we arrived.

In many cases this belief holds true. For example, as other diners finish eating, our expected waiting time for a table at a restaurant drops. Similarly, subway trains follow a schedule with (nearly) deterministic gaps between trains and thus, as long as the train is on schedule, our expected remaining waiting time decreases as we wait. However, a startling aspect of heavy-tailed distributions is that this is not always true. For example, if you have been waiting for a long time after the scheduled arrival time for a subway train, then it is very likely that there was some failure and the train may take an extremely long time to arrive, and so your expected remaining waiting time has actually increased while you waited. Similarly, if you are waiting for a response to an email and have not heard for a few days, it is likely to be a very long time until a response comes (if it ever does).

These examples highlight another fundamental distinction between light-tailed distributions (e.g., restaurant waiting times) and heavy-tailed distributions (e.g., email waiting times). To make the contrast even clearer, we can illustrate the same distinction using our classic examples of heavy-tailed and light-tailed distributions: wealth and heights. If we know someone is taller than 6 feet tall, then it is most likely that they are only a few inches taller; but if we know someone has more than $1 million, then it is much more likely that they are multimillionaires than that they are just barely millionaires.

All these examples suggest that, as with scale-invariance and the catastrophe principle, the behavior we expect to see is aligned with what happens under light-tailed distributions and so, upon first encounter, the behavior of heavy-tailed distributions is mysterious. We expect that if we have waited a long time, the remaining waiting time (i.e., the *residual life*) should have decreased, and so it is particularly jarring that under heavy-tailed distributions the residual life will likely have increased dramatically.

In this chapter we explore the residual life of heavy-tailed distributions in order to build intuition for the counterintuitive phenomena just described. To do this, we start by exploring the distribution of residual life via two common measures: the hazard rate function and the mean residual life function. We then study the relationship between heavy-tailed distributions and properties of the hazard rate and the mean residual life, which leads us to the formalization of a subclass of heavy-tailed distributions, termed *long-tailed distributions*, that we explore in depth.

4.1 Residual Lives and Hazard Rates

The foundation of this chapter is the concept of the residual life of a distribution. The term "residual life" refers to the remaining waiting time, given that you have already been waiting for some amount of time. Clearly, the residual life crucially depends on how long you have waited already, and so it is a conditional concept. Formally, we define the residual life distribution as follows.

Definition 4.1 For a nonnegative random variable X with distribution function F, the residual life distribution $R_x(t)$ equals

$$R_x(t) = 1 - \Pr\left(X > x + t | X > x\right) = 1 - \frac{\bar{F}(x+t)}{\bar{F}(x)} \qquad (\bar{F}(x) > 0,\ t \geq 0).$$

The complementary residual life distribution $\bar{R}_x(t)$ is defined as $\bar{R}_x(t) = 1 - R_x(t)$.

The residual life distribution $R_x(t)$ is the distribution of the waiting time, given that you have already waited for x time, or in different terminology the size of the *excess* beyond the threshold t, given that the random variable exceeds the threshold. The residual life distribution is a foundational concept that is found in widely varying applications, ranging from the insurance assessment and reliability theory to the social sciences, where it has found use in studying the lifetimes of everything from the length of wars to human life expectancies. In this book, we make use of properties of the residual life distribution when studying the emergence of heavy-tailed distributions under multiplicative and extremal processes in Chapters 6 and 7; and also in the design of statistical tools for estimating heavy-tailed phenomena in Chapter 9.

To begin to get a feel for the residual life distribution, let us consider some examples. Conveniently, it tends to be quite straightforward to calculate $\bar{R}_x(t)$ for common distributions. For example, $\bar{R}_x(t)$ for the Pareto distribution can be computed as follows:

$$\text{Pareto: } \bar{R}_x(t) = \frac{\left(\frac{x_m}{x+t}\right)^{\alpha}}{\left(\frac{x_m}{x}\right)^{\alpha}} = \left(1 + \frac{t}{x}\right)^{-\alpha}. \tag{4.1}$$

This is shown in Figure 4.1. Interestingly, for any x, the residual life distribution under the Pareto distribution, $\bar{R}_x(t)$, follows a Burr distribution. Thus, the residual life distribution has a regularly varying tail.

Similarly, the residual life distribution of the exponential can be calculated easily.

$$\text{Exponential: } \bar{R}_x(t) = \frac{e^{-\mu(x+t)}}{e^{-\mu x}} = e^{-\mu t}. \tag{4.2}$$

A striking aspect of Equation (4.2) is that $\bar{R}_x(t) = \bar{F}(t)$. This is a restatement of the "memoryless" property of the exponential distribution, which says that regardless of how long you have waited so far, the distribution of the remaining time you have to wait is exactly the

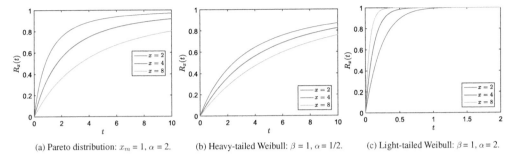

(a) Pareto distribution: $x_m = 1$, $\alpha = 2$. (b) Heavy-tailed Weibull: $\beta = 1$, $\alpha = 1/2$. (c) Light-tailed Weibull: $\beta = 1$, $\alpha = 2$.

Figure 4.1 Illustration of the residual life distribution for different choices of F. Note that when F is heavy-tailed (see (a) and (b)), the residual life distribution "grows" stochastically with increasing age x; whereas the residual life distribution "shrinks" stochastically with increasing x in the light-tailed example (see (c)).

same as if you just arrived. This is a particularly special property of exponential distributions, which are the only continuous distributions that are memoryless (see Exercise 1).

Finally, let us consider the residual life of the Weibull distribution. Again, it is simple to calculate.

$$\text{Weibull: } \bar{R}_x(t) = \frac{e^{-\beta(x+t)^\alpha}}{e^{-\beta x^\alpha}} = e^{-\beta[(x+t)^\alpha - x^\alpha]}. \quad (4.3)$$

However, though it is simple to calculate $\bar{R}_x(t)$ for the Weibull, the resulting form is not particularly informative. It is hard to get much insight from it directly. However, Figure 4.1 highlights the shape of the residual life distribution for both light-tailed and heavy-tailed Weibull distributions.

More generally, despite the fact that the residual life distribution is easy to derive for many distributions, it is often useful (or even necessary) to look at statistics of the residual life distribution in order to obtain insight into its behavior. There are two statistics that are most commonly used: the *mean residual life* and the *hazard rate*. These are the focus of the next two sections.

4.1.1 The Mean Residual Life

Whenever we consider a distribution, it is natural to use its mean in order to obtain insight. In this case, because residual life is a conditional concept that depends on how long you have waited so far, the mean of the residual life distribution is a function of the time you have waited. More specifically, the mean residual life function is defined as follows.

Definition 4.2 For a nonnegative random variable X with distribution function F, the mean residual life (MRL) function equals $m(x) = \mathbb{E}\left[X - x \mid X > x\right]$ (for $x \geq 0$ satisfying $\bar{F}(x) > 0$). Equivalently,

$$m(x) = \int_0^\infty \bar{R}_x(t)dt = \int_0^\infty \frac{\bar{F}(x+t)}{\bar{F}(x)}dt. \quad (4.4)$$

Note that the MRL function $m(x)$ is well defined over the interval $\{x \geq 0 \colon \bar{F}(x) > 0\}$. Over this interval, the MRL is bounded if and only if the distribution has finite mean.

The MRL function $m(x)$ appears far less frequently than the density function $f(x)$, the distribution function $F(x)$, or moment generating function $M(s)$; however, $m(x)$ also completely determines the distribution when the distribution has a finite mean. Thus, for example, it is possible to "invert" $m(x)$ to calculate the distribution function $F(x)$ [106]. Further, as we show in Chapter 9, the mean residual life also can be useful in statistical exploration of heavy-tailed phenomena.

It is typically not hard to compute $m(x)$ for common distributions. To illustrate the behavior of $m(x)$ let us return to the examples of the Pareto, Exponential, and Weibull distributions. For the Pareto distribution, it is quite straightforward to calculate $m(x)$. In particular, assuming that the mean is finite ($\alpha > 1$), we have

$$\text{Pareto: } m(x) = \int_0^\infty \bar{R}_x(t)dt = \int_0^\infty \left(1 + \frac{t}{x}\right)^{-\alpha} dt = \frac{x}{\alpha - 1}. \qquad (4.5)$$

Interestingly, from Equation (4.5) we see that the mean residual life of the Pareto distribution is increasing and grows unboundedly with x. In particular, under a Pareto distribution, the expected remaining waiting time grows linearly with the amount of time you have waited so far.

The calculation of the mean residual life is also straightforward for the Exponential distribution:

$$\text{Exponential: } m(x) = \int_0^\infty \bar{R}_x(t)dt = \frac{1}{\mu} \int_0^\infty \mu e^{-\mu t}dt = \frac{1}{\mu}.$$

This derivation highlights a consequence of the memoryless property of Exponential distributions – the mean residual life is constant with respect to x, specifically $m(x) = 1/\mu = \mathbb{E}[X]$. That is, the expected remaining waiting time is the same as when you first arrived, regardless of how long you have waited.

Though it is straightforward to calculate the mean residual life function $m(x)$ for the Pareto and the Exponential, it is not always so easy. In fact, it is difficult to derive an explicit formula for $m(x)$ for the Weibull distribution. However, we can numerically compute the mean residual life; see Figures 4.2(a)–(c). The figure shows that when $\alpha < 1$, the mean residual life is increasing, and thus one should expect the remaining waiting time to grow as you wait longer, but that the opposite is true when $\alpha > 1$. In that case, the mean residual life is decreasing, which means the expected remaining waiting time decreases as you wait.

4.1.2 The Hazard Rate

A second important statistic of the residual life distribution is the hazard rate. We have already mentioned the hazard rate a few times in the book, but here we introduce it formally and study it in detail, since it is fundamentally related to the residual life distribution.

The residual life distribution $\bar{R}_x(t)$ looks at the remaining waiting time given that you have already waited for a certain amount of time x, while the hazard rate can be thought of as the likelihood of the wait time ending now, given that you have waited x time already. Formally, the hazard rate is defined as follows.

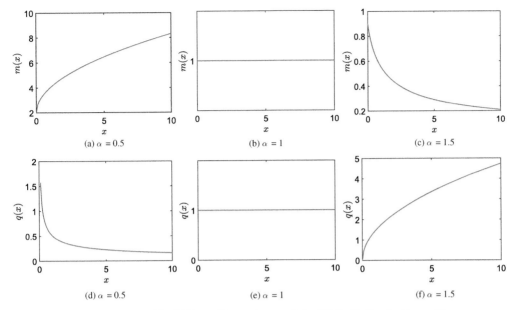

Figure 4.2 Mean residual life and hazard rate of the Weibull distribution with $\beta = 1$ and different values of α. Recall that $\alpha = 1$ corresponds to the Exponential distribution.

Definition 4.3   For a nonnegative random variable X with distribution function F and density function f, define the hazard rate, aka the failure rate, as $q(x) = f(x)/\bar{F}(x)$ (for $x \geq 0$ satisfying $\bar{F}(x) > 0$).[1] Further, define the cumulative hazard as $Q(y) = \int_0^y q(x)dx$.

It is easy to see that the hazard rate and the residual life are intimately related. The hazard rate is the density of the residual life distribution evaluated at zero.

$$R'_x(0) = \frac{d}{dt}\left(1 - \frac{\bar{F}(x+t)}{\bar{F}(x)}\right)\Big|_{t=0} = \frac{f(x+t)}{\bar{F}(x)}\Big|_{t=0} = \frac{f(x)}{\bar{F}(x)} = q(x).$$

Further, the hazard rate is intrinsically tied to the tail of the distribution. To see this, note that

$$q(t) = \frac{f(t)}{\bar{F}(t)} = -\frac{d}{dt}\log \bar{F}(t),$$

and so

$$Q(x) = -\log \bar{F}(x),$$

which gives

$$\bar{F}(x) = e^{-Q(x)} = e^{-\int_0^x q(t)dt}. \tag{4.6}$$

---

[1] Unlike the MRL, the hazard rate is only defined for continuous random variables. Moreover, the domain of the hazard rate function is the same as that of the residual life distribution and the MRL, i.e., $\{x \geq 0 : \bar{F}(x) > 0\}$.

As a consequence, it is easy to see that the hazard rate and the mean residual life are also closely related. In particular, as long as both exist, we have

$$m(x) = \int_0^\infty e^{-\int_x^{x+t} q(y)dy} dt, \qquad (4.7)$$

which implies that if the hazard rate is monotonically decreasing (increasing), then the mean residual life will be monotonically increasing (decreasing). Further, it is possible to show (see Exercise 3) that

$$m'(x) = m(x)q(x) - 1. \qquad (4.8)$$

The interested reader is referred to [106] and the references therein for more details.

To get a feeling for the behavior of the hazard rate, let us return again to our examples of the Pareto, Exponential, and Weibull distributions. Either by computing directly or by using the above relationships, it is straightforward to see that the hazard rates of the Pareto and the Exponential are as follows:

$$\text{Pareto: } q(t) = \frac{\frac{\alpha}{x_m}\left(\frac{x_m}{t}\right)^{\alpha+1}}{\left(\frac{x_m}{t}\right)^{\alpha}} = \frac{\alpha}{t},$$

$$\text{Exponential: } q(t) = \frac{\mu e^{-\mu t}}{e^{-\mu t}} = \mu.$$

Thus, the Pareto has a hazard rate that decreases to zero, while the Exponential has a constant hazard rate. This contrast is interesting: the memoryless property of the Exponential distribution means that the likelihood that your waiting time ends is unchanging as you wait, while under the Pareto distribution the likelihood that your waiting time ends decreases to zero as you wait.

It is also straightforward to compute the hazard rate of the Weibull distribution, which is in contrast to the difficulty of computing $m(x)$ in this case. In particular, the hazard rate of the Weibull is:

$$\text{Weibull: } q(t) = \frac{\alpha \beta t^{\alpha-1} e^{-\beta t^\alpha}}{e^{-\beta t^\alpha}} = \alpha \beta t^{\alpha-1}.$$

The form of the hazard rate under the Weibull distribution highlights a similar contrast to what we saw between the Pareto and the Exponential, only more extreme (see Figures 4.2 (d)–(f)). When $\alpha > 1$, the hazard rate is increasing (and thus the mean residual life is decreasing), which means that the likelihood your wait ends increases as you wait, while when $\alpha < 1$, the hazard rate is decreasing (and thus the mean residual life is increasing), similarly to that of the Pareto distribution. Of course, $\alpha = 1$ corresponds to the case of the Exponential distribution, and so the hazard rate is constant.

4.2 Heavy Tails and Residual Lives

The simple examples of the Pareto, Exponential, and Weibull that we have used so far in the chapter illustrate the contrast between light-tailed and heavy-tailed distributions that we have discussed informally in the introduction to the chapter: if we have waited a long time, then under light-tailed Weibull distributions the expected remaining waiting time will have decreased, while under the heavy-tailed Weibull and Pareto distributions the expected remaining waiting time will have increased dramatically.

In particular, we have seen that under the light-tailed Weibull the mean residual life is decreasing and the hazard rate is increasing, while under the heavy-tailed Weibull and Pareto distributions the mean residual life is increasing unboundedly and the hazard rate is decreasing to zero. The fact that these three distributions all have monotonic hazard rates and mean residual lives points us toward the importance of this property, and in particular, motivates the definition of the following four classes of distributions.

Definition 4.4 A nonnegative distribution F with mean residual life function m is said to have increasing/decreasing mean residual life (IMRL/DMRL) if $m(x)$ is increasing/decreasing in x for all x such that $F(x) \in (0, 1)$.[2]

Definition 4.5 A nonnegative distribution F with hazard rate q is said to have increasing/decreasing hazard rate (IHR/DHR) if $q(x)$ is increasing/decreasing in x for all x such that $F(x) \in (0, 1)$.[3]

Clearly, heavy-tailed Weibull and Pareto distributions are IMRL and DHR; while light-tailed Weibull distributions are DMRL and IHR. Given these examples and the relationship between the hazard rate and the mean residual life, one would expect a strong connection between the DMRL/IMRL and IHR/DHR, and this is indeed the case. In fact, it follows immediately from (4.7) that the IHR class is contained within the DMRL class and the DHR class is contained within the IMRL class.

Theorem 4.6 *All distributions with an increasing (decreasing) hazard rate have a decreasing (increasing) mean residual life, that is, IHR $\subseteq$ DMRL and DHR $\subseteq$ IMRL.*

At this point it is natural to notice that, because the Exponential distribution has constant mean residual life and hazard rate, it is, in some sense, the boundary between the IHR and DHR classes and between the IMRL and DMRL classes. Of course, Exponential distributions also serve as the boundary between light-tailed and heavy-tailed distributions, and so it is quite tempting to think of IMRL/DHR distributions as "heavy-tailed" and DMRL/IHR distributions as "light-tailed." In fact, the temptation is so strong that in some disciplines, IMRL is used as a defining property of "heavy-tailed" distributions. However, heavy-tailed and IMRL/DHR are actually quite different concepts, as are light-tailed and DMRL/IHR.

Specifically, it is easy to construct examples of IMRL and DHR distributions that are not heavy-tailed, and it is easy to construct examples of heavy-tailed distributions that are not

---

[2] A random variable X with distribution F takes values in the interval $\{x : \bar{F}(x) < 1\}$ with probability 1. Moreover, note that the mean residual life is defined for x such that $\bar{F}(x) > 0$. Thus, for defining whether or not a distribution is IMRL/DMRL, we check for monotonicity of $m(\cdot)$ only over the interval $\{x : \bar{F}(x) \in (0, 1)\}$.

[3] Note that a continuous random variable X with distribution F takes values in the interval $\{x : \bar{F}(x) \in (0, 1)\}$ with probability 1. Thus, to define a distribution as IHR/DHR, we only check for monotonicity of $q(\cdot)$ over this interval.

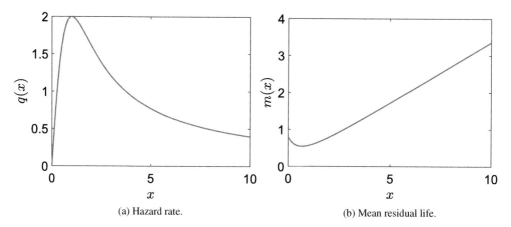

(a) Hazard rate.   (b) Mean residual life.

Figure 4.3 Hazard rate and mean residual life for the Burr distribution, with $\lambda = 1$, $c = k = 2$. Note that the distribution, though heavy-tailed, is neither IMRL not DHR.

IMRL or DHR. For example, a heavy-tailed distribution that is not IMRL or DHR is the Burr distribution. Recall that the Burr distribution is defined by $\bar{F}(x) = (1 + \lambda x^c)^{-k}$ for $x \geq 0$, where λ, c, k are positive parameters. It is easy to check that the Burr is regularly varying (with index $-ck$) and thus heavy-tailed. A simple calculation shows that the hazard rate function for the Burr distribution is given by

$$q(x) = \frac{k\lambda c x^{c-1}}{1 + \lambda x^c}.$$

Clearly, q eventually decreases, with $\lim_{x\to\infty} q(x) = 0$. However, for $c > 1$, note that $q(0) = 0$. This means that when $c > 1$, the Burr distribution is not DHR. Next, we turn to the mean residual life. Note that the expectation of the Burr is finite if $ck > 1$. While it is difficult to come up with an explicit formula for the mean residual life of the Burr distribution, we can invoke (4.8) to argue that $m'(0) = -1$ whenever $c > 1$. Thus, we may conclude that the Burr distribution is not IMRL for $c > 1$. This means that, though $m(x)$ is eventually increasing and $q(x)$ is eventually decreasing for the Burr distribution, they are not monotonic over the entire support of the distribution; see Figure 4.3.

Similarly, it is easy to construct examples of light-tailed distributions that are not DMRL and IHR. The Hyperexponential distribution is one such example. The Hyperexponential distribution is a mixture of n exponential distributions where, with probability p_i, a sample is drawn from an Exponential distribution with rate μ_i, for $i = 1, \ldots, n$ with $\sum p_i = 1$. Note that the Hyperexponential distribution is defined by the c.c.d.f. $\bar{F}(x) = \sum_{i=1}^n p_i e^{-\mu_i x}$ for $x \geq 0$. The hazard rate is now easily seen to be, for $x \geq 0$,

$$q(x) = \frac{\sum_{i=1}^n p_i \mu_i e^{-\mu_i x}}{\sum_{i=1}^n p_i e^{-\mu_i x}}.$$

We now differentiate q to see that the Hyperexponential is DHR:

$$q'(x) = \frac{\left(\sum_{i=1}^n p_i \mu_i e^{-\mu_i x}\right)^2 - \left(\sum_{i=1}^n p_i e^{-\mu_i x}\right)\left(\sum_{i=1}^n p_i \mu_i^2 e^{-\mu_i x}\right)}{\left(\sum_{i=1}^n p_i e^{-\mu_i x}\right)^2} \leq 0,$$

where the inequality is a consequence of the Cauchy–Schwarz inequality.[4] Since the Hyperexponential is DHR, it follows from Theorem 4.6 that it is also IMRL.

4.3 Long-Tailed Distributions

The discussion in the previous section encourages caution in connecting "heavy-tailed" with the concepts of "increasing mean residual life" and "decreasing hazard rate." However, if we think again about the informal examples of waiting times that we discussed at the beginning of the chapter, it becomes clear that IMRL and DHR are too "precise" to capture the phenomena we were describing. For example, consider the case of waiting for a response to an email. It is not that we expect our average remaining waiting time to monotonically increase as we wait. In fact, we are very likely to get a response quickly, and so the expected waiting time should drop initially (and the hazard rate should increase initially). It is only after we have waited a "long" time already, in this case a few days, that we expect to see a dramatic increase in our residual life. Further, in the extreme, if we have not received a response in a month, we can reasonably expect that we may never receive a response, and so the mean residual life is, in some sense, growing unboundedly, or equivalently the hazard rate is decreasing to zero. The example of waiting for a subway train highlights the same issues. Initially, we expect that the mean residual life should decrease, because if the train is on schedule, things are very predictable. However, once we have waited a long time beyond when the train was supposed to arrive, it likely means something went wrong, and could mean the train has had some sort of mechanical problem and will never arrive.

These examples illustrate two important aspects that need to be captured in a formalization of these phenomena. First, they show that strict monotonicity of the residual life is not crucial (or desirable), and that we should instead focus on the behavior of the tail. This is similar to the way we relaxed scale-invariance to asymptotic scale-invariance in Chapter 2. Second, they show that the phenomena we would like to capture include the fact that the residual life distribution "blows up," in the sense that if we have waited a very long time, we should expect to wait forever. Note that this property is true of heavy-tailed Weibull and Pareto distributions but is not true of the light-tailed distributions that are a part of the IMRL and DHR classes. For example, the Hyperexponential distribution with rate parameters $\mu_1, \ldots, \mu_n$ has a mean residual life that is upper bounded by $\max_i(1/\mu_i)$ (see Exercise 4).

These two observations lead us to the following definition of the class of long-tailed distributions.

Definition 4.7 A distribution F over the nonnegative reals is said to be long-tailed, denoted by $F \in \mathcal{L}$, if

$$\lim_{x \to \infty} \bar{R}_x(t) = \lim_{x \to \infty} \frac{\bar{F}(x+t)}{\bar{F}(x)} = 1 \tag{4.9}$$

for all $t > 0$, that is, $\bar{F}(x+t) \sim \bar{F}(x)$ as $x \to \infty$. A nonnegative random variable is said to be long-tailed if its distribution function is long-tailed.

---

[4] The Cauchy–Schwarz inequality states that for real n-dimensional vectors x and y, $(\sum_{i=1}^n x_i y_i)^2 \leq (\sum_{i=1}^n x_i^2)(\sum_{i=1}^n y_i^2)$.

The definition of long-tailed distributions exactly parallels our discussion above. In particular, long-tailed distributions are those where the distribution of residual life "blows up," that is, for any finite t, the probability the residual life is larger than t goes to 1 as $x \to \infty$. Naturally, this leads to the immediate consequence that the mean residual life grows unboundedly. We leave the proof of this result to the reader (see Exercise 8).

Lemma 4.8 *Suppose that the distribution F is long-tailed. Then, $\lim_{x \to \infty} m(x) = \infty$.*[5]

The connection between long-tailed distributions and the asymptotic behavior of the mean residual life and the hazard rate can be made more precise if we restrict attention to continuous distributions that are "well-behaved;" specifically, distributions having the property that the limit of the hazard rate function $q(x)$ exists as $x \to \infty$. Note that this assumption is not restrictive and captures all nonpathological continuous distributions.

Theorem 4.9 *Suppose that the distribution F over the nonnegative reals is associated with a density function f. Assuming $\lim_{x \to \infty} q(x)$ exists,*

$$F \in \mathcal{L} \iff \lim_{x \to \infty} q(x) = 0 \iff \lim_{x \to \infty} m(x) = \infty.[5]$$

Proof Note that we have made the assumption that the hazard rate $q(x)$ has a limit as $x \to \infty$. Let $c = \lim_{x \to \infty} q(x)$. Clearly, $c \geq 0$.

We first prove that $F \in \mathcal{L} \iff c = 0$. Recall that the c.c.d.f. can be represented in terms of the hazard rate as follows: $\bar{F}(x) = e^{-\int_0^x q(s)ds}$. It follows from the preceding representation that for $t \geq 0$,

$$\frac{\bar{F}(x+t)}{\bar{F}(x)} = e^{-\int_x^{x+t} q(s)ds}.$$

It is now not hard to see that

$$\lim_{x \to \infty} \frac{\bar{F}(x+t)}{\bar{F}(x)} = 1 \iff \lim_{x \to \infty} \int_x^{x+t} q(s)ds = 0 \iff c = 0.$$

Next, we show that $c = 0 \iff \lim_{x \to \infty} m(x) = \infty$. If $c = 0$, we have already proved that $F \in \mathcal{L}$, which implies, from Lemma 4.8, that $\lim_{x \to \infty} m(x) = \infty$. It thus suffices to show that if $c > 0$, then $m(x)$ is bounded. To see this, recall that

$$m(x) = \int_0^{\infty} e^{-\int_x^{x+t} q(s)ds} dt.$$

Assuming $c > 0$, there exists $x_0 > 0$ such that $q(x) \geq c/2$ for $x \geq x_0$. Therefore, for $x \geq x_0$,

$$m(x) \leq \int_0^{\infty} e^{-ct/2} dt = \frac{2}{c} < \infty.$$

This completes the proof.   □

---

[5] Note that if the expectation corresponding to the distribution F is ∞, then $m(x) = \infty$ identically. In this case, we follow the convention that $\lim_{x \to \infty} m(x) = \infty$ holds.

These two results demonstrate that the definition of long-tailed is both "weaker" than IMRL/DHR in that it focuses only on the tail, and "stronger" than IMRL/DHR in that it requires the mean residual life to grow unboundedly and the hazard rate to decrease to zero asymptotically.

Though the name "long-tailed" may initially seem strange, since it has no connection to the idea of residual life, it is actually natural to see from the definition why the name "long-tailed" is appropriate. In particular, $\bar{F}(x + t) \sim \bar{F}(x)$ as $x \to \infty$ for all t means that the tail stretches out with seemingly no decay over any finite range t. Thus, the tail of the distribution is indeed quite long. As this suggests, long-tailed distributions are a subclass of heavy-tailed distributions. In fact, if you recall, the definition of long-tailed distributions came up organically in our proof that subexponential distributions are heavy-tailed (Lemma 3.8), which, in retrospect, consisted of first showing that subexponential distributions are a subclass of long-tailed distributions and then showing that long-tailed distributions are heavy-tailed. Thus, we have already proven the following theorem in Chapter 3.

Theorem 4.10 *All long-tailed distributions are heavy-tailed.*

The class of long-tailed distributions is an extremely broad subclass of heavy-tailed distributions. As we just mentioned, the class of long-tailed distributions contains the class of subexponential distributions, which in turn contains the class of regularly varying distributions. As a result, all common heavy-tailed distributions are long-tailed, for example, the Pareto, the Weibull (with $\alpha < 1$), the LogNormal, the Burr, etc. Figure 4.4 illustrates how the classes we discuss in this chapter fit into those we have discussed previously. As the figure shows, there is a complex zoo of terminology surrounding subclasses of heavy-tailed distributions but, by this point in the book, each of the terms should mean something precise to you.[6]

In particular, to distinguish long-tailed distributions from heavy-tailed distributions, we recall one of our equivalent definitions of heavy-tailed distributions (Lemma 1.2). A distribution F over the nonnegative reals is heavy-tailed if

$$\liminf_{x \to \infty} \frac{-\log \bar{F}(x)}{x} = 0.$$

In contrast, for long-tailed distributions, we have the following, which we proved in the process of proving that subexponential distributions are heavy-tailed (Lemma 3.8):

Lemma 4.11 *If $F \in \mathcal{L}$, then $\lim_{x \to \infty} \frac{-\log \bar{F}(x)}{x} = 0$.*

---

[6] Since our descriptions of residual life, hazard rate, and MRL are essentially applicable to only nonnegative random variables, we mark the Gaussian and the Gumbel distributions as light-tailed but not DMRL in Figure 4.4. Similarly, we mark the Cauchy distribution as regularly varying but not IMRL. Proving the placement of the Erlang, the LogNormal, the Fréchet, and the Lévy distributions in Figure 4.4 is the goal of Exercises 5, 6, 7, and 8, respectively. Finally, that all DMRL distributions are light-tailed follows from the following property of DMRL distributions (see Prop. 6.1.2 in [202]): If a random variable X having mean $1/\mu$ is DMRL, then for $Y \sim \mathrm{Exp}(\mu)$, $\mathbb{E}[g(X)] \leq \mathbb{E}[g(Y)]$ for all convex functions $g(\cdot)$. Taking $g(x) = e^{\epsilon x}$ for $\epsilon \in (0, \mu)$, it then follows that $\mathbb{E}\left[e^{\epsilon X}\right] < \infty$, which implies that X is light-tailed.

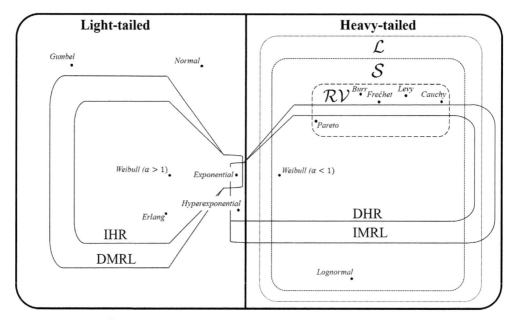

Figure 4.4 Illustration of various classes of heavy-tailed distributions, along with common distributions.

Thus, in order to construct a heavy-tailed distribution that is not long-tailed, all that is necessary is to ensure that the limit of $\frac{-\log \bar{F}(x)}{x}$ does not exist as $x \to \infty$, but $\liminf_{x \to \infty} \frac{-\log \bar{F}(x)}{x} = 0$ (see Exercise 8).

To understand the contrast between long-tailed distributions and subexponential distributions, let us go back to our proof of Lemma 3.8 in Chapter 3. In particular, in that proof, we showed that F is subexponential if and only if

$$\lim_{t \to \infty} \int_0^t \frac{\bar{F}(t-u)}{\bar{F}(t)} dF(u) = 1.$$

Correspondingly, if F is long-tailed, then it satisfies the above condition so long as we may interchange the limit and the integral. The implication is that the class of subexponential distributions corresponds to those long-tailed distributions that are "well-behaved" enough for this interchange to be valid.

The distinction between long-tailed and subexponential distributions can also be understood in terms of the hazard rate. We have already seen a sufficient condition in terms of the hazard rate for a distribution to be subexponential in Lemma 3.7. In fact, Pitman [177] proved the following (stronger) necessary and sufficient condition for a distribution to be subexponential.

Lemma 4.12 *Suppose that the hazard rate $q(\cdot)$ associated with the distribution F (over the nonnegative reals) is eventually decreasing, with $\lim_{x \to \infty} q(x) = 0$. Then $F \in \mathcal{S}$ if and only if*

$$\lim_{x \to \infty} \int_0^x e^{yq(x)} f(y) dy = 1. \tag{4.10}$$

Now, from Theorem 4.9, we see that a distribution F having a hazard rate eventually decreasing to zero, but that does not satisfy (4.10), is long-tailed but not subexponential. The interested reader is referred to [177] for an example of such a construction.

4.4 An Example: Random Extrema

Since the class of long-tailed distributions can be defined in terms of the limiting behavior of the hazard rate and the residual life, it is natural that it finds application most readily in the study of extremes, that is, the maximum and minimum of a set of random variables. Of course, the study of extremes is crucial to a wide variety of areas such as identifying outliers in statistics; determining the likelihood of extreme events in risk management; and many more, including applications in both the physical and social sciences. In this book, we focus on extremal processes in Chapter 7 when discussing the emergence of heavy-tailed distributions and then again in Part III, where extreme value theory plays a crucial role in the statistical tools we develop for estimating heavy-tailed distributions from data.

In settings where extremes are of interest, the core of the analysis often relies on understanding a very simple process – the maximum or minimum of a random number of independently and identically distributed random variables. For example, if one wants to predict the size of the maximal earthquake damage in the US in a particular year, then there is a random number of earthquakes in a year, and their sizes may be assumed to be independent and identically distributed. Of course, either (or both) the distribution of the number of events and the payout of each event could be heavy-tailed.

More formally, let us consider the following setting, which parallels the example of random sums considered in Chapter 3. Suppose $\{X_i\}_{i \geq 1}$ is a sequence of independent and identically distributed random variables with mean $\mathbb{E}[X]$ and the random variable N takes values in $\mathbb{N}$ and is independent of $\{X_i\}_{i \geq 1}$. Our goal is to characterize

$$M_N = \max(X_1, \ldots, X_N) \text{ and } m_N = \min(X_1, \ldots, X_N).$$

While you may not have studied the random extrema M_N and m_N before, you have almost certainly studied the version where the number of random variables is fixed at n (i.e., M_n and m_n). In particular, for each of these, it is simple to characterize the distribution function

$$\bar{F}_{m_n}(x) = \Pr(X_1 > x, \ldots, X_n > x) = \bar{F}(x)^n,$$
$$F_{M_n}(x) = \Pr(X_i < x, \ldots, X_n < x) = F(x)^n.$$

Using this, it is not difficult to see that the class of long-tailed distributions is well-behaved with respect to extrema. For example, we show in the following that the class is closed with respect to max and min.

Lemma 4.13 *For $n \geq 2$, suppose that $X_1, X_2, \ldots, X_n$ are i.i.d., long-tailed random variables. Then*

(i) $\max(X_1, X_2, \ldots, X_n)$ *is long-tailed, and*
(ii) $\min(X_1, X_2, \ldots, X_n)$ *is long-tailed.*

Proof Let us denote the distribution of each of the i.i.d. X_i by F. We start by focusing on the first claim: that $M_n = \max(X_1, X_2, \ldots, X_n)$ is long-tailed. To begin, note that the tail of M_n satisfies $\Pr(M_n > x) \sim n\bar{F}(x)$, since

$$
\begin{aligned}
\lim_{x \to \infty} \frac{\Pr(M_n > x)}{\bar{F}(x)} &= \lim_{x \to \infty} \frac{1 - (1 - \bar{F}(x))^n}{\bar{F}(x)} \\
&= \lim_{x \to \infty} \frac{1 - (1 - n\bar{F}(x) + \binom{n}{2}\bar{F}(x)^2 - \cdots)}{\bar{F}(x)} \\
&= \lim_{x \to \infty} \frac{n\bar{F}(x) + o(\bar{F}(x))}{\bar{F}(x)} = n.
\end{aligned}
$$

Next, for any $y > 0$, we can write

$$
\frac{\Pr(M_n > x + y)}{\Pr(M_n > x)} = \left(\frac{\Pr(M_n > x + y)}{n\bar{F}(x + y)} \right) \cdot \left(\frac{n\bar{F}(x)}{\Pr(M_n > x)} \right) \cdot \left(\frac{\bar{F}(x + y)}{\bar{F}(x)} \right).
$$

Each of the three functions in brackets above approaches 1 as $x \to \infty$, where the third is a result of the fact that X_i is long-tailed. Thus, it follows that

$$
\lim_{x \to \infty} \frac{\Pr(M_n > x + y)}{\Pr(M_n > x)} = 1,
$$

which completes the proof of the first claim in the lemma.

The second claim, that $m_n = \min(X_1, X_2, \ldots, X_n)$ is long-tailed, is easy to verify too. Given that X_i is long-tailed, we have $\lim_{x \to \infty} \frac{\bar{F}(x+y)}{\bar{F}(x)} = 1$. Thus, for any $y > 0$,

$$
\lim_{x \to \infty} \frac{\Pr(m_n > x + y)}{\Pr(m_n > x)} = \lim_{x \to \infty} \frac{\bar{F}(x + y)^n}{\bar{F}(x)^n} = 1,
$$

which completes the proof.   □

While the lemma above is useful, it is concerned only with the minimum and maximum of a fixed number of random variables. Our goal in this section is to study *random extrema*. The fact that long-tailed distributions are closed with respect to max/min suggest that the same may be true for random extrema. In fact, it is possible to say quite a bit more about the behavior of the tail of random extrema. In particular, it is possible to derive precise characterizations of the tail behavior, in quite general settings, with elementary analytic techniques.

Theorem 4.14 *Consider an infinite i.i.d. sequence of random variables $X_1, X_2, \ldots$, and a random variable $N \in \mathbb{N}$ that is independent of $\{X_i\}_{i \geq 1}$ and has $\mathbb{E}[N] < \infty$. Define n_0 such that $\Pr(N \geq n_0) = 1$ and $\Pr(N = n_0) > 0$. Then,*

$$
\Pr(\max(X_1, X_2, \ldots, X_N) > t) \sim \mathbb{E}[N]\Pr(X_1 > t)
$$

and

$$
\Pr(\min(X_1, X_2, \ldots, X_N) > t) \sim \Pr(N = n_0)\Pr(X_1 > t)^{n_0}.
$$

Thus, if X_i are long-tailed, then both $\max(X_1, X_2, \ldots, X_N)$ and $\min(X_1, X_2, \ldots, X_N)$ are also long-tailed.

To get intuition for this theorem, an interesting special case to consider is when $N \sim$ Geometric(p). In this case, $\Pr(N = i) = p(1 - p)^{i-1}$ for $i \geq 1$ and $\mathbb{E}[N] = 1/p$. So, to apply Theorem 4.14 we set $n = 1$ and $p_n = p$, and thus we obtain

$$\Pr(\max(X_1, X_2, \ldots, X_N) > t) \sim \frac{1}{p}\Pr(X_1 > t),$$

$$\Pr(\min(X_1, X_2, \ldots, X_N) > t) \sim p\Pr(X_1 > t),$$

which gives

$$\lim_{t \to \infty} \frac{\Pr(\max(X_1, X_2, \ldots, X_N) > t)}{\Pr(\min(X_1, X_2, \ldots, X_N) > t)} = \frac{1}{p^2} = \mathbb{E}[N]^2.$$

Interestingly, this does not depend on the distribution of X_i, just on the distribution of N.

Beyond the example of a Geometric distribution, note that the form of Theorem 4.14 for the random maximum exactly parallels the form of the maximum of a deterministic number of samples (i.e., $N = n$). In particular, the tail of the max of n random variables satisfies

$$\Pr(\max(X_1, \ldots, X_n) > t) \sim n\Pr(X_1 > t). \tag{4.11}$$

The parallel form of (4.11) to that of Theorem 4.14 indicates that, with respect to the tail, we can basically ignore the fact that N is random when studying random maxima. Interestingly, this is similar to the insight provided by Wald's equation for random sums, as we discussed in Section 3.3.

Given this observation, it is natural to further contrast behavior of random extrema (Theorem 4.14) with the behavior of random sums (Theorems 3.9 and 3.10). For random sums, the behavior of the tail depends crucially on whether the tail of N is heavier or lighter than the tail of the X_i. In contrast, for random extrema, the tail of N is unimportant. Further, note that, combining Theorem 4.14 with Theorem 3.9, we see that, when X_i are subexponential,

$$\Pr\left(\sum_{i=1}^{N} X_i > t\right) \sim \mathbb{E}[N]\Pr(X_1 > t) \sim \Pr(\max(X_1, X_2, \ldots, X_N) > t).$$

Thus, the random sum is tightly coupled to the random max in this case, as should be expected for subexponential distributions, given the catastrophe principle.

Now, with this intuition in hand, let us move to the proof of Theorem 4.14.

Proof We begin with the following representation of the tail of $M_N = \max(X_1, X_2, \ldots, X_N)$:

$$\Pr(M_N > t) = \sum_{i \in \mathbb{N}} \Pr(N = i)\Pr(\max(X_1, X_2, \ldots, X_i) > t).$$

Therefore,

$$\lim_{t \to \infty} \frac{\Pr(M_N > t)}{\Pr(X_1 > t)} = \lim_{t \to \infty} \sum_{i \in \mathbb{N}} \Pr(N = i)\frac{\Pr(\max(X_1, X_2, \ldots, X_i) > t)}{\Pr(X_1 > t)}.$$

Now, since $\lim_{t \to \infty} \frac{\Pr(\max(X_1, X_2, \ldots, X_i) > t)}{\Pr(X_1 > t)} = i$, the statement of the theorem follows once we justify interchanging the order of the limit and the summation in the preceding equation. As we show below, this interchange can be justified using the dominated convergence theorem (DCT).

The dominated convergence theorem states that, if we can find an upper bound β_i of $\frac{\Pr(\max(X_1,X_2,\ldots,X_i)>t)}{\Pr(X_1>t)}$ that is independent of t, such that $\sum_{i\in\mathbb{N}} \Pr(N=i)\beta_i < \infty$, then the order of the limit and the summation can be interchanged. In this case, we may take $\beta_i = i$ since

$$\frac{\Pr\left(\max(X_1,X_2,\ldots,X_i)>t\right)}{\Pr\left(X_1>t\right)} = \frac{1-(1-\Pr\left(X_1>t\right))^i}{\Pr\left(X_1>t\right)} \le i,$$

where we have used the inequality $(1-x)^i \ge 1 - ix$ for $x \ge 0$. The application of the DCT is now justified, noting that $\sum_{i\in\mathbb{N}} \Pr(N=i)\beta_i = \mathbb{E}[N] < \infty$. This completes the proof of the tail characterization of M_n.

We now turn our attention to the tail characterization of $m_N = \min(X_1, X_2, \ldots, X_N)$. As before, we may represent the tail of m_N by conditioning with respect to N:

$$\Pr\left(m_N > t\right) = \sum_{i\in\mathbb{N}} \Pr\left(N=i\right)\Pr\left(\min(X_1,X_2,\ldots,X_i)>t\right) = \sum_{i\in\mathbb{N}} \Pr\left(N=i\right)\Pr\left(X_1>t\right)^i.$$

Noting that n_0 is the smallest value taken by N with positive probability, we have

$$\Pr\left(m_N > t\right) = \sum_{i\ge n_o} \Pr\left(N=i\right)\Pr\left(X_1>t\right)^i$$

$$= \Pr\left(X_1>t\right)^{n_o}\left(\Pr\left(N=n_0\right) + \Pr\left(N=n_0+1\right)\Pr\left(X_1>t\right)\right.$$

$$\left. + \Pr\left(N=n_0+2\right)\Pr\left(X_1>t\right)^2 + \cdots\right)$$

$$= \Pr\left(X_1>t\right)^{n_o} g(\Pr\left(X_1>t\right)), \tag{4.12}$$

where the function g is defined via the power series

$$g(x) = \sum_{i=0}^{\infty} \Pr\left(N=n_0+i\right)x^i.$$

Suppose that we can show that g is continuous at 0, that is, $\lim_{x\to 0} g(x) = g(0) = \Pr\left(N=n_0\right)$. Then it follows from (4.12) that

$$\lim_{t\to\infty} \frac{\Pr\left(m_N>t\right)}{\Pr\left(X_i>t\right)^{n_o}} = \lim_{t\to\infty} g(\Pr\left(X_1>t\right)) = \Pr\left(N=n_0\right),$$

which is our desired characterization of the tail of m_N.

It remains then to argue that g is continuous at 0. Note that $g(1) = \sum_{i\ge n_o} \Pr\left(N=i\right) = 1$, that is, the power series defined by $g(x)$ converges at $x = 1$. It therefore follows that the power series is continuous over $|x| < 1$, which, of course, implies the continuity of g at 0. □

4.5 Additional Notes

In this chapter we introduced IHR/DHR and IMRL/DMRL in the context of contrasting heavy-tailed and light-tailed distributions; however, it is important to note that these classes are useful in much broader contexts. In particular, monotonicity properties of the hazard rate and the mean residual life yield crucial tools in the area of stochastic comparisons and stochastic orderings, which are fundamental tools for applied probability. For an introduction to the connections between these tools and stochastic orderings we point the interested reader to [45, 156, 193].

Our introduction to long-tailed distributions focused on providing intuitive proofs of some important properties of the class. However, readers who want to go into more depth in the study of long-tailed distributions should note that most elementary properties of long-tailed distributions can be shown to quickly follow from their connection with slowly varying functions (as introduced in Chapter 2). This connection is highlighted in [89], which also includes a comprehensive treatment of long-tailed distributions, including properties of long-tailed distributions on the real line, and so-called locally long-tailed distributions.

Unlike the classes of regularly varying and subexponential distributions, it is usually too ambitious to expect asymptotic approximations to hold for the full class of long-tailed distributions. This makes them more difficult to work with in applications. However, in several contexts it is possible to develop asymptotic lower bounds without additional assumptions, thus making the class more easily applicable. For example, in [226] the following is shown. If $S_n = X_1 + \cdots + X_n$ is a random walk with negative drift, with $P(X_1 > x + y) \sim P(X_1 > x)$, then $\lim \inf_{x \to \infty} \frac{\Pr(\sup_n S_n > x)}{\int_x^\infty \Pr(X_1 > u) du} \geq -E[X_1]$.

A second, more advanced application where the class of long-tailed distributions is used is in the analysis of Brownian motion. Specifically, let $B(t)$ be a Brownian motion with strictly positive drift, $M(t) = \sup_{s<t} B(s)$ and T an independent long-tailed random variable. Then, $P(M(T) > x) \sim P(B(T) > x)$. For a proof and a discussion of some applications of this result, see [221] and [231].

4.6 Exercises

1. The goal of this exercise is to prove that the exponential is the only continuous distribution that is memoryless. Specifically, suppose that the distribution F corresponding to a continuous nonnegative random variable satisfies the following property:

$$\frac{\bar{F}(x+t)}{\bar{F}(x)} = \bar{F}(t) \qquad \forall \, s, t \geq 0.$$

    Prove that F is necessarily exponential.
    Hint: Recall our discussion on Cauchy's functional equation in Chapter 2.

2. Consider a distribution F corresponding to a nonnegative, continuous random variable having finite mean. Show that if the hazard rate function $q(\cdot)$ corresponding to F is eventually decreasing, then the mean residual life function $m(\cdot)$ corresponding to F is eventually increasing.

3. Consider a distribution F corresponding to a nonnegative, continuous random variable having finite mean. Let $m(\cdot)$ and $q(\cdot)$ denote, respectively, the MRL and hazard rate corresponding to F.
    Prove that $m'(x) = m(x)q(x) - 1$.

4. Recall that the hyperexponential distribution is the mixture of independent exponentials. Specifically, consider the hyperexponential distribution F defined by the c.c.d.f.

$$\bar{F}(x) = \sum_{i=1}^{n} p_i e^{-\mu_i x} \qquad (x \geq 0).$$

Here, $\mu_i > 0$ and $p_i > 0$ for $1 \leq i \leq n$, with $\sum_{i=1}^{n} p_i = 1$. Let $m(\cdot)$ denote the mean residual life function corresponding to F.

Prove that $\lim_{x \to \infty} m(x) = 1/\hat{\mu}$, where $\hat{\mu} = \min_i \mu_i$.

Note: Since the hyperexponential is IMRL, this statement also implies that $m(x) \leq 1/\hat{\mu}$ for all x.

5. Recall that the Erlang distribution with parameters (k, μ), where $k \in \mathbb{N}$ and $\mu > 0$, is associated with the c.c.d.f.

$$\bar{F}(x) = \begin{cases} e^{-\mu x} \sum_{i=0}^{k-1} \dfrac{(\mu x)^i}{i!} & \text{for } x \geq 0, \\ 1 & \text{for } x < 0. \end{cases}$$

Prove that the Erlang distribution is IHR.

6. Prove that the LogNormal distribution is neither IMRL nor DMRL.

    Hint: It suffices to establish the following properties of the LogNormal hazard rate function $q(\cdot)$: (i) $\lim_{x \downarrow 0} q(x) = 0$, and (ii) $\lim_{x \to \infty} q(x) = 0$. While Property (i) implies that $\lim_{x \downarrow 0} m'(x) < 0$, Property (ii) implies that $\lim_{x \to \infty} m(x) = \infty$. To prove Property (ii), you may use the following property of the hazard rate function q_N of the standard Gaussian: $q_N(x) \sim x$ as $x \to \infty$ (this follows from Theorem 1.2.3 in [73]).

7. Recall that the Fréchet distribution with shape parameter $\alpha > 0$ is characterized by the c.d.f.

$$F(x) = \begin{cases} e^{-x^{-\alpha}} & \text{for } x \geq 0, \\ 0 & \text{for } x < 0. \end{cases}$$

Prove that the Fréchet distribution is neither DHR nor IHR. For $\alpha > 1$, prove that the Fréchet distribution is neither DMRL nor IMRL.

    Hint: Once again, it suffices to show that the hazard rate function $q(\cdot)$ of the Fréchet distribution satisfies: (i) $\lim_{x \downarrow 0} q(x) = 0$, and (ii) $\lim_{x \to \infty} q(x) = 0$.

8. Recall that the Lévy distribution with scale parameter $c > 0$ is characterized by the density function

$$f(x) = \begin{cases} \sqrt{\dfrac{c}{2\pi}} \dfrac{e^{-\frac{c}{2x}}}{x^{\frac{3}{2}}} & \text{for } x > 0, \\ 0 & \text{for } x \leq 0. \end{cases}$$

Prove that the Lévy distribution is neither DMRL nor IMRL.

9. Prove that, to check that a distribution F is long-tailed, it suffices to check that (4.9) holds for *some* $t > 0$.

10. Prove that a nonnegative random variable X is long-tailed if and only if $\lfloor X \rfloor$ is long-tailed.

11. Let $m(\cdot)$ denote the mean residual life function corresponding to distribution F over the nonnegative reals. If F is long-tailed, prove that $\lim_{x \to \infty} m(x) = \infty$.

12. Consider a distribution F over $\mathbb{R}_+$ with finite mean μ. Recall that the *excess* distribution corresponding to F is defined as

$$\bar{F}_e(x) = \frac{1}{\mu} \int_x^\infty \bar{F}(y)dy.$$

Prove that if F is long-tailed, then F_e is long-tailed.

13. The goal of this exercise is to come up with a representation theorem for long-tailed distributions, analogous to the Karamata representation theorem for regularly varying functions (Theorem 2.12 in Chapter 2). Indeed, the representation theorem for long-tailed distributions turns out to be a direct consequence of the Karamata representation theorem for slowly varying functions.

Consider a distribution F over the nonnegative reals.

(a) Show that $F \in \mathcal{L}$ if and only if $\bar{F}(\log(\cdot))$ is slowly varying.

(b) Use the representation theorem for slowly varying functions (Theorem 2.12) to establish the following representation theorem for long-tailed distributions.

   F is long-tailed if and only if $\bar{F}$ can be represented as a monotonically decreasing function of the form

$$\bar{F}(x) = \bar{c}(x)e^{\int_0^x \bar{\beta}(t)dt},$$

   where $\lim_{x \to \infty} \bar{c}(x) = c \in (0, \infty)$ *and* $\lim_{x \to \infty} \bar{\beta}(x) = 0$.

(c) Deduce from this representation theorem that long-tailed distributions are heavy-tailed distributions. Specifically, show that if $F \in \mathcal{L}$, then

$$\lim_{x \to \infty} \frac{\bar{F}(x)}{e^{-\mu x}} = \infty \qquad \forall \, \mu > 0.$$

14. Construct a distribution that is heavy-tailed but not long-tailed.

15. Recall that a nonnegative random variable X is square-root insensitive if $\Pr(X > x) \sim \Pr(X > x - \sqrt{x})$. We discussed these distributions briefly in Chapter 3. Prove that if X is square-root insensitive, then $\sqrt{X}$ is long-tailed.

Part II

Emergence

When heavy tails are found in the world around us they are often framed as surprising curiosities, something in stark contrast to the Gaussian distribution that the central limit theorem teaches us to expect. Of course, the Gaussian distribution *is* commonly encountered in the world around us; but it is also true that heavy-tailed distributions are more than a curiosity – they arise frequently and in many disparate contexts. This prompts the question: *Why?* Are there simple laws that can "explain" the emergence of heavy-tailed distributions in the same way that the central limit theorem "explains" the prominence of the Gaussian distribution?

In Part II of this book we focus on this question. We study three generic, foundational stochastic processes in order to understand when one should expect the emergence of heavy-tailed distributions as opposed to light-tailed distributions. In particular, we study additive processes (Chapter 5), multiplicative processes (Chapter 6), and extremal processes (Chapter 7). Additive processes are likely the most familiar, given their connection to the central limit theorem, but multiplicative and extremal processes are nearly as ubiquitous. Multiplicative processes arise in situations where growth happens proportionally to the current size, such as growth in investments, incomes, or populations; whereas extremal processes are crucial for characterizing extreme events such as large earthquakes, floods, or world records.

The unifying thread in the chapters that follow is that heavy-tailed distributions should *not* be viewed as anomalies. In fact, heavy tails should not be surprising at all; in many cases they should be *expected*. Both additive processes and extremal processes can lead to both light-tailed and heavy-tailed distributions, in some cases *creating* heavy-tailed distributions from light-tailed inputs. Further, multiplicative processes are almost guaranteed to lead to heavy-tailed distributions, regardless of the inputs to the process. Taken together, these chapters highlight the reality that the emergence of heavy-tailed distributions is as natural as, if not more natural than, the emergence of the Gaussian distribution.

5

Additive Processes

Discoveries of heavy-tailed phenomena are often greeted with surprise, as if heavy-tailed distributions are merely a probabilistic curiosity. In large part, this is a consequence of the prominence and beauty of the central limit theorem. The central limit theorem invites the belief that the emergence of the Gaussian (Normal) distribution is almost a "rule of nature," since processes in nature can often be viewed as being formed by a sum of many random events, that is, via some *additive process*. Consequently, the Gaussian distribution becomes our expectation about how the world around us will behave.

Of course, the Gaussian distribution *is* prominent in our lives. Many aspects of human growth and behavior are approximately Gaussian – heights, weights, test scores. However, heavy-tailed distributions are also very prominent in the world around us, as we see throughout this book. The ubiquity of heavy-tailed phenomena seems to fly in the face of the "explanation" provided by the central limit theorem for the naturalness of the Gaussian distribution. Thus, the question emerges:

Given that the central limit theorem predicts the emergence of the Gaussian distribution, why are heavy-tailed distributions so common?

To answer this question, it is reasonable to first think about where the central limit theorem applies, and where it does not. The most obvious limitation of the central limit theorem is that it only applies to additive processes. Of course, there are many other ways things can evolve besides additive processes and other processes could certainly lead to other, possibly heavy-tailed, distributions. In fact, other types of processes do lead to distributions besides the Gaussian, as we demonstrate in the following two chapters where we consider two other general processes – multiplicative processes and extremal processes.

But, even in the case when things evolve according to an additive process, it turns out that there is more to the story than the version of the central limit theorem typically taught in introductory probability and statistics courses. In particular, the typical statement of the central limit theorem that we learn in these courses is not the full version of the central limit theorem. There is a generalized central limit theorem that highlights that one should not expect additive processes to *always* yield distributions that are approximately Gaussian. The generalized central limit theorem tells us that a broad class of *stable distributions* can emerge from additive processes, and that these distributions often have heavy tails, specifically regularly varying tails.

Thus, in some sense, intuitive expectations about the universality of the Gaussian distribution are skewed because people only have a partial view of the central limit theorem. In this chapter we seek to remedy this by introducing the generalized central limit theorem and the class of stable distributions. Note that the typical treatment of the generalized central limit theorem and the class of stable distributions is extremely technical, but we do our best to provide an elementary account here. Of course, this means that we do not prove results in their full generality. However, throughout the chapter, we point interested readers to references where the full details can be found.

5.1 The Central Limit Theorem

Our focus in this chapter is on additive processes. As the name implies, additive processes are processes that evolve as the sum of random events. A simple, general class of additive processes, which is our focus in this chapter, is the following:

$$S_n = X_1 + X_2 + \cdots + X_n, \text{ where } X_i \text{ are i.i.d.}$$

The study of additive processes is a classical and important area, and two of the most celebrated results in probability provide us insight into the behavior of S_n: (i) *the law of large numbers* and (ii) *the central limit theorem*.

While you have almost certainly seen both of these results before, we present them here in a non-standard way in order to highlight a perspective on these results that is not often taught in introductory probability courses. In particular, in the treatment that follows, we show that the law of large numbers and the central limit theorem can be interpreted as the *first- and second-order approximations* of the additive process S_n. This "engineers" view is particularly useful for applications in algorithms and computer networking, where sums of i.i.d. random variables are often key to analysis.

To begin to understand this view, let us first recall the statement of the law of large numbers.

Theorem 5.1 (The strong law of large numbers) *Consider an infinite sequence of i.i.d. random variables $X_1, X_2, \ldots$ having $\mathbb{E}[X_i] = \mathbb{E}[X] \in (-\infty, \infty)$. Then,*

$$\frac{S_n}{n} \overset{a.s.}{\to} \mathbb{E}[X] \text{ as } n \to \infty.^1$$

Informally, the law of large numbers says that, up to a first-order approximation, S_n is well approximated by its mean, and is thus linear in n, that is,

$$S_n \overset{a.s.}{=} \mathbb{E}[X] n + o(n).$$

This is a particularly nice approximation because it is deterministic; all the randomness of the S_n has disappeared, and the approximation only depends on n and $\mathbb{E}[X]$. Figure 5.1 highlights this view of the law of large numbers, and the accuracy of the first-order approximation the law of large numbers provides. In particular, Figure 5.1 shows that it does not take

---

[1] Almost sure (a.s.) convergence of a sequence of random variables $\{X_n\}_{n\geq 1}$ to a (possibly random) limit X implies that with probability 1, the sequence of values taken by the random variables $\{X_n\}_{n\geq 1}$ converge to the value taken by the random variable X [111, Chapter 2]. Almost sure convergence is also commonly referred to as convergence *with probability 1* (w.p.1).

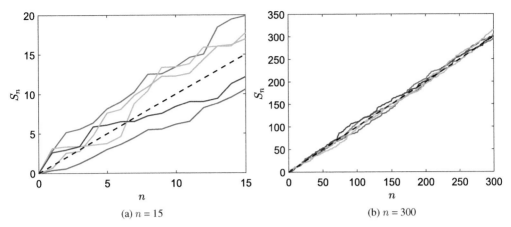

Figure 5.1 Illustration of the first-order approximation of S_n provided by the law of large numbers. The plots illustrate five realizations of S_n with $X_i \sim \text{Exp}(1)$. The dotted line shows the law of large numbers approximation.

very large n before the linear, deterministic, first-order approximation of S_n becomes accurate: (a) shows that the approximation is not particularly useful at $n = 15$, but by $n = 300$ the approximation is quite predictive.

Given that the first-order approximation of S_n from the law of large numbers is deterministic, it is natural to ask if it can be made more precise, that is, what is the second-order correction term? One would hope to have a correction term that captures the randomness of S_n, and this is exactly what is provided by the central limit theorem.

Theorem 5.2 (The central limit theorem) *Consider an infinite sequence of i.i.d. random variables $X_1, X_2, \ldots$ having $\mathbb{E}[X_i] = \mathbb{E}[X] \in (-\infty, \infty)$ and $\text{Var}[X_i] = \sigma^2 < \infty$. Then,*

$$\frac{S_n - n\mathbb{E}[X]}{\sqrt{n}} \xrightarrow{d} Z \text{ as } n \to \infty, \text{ where } Z \sim Gaussian(0, \sigma^2).^2$$

The statement of the central limit theorem can be viewed as providing a correction term for the law of large numbers. Specifically, the term $S_n - n\mathbb{E}[X]$ is the error of the approximation provided by the law of large numbers, and so the central limit theorem states that this error is on the order of $\sqrt{n}$. Thus, by combining the central limit theorem and the law of large numbers, we get a second-order approximation of the additive process S_n, where the second term is of order $\sqrt{n}$:

$$S_n \stackrel{d}{=} \mathbb{E}[X]n + Z\sqrt{n} + o(\sqrt{n}), \text{ where } Z \sim Gaussian(0, \sigma^2).^3$$

---

[2] A sequence $\{X_n\}_{n \geq 1}$ random variables converges *in distribution* to a (possibly random) limit X (denoted $X_n \xrightarrow{d} X$) if the distributions of X_n converge to the distribution of X as $n \to \infty$, i.e., $\lim_{n \to \infty} \Pr(X_n \leq x) = \Pr(X \leq x)$ for all x. To be more precise, if the c.d.f. $\Pr(X \leq x)$ has discontinuities, then we require that the preceding convergence hold at all points x where $\Pr(X \leq x)$ is continuous [111, Chapter 2].

[3] Recall that $X \stackrel{d}{=} Y$ if the random variables X and Y have the same distribution, i.e., $\Pr(X \leq x) = \Pr(Y \leq x)$ for all x. Thus, the central limit theorem gives us a distributional description of S_n for large n.

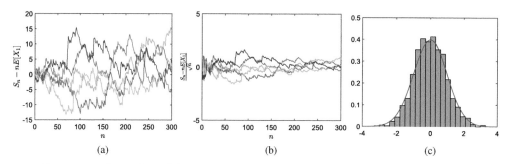

Figure 5.2 Illustration of the second-order approximation of S_n provided by the central limit theorem. The plots illustrate five realizations of S_n with $X_i \sim \mathrm{Exp}(1)$ for $n = 300$. (a) shows the error of the first-order approximation provided by the law of large numbers, that is, $S_n - n\mathbb{E}[X_i]$. (b) shows the normalized error of the first-order approximation, that is, $(S_n - n\mathbb{E}[X_i])/\sqrt{n}$. (c) shows the histogram of the normalized error of the first-order approximation for $n = 300$ over 1,000 sample paths. The red curve is the probability density function of the standard Gaussian distribution.

Of course, since the central limit theorem does not have a deterministic limit, the approximation is no longer deterministic. However, this is natural (and even desirable) given the randomness of S_n.

Figure 5.2 highlights the view of the central limit theorem as a "correction term" for the approximation of S_n. In particular, the y-axis of these plots can be interpreted as the error of the first-order approximation provided by the law of large numbers. Figure 5.2(a) shows the error without normalization, which is growing with n on the order of $\sqrt{n}$. Figure 5.2(b) shows the error normalized by $\sqrt{n}$. The normalization prevents the correction term from blowing up but, as the figure reveals, significant randomness remains (see the contrast between Figure 5.2(b) and Figure 5.1(b)). However, the central limit theorem states that this randomness has structure – it follows a Gaussian distribution. This is illustrated in Figure 5.1(c).

Since understanding and generalizing the central limit theorem is core to our goals in this chapter, it is important for us to consider the proof. In fact, when we move to the generalized central limit theorem, the proof is more technical, but the structure mimics the approach we use in the proof of the classical version of the central limit theorem in what follows.

Our proof of the central limit theorem relies on manipulating the characteristic function of the sum, and so let us first recall some important properties of the characteristic function. Recall that the characteristic function of a random variable X is defined as

$$\phi_X(t) = \mathbb{E}\left[e^{itX}\right] = \mathbb{E}\left[\cos(tX) + i\sin(tX)\right].$$

The characteristic function is a particularly useful quantity because it completely determines the distribution of the random variable (i.e., it can be inverted to find the distribution function) and it is often easier than the distribution function to work with analytically. For example, an important property that highlights the value of the characteristic function is the fact that the characteristic function of a linear combination of independent random variables can be expressed as the product of the characteristic functions of the random variables, that is,

$$\phi_{a_1 X_1 + a_2 X_2 + \cdots + a_n X_n}(t) = \phi_{X_1}(a_1 t)\phi_{X_2}(a_2 t) \cdots \phi_{X_n}(a_n t).$$

As you can imagine, this property makes the characteristic function an appealing tool for studying additive processes in general and, specifically, for proving the central limit theorem.

Proof of Theorem 5.2   To begin, we shift the X_i to obtain a zero mean random variable $Y_i = (X_i - \mathbb{E}[X])$ and denote the characteristic function of the i.i.d. Y_i by ϕ_Y. Though we do not know much about the form of ϕ_Y, we do know that $\phi_Y(0) = 1$ and that $\phi_Y'(0) = i\mathbb{E}[Y] = 0$. Further, we know that the variance of X_i is finite, and thus the variance of Y is finite as well. It follows that the characteristic function of Y is twice differentiable and that $\phi_Y''(0) = -\mathbb{E}[Y^2] = -\text{Var}[Y] = -\sigma^2$. This allows us to write a Taylor expansion of ϕ_Y around 0 as follows:

$$\phi_Y(t) = \phi_Y(0) + \phi_Y'(0)t + \phi_Y''(0)\frac{t^2}{2} + o(t^2) \text{ as } t \downarrow 0$$

$$= 1 - \frac{\sigma^2 t^2}{2} + o(t^2), \text{ as } t \downarrow 0.$$

Next, we use $Z_n = \frac{\sum_{i=1}^{n} Y_i}{\sqrt{n}}$ to denote the normalized additive process we are interested in characterizing. We can write the characteristic function of Z_n as follows:

$$\phi_{Z_n}(t) = \left(\phi_Y(t/\sqrt{n})\right)^n = \left(1 - \frac{\sigma^2 t^2}{2n} + o(t^2/n)\right)^n \to e^{-\sigma^2 t^2/2} \text{ as } n \to \infty.$$

The key of the proof is to recognize that this limit is exactly the characteristic function of a Gaussian distribution with mean zero and variance σ^2. This completes the proof because, by Lévy's continuity theorem [82], convergence of the characteristic functions implies convergence in distribution.   □

While the central limit theorem is likely second nature for anyone who has taken an introductory probability course, there are some surprises when considering it in the context of heavy-tailed distributions. In particular, note that the central limit theorem applies even when the X_is are heavy-tailed, as long as the variance is finite. So, for example, if one considers the Pareto distribution (or, more generally, a regularly varying distribution) with index $\alpha > 2$, the variance is finite and so $\frac{1}{\sqrt{n}}(S_n - nE[X])$ converges in distribution to a Gaussian random variable.

This initially seems natural, since the central limit theorem is so familiar. However, upon more careful examination it is a bit surprising. Recall that if X_1 and X_2 are regularly varying with index $-\alpha$, then $X_1 + X_2$ is also regularly varying with index $-\alpha$ (see Lemma 2.18). Thus, by induction, any finite sum $X_1 + \cdots + X_n$ is regularly varying with index $-\alpha$. So, for all n, $S_n = X_1 + \cdots + X_n$ is heavy-tailed; yet as $n \to \infty$, the central limit theorem gives a limiting distribution that is light-tailed – the Gaussian. Even more surprisingly, while all moments of the Gaussian distribution are finite, S_n will have infinite higher-order moments for all finite n! Specifically, $E[S_n^k] = \infty$ for all n and $k > \alpha$.

This seems contradictory at first; however, the reason for the apparent contradiction is simple. For any large but finite n, the approximation of S_n provided by the Gaussian distribution is only accurate up to a particular point in the tail, say for $t < t_n$. Since the Gaussian is light-tailed, it cannot approximate the full tail of a heavy-tailed distribution but, as $n \to \infty$, $t_n \to \infty$, and so convergence in distribution is achieved. For characterizations of such t_n, we refer to the survey [155].

5.2 Generalizing the Central Limit Theorem

The classical statement of the central limit theorem gives a convincing explanation for the emergence of the Gaussian distribution in the world around us: if a process grows additively, then it can be approximated via the Gaussian distribution. But, of course, there are conditions on when this may happen. Relaxing these conditions reveals that the Gaussian is not as special as the classical statement of the central limit theorem makes it appear.

In particular, there are four key assumptions in the classical central limit theorem: (i) X_i are identically distributed; (ii) X_i are independent; (iii) X_i have finite mean; and (iv) X_i have finite variance. It turns out that (i) and (ii) are not particularly crucial. Specifically, if X_i are not identically distributed, then it is still possible to obtain versions of the central limit theorem where S_n converges to a Gaussian distribution (e.g., the Lyapunov central limit theorem [30, Section 27]). Similarly, if X_i are dependent, it is still possible to prove versions of the central limit theorem where S_n converges to a Gaussian distribution, as long as the dependence is "local" (e.g., see Theorem 27.4 of [30]). Of course, if the dependence is extreme, then it can be constructed to create an arbitrary limiting distribution for S_n.

Thus, we are led to considering assumptions (iii) and (iv) – that the X_i have finite mean and variance. If one of these assumptions does not hold, we can immediately see that S_n cannot have a limit that is Gaussian since the Gaussian distribution is defined via a finite mean and variance. However, this does not mean that a version of the central limit theorem does not apply here.

In particular, we might still hope to have a result of the form

$$\frac{(X_1 + X_2 + \cdots + X_n) - b_n}{a_n} \xrightarrow{d} G, \tag{5.1}$$

for some sequences of scaling parameters a_n and translation parameters b_n, and some random variable G. This form naturally generalizes both the law of large numbers and the classical central limit theorem. In the case of the law of large numbers, $a_n = n$ and $b_n = 0$, and in the case of the standard central limit theorem, $a_n = \sqrt{n}$ and $b_n = n\mathbb{E}[X]$.

If we hope to obtain a generalized form of the central limit theorem such as the one in (5.1), then a first step is to understand what distributions may serve as limits, that is,

What distributions might the limiting distribution, G, follow?

Clearly, the Gaussian distribution is a candidate. In fact, the Gaussian distribution has some very useful properties that make it natural in the context of the central limit theorem. The most important is that, if X_1, X_2 are independent and Gaussian with mean μ and variance σ^2, then $X_1 + X_2$ is Gaussian with mean 2μ and variance $2\sigma^2$. More generally, for any $a_1, a_2 > 0$, $a_1 X_1 + a_2 X_2$ is Gaussian distributed with mean $(a_1 + a_2)\mu$ and variance $(a_1^2 + a_2^2)\sigma^2$, which implies that

$$a_1 X_1 + a_2 X_2 \stackrel{d}{=} \left(\sqrt{a_1^2 + a_2^2}\right) X_1 + \left(a_1 + a_2 - \sqrt{a_1^2 + a_2^2}\right)\mu.$$

Thus, the distribution of the sum of two Gaussian random variables yields a simple linear scaling of one of the original random variables. So, the Gaussian is, in a sense, "stable" with respect to addition and scalar multiplication.

This notion of "stability" turns out to be interesting beyond the Gaussian distribution. In particular, it is the foundation of the class of *stable distributions*, which is defined as follows.

Definition 5.3 A distribution F is stable if, for any $n \geq 2$ i.i.d. random variables $X_1, X_2, \ldots, X_n$ with distribution F, there exist constants $c_n > 0, d_n \in \mathbb{R}$ such that

$$X_1 + X_2 + \cdots + X_n \stackrel{d}{=} c_n X_1 + d_n.$$

We have already seen that the Gaussian distribution is an example of a stable distribution. Another (trivial) example of a stable distribution is a degenerate point mass where $X_i = c$ with probability 1. The Cauchy and Lévy distributions provide two other common examples of a stable distribution. However, beyond these examples, it is difficult to understand how broad the class of stable distributions is and what properties stable distributions have from the definition alone.

To better understand the definition of stable distributions, note that it is tightly coupled to the form of the central limit theorem, and to its generalized form in (5.1). In particular, the class of stable distributions is of foundational importance because stable distributions are precisely the distributions that can serve as the limiting distribution G in (5.1).

Theorem 5.4 *A random variable Z has a stable distribution if and only if there exists an infinite sequence of i.i.d. random variables $X_1, X_2, \ldots$ and deterministic sequences $\{a_n\}, \{b_n\}$ ($a_n > 0$), such that*

$$\frac{(X_1 + X_2 + \cdots + X_n) - b_n}{a_n} \stackrel{d}{\to} Z.$$

In a sense, Theorem 5.4 provides a first-cut at a generalized central limit theorem. In particular, it shows that there is an entire class of distributions, stable distributions, that serve as candidate limiting distributions for additive processes. So, the Gaussian distribution is not as "special" as the standard statement of the central limit theorem makes it appear, and the fact that it is the limiting distribution can be "explained" simply by the fact that it is a stable distribution. However, Theorem 5.4 is not particularly satisfying when interpreted as a generalized central limit theorem since it says very little about the form of the distributions that can emerge as limiting distributions or about what the limiting distribution will be for specific X_i. The next two sections fill in these holes as we continue to move toward the full statement of the generalized central limit theorem.

Proof of Theorem 5.4 First, we show that if F is a stable distribution, then it is the limit, in distribution, of a centered, normalized additive process. Let $\{X_i\}_{i \geq 1}$ denote an i.i.d. sequence of random variables with distribution F. By Definition 5.3, for any $n \geq 2$, there exist constants $c_n > 0, d_n \in \mathbb{R}$ such that

$$X_1 + X_2 + \cdots + X_n \stackrel{d}{=} c_n X_1 + d_n.$$

In other words, for any $n \geq 2$,

$$\frac{(X_1 + X_2 + \cdots + X_n) - d_n}{c_n} \stackrel{d}{=} X_1.$$

It therefore follows trivially that as $n \to \infty$,

$$\frac{(X_1 + X_2 + \cdots + X_n) - d_n}{c_n} \overset{d}{\to} F.$$

Next, we show that if the distribution F is the limit in distribution of a centered, normalized additive process, then F is stable. Accordingly, suppose that

$$\frac{(X_1 + X_2 + \cdots + X_n) - b_n}{a_n} \overset{d}{\to} F,$$

where $\{X_i\}_{i \geq 1}$ is an i.i.d. sequence of random variables, and $\{a_n\}$, $\{b_n\}$ are deterministic sequences satisfying $a_n > 0$. Now, fix integer $k \geq 2$, and define, for $m = jk$, $j \in \mathbb{N}$,

$$Y_m = \frac{(X_1 + X_2 + \cdots X_m) - b_m}{a_m},$$

$$Z_m = \frac{(X_1 + X_2 + \cdots X_m) - kb_j}{a_j}.$$

Consider the limit now as $m \to \infty$ by taking $j \to \infty$. Clearly, $Y_m \overset{d}{\to} F$. On the other hand, note that Z_m is the sum of k i.i.d. random variables, each distributed as $\frac{(X_1 + X_2 + \cdots + X_j) - b_j}{a_j}$. Therefore, $Z_m \overset{d}{\to} F^{*k}$, where F^{*k} is the distribution corresponding to the sum of k i.i.d. random variables having distribution F. Moreover, since Y_m and Z_m differ only via translation and scaling parameters, it can be shown that their limiting distributions also only differ via translation and scaling parameters. In other words, F^{*k} and F differ only via translation and scaling parameters. Since this is true for all $k \geq 2$, it then follows from Definition 5.3 that F is stable.

The above argument can be formalized by invoking the following classical result: Suppose that a random sequence $\{Z_n\}_{n \geq 1}$ converges in distribution to a (nondegenerate) limit Z. Then for a deterministic real sequence $\{\beta_n\}_{n \geq 1}$ and a deterministic positive sequence $\{\alpha_n\}_{n \geq 1}$, if the sequence $\frac{Z_n - \beta_n}{\alpha_n}$ has a (nondegenerate) limit, then this limit must be a translated and scaled version of Z (see [100, Section 10] or [75, Appendix A1.5]). $\qquad \square$

5.3 Understanding Stable Distributions

The discussion in the previous section highlights the importance of the class of stable distributions in the context of generalizing the central limit theorem; however, the results we have discussed so far are not particularly satisfying because they provide little information about the form of the limiting distributions or about when different limiting distributions will emerge (e.g., Theorem 5.4). To address these holes it is necessary to develop a deeper understanding of the class of stable distributions.

To this point in the chapter, we have noted that the class of stable distributions includes a few common distributions: the Gaussian distribution, deterministic distributions, the Lévy distribution, and the Cauchy distribution. But, beyond these distributions we have said very little. The reason is that it is difficult to understand the generality of the class from the definition directly. However, it turns out that the class of stable distributions can be characterized quite concisely. In particular, the following "representation theorem" characterizes all stable distributions via their characteristic functions.[4]

Theorem 5.5 (Representation theorem) *A nondegenerate random variable X is α-stable if and only if $X \overset{d}{=} aZ + b$, where $a > 0$, $b \in \mathbb{R}$, and the random variable Z has a characteristic function of the following form, parameterized by $\alpha \in (0, 2]$, and $\beta \in [-1, 1]$.*

$$\phi_Z(t) = \exp\left\{-|t|^\alpha \left(1 - i\beta\mathrm{sign}(t)\gamma(t, \alpha)\right)\right\}, \tag{5.2}$$

where

$$\gamma(t, \alpha) = \begin{cases} \tan\left(\frac{\pi\alpha}{2}\right) & \text{for } \alpha \neq 1, \\ -\frac{2}{\pi}\log|t| & \text{for } \alpha = 1 \end{cases}$$

and

$$\mathrm{sign}(x) = \begin{cases} -1 & \text{if } x < 0, \\ 0 & \text{if } x = 0, \\ 1 & \text{if } x > 0. \end{cases}$$

This representation theorem gives a concrete characterization of stable distributions, which is crucial for understanding the generality of the class. In particular, the representation theorem implies that the class of stable distributions can be parameterized via two parameters, $\alpha \in (0, 2]$, and $\beta \in [-1, 1]$. Typically, α is referred to as the stability parameter, and stable distributions with stability parameter α are referred to as α-stable distributions. It turns out that α is closely tied to the tail of the distribution. On the other hand, β is the skewness parameter: $\beta = 0$ yields distributions symmetric around zero, while $\beta > 0$ ($\beta < 0$) yields distributions skewed to the right (left) of zero. Note that $\beta = 0$ corresponds to a real-valued ϕ, which implies a symmetric distribution.

While the representation theorem provides a concise characterization of the class of stable distributions, the fact that it is stated in terms of the characteristic function makes it difficult to derive insight about the properties of stable distribution functions from it directly. It would be more convenient to have a characterization of the class in terms of the distribution function itself, as we had for the class of regularly varying distributions (Chapter 2), subexponential distributions (Chapter 3), and long-tailed distributions (Chapter 4). However, while the characteristic functions of stable distributions have closed forms, only in a handful of cases can these distributions be described via their density function or distribution function in closed form.

We have already discussed a few of these. The case of $\alpha = 2$, $\beta = 0$ corresponds to the Gaussian distribution. This is easy to observe by substituting into (5.2), which yields $\phi_Z(t) = e^{-t^2}$, which, of course, means Z is Gaussian distributed with mean 0 and variance 2. Similarly, it is easy to see that $\alpha = 1$, $\beta = 0$ corresponds to the Cauchy distribution. Another well-known stable distribution is the Lévy distribution, which corresponds to $\alpha = 1/2$, $\beta = 1$.

---

[4] Proofs of this result tend to be quite technical, and so we do not prove them here. The interested reader can refer to [82, Chapter 17] for the proofs.

In these three cases, the distribution function can be written in closed form, but it is difficult to explicitly write the distribution and density functions for stable distributions in general. However, there are some properties of the distribution functions of stable distributions that can be characterized. One that is of particular interest in the context of this book is the tail of stable distributions. Specifically, the following theorem characterizes the tail behavior of α-stable distributions, for $\alpha \in (0, 2)$. Given that the case of $\alpha = 2$ corresponds to the Gaussian distribution, this gives a complete characterization of the tail behavior of nondegenerate stable distributions.

Theorem 5.6   *If X is α-stable for $\alpha \in (0, 2)$, there exist $p, q \geq 0$ with $p + q > 0$ such that, as $x \to \infty$,*

$$\Pr\left(X > x\right) = (p + o(1))x^{-\alpha}, \quad \Pr\left(X < -x\right) = (q + o(1))x^{-\alpha}.$$

Thus, either the left tail, the right tail, or both tails of X are regularly varying with index $-\alpha$.

This theorem has deep consequences. It implies that, within the class of stable distributions, the only nondegenerate light-tailed distribution is the Gaussian distribution. Further, it implies that the only stable distribution with finite variance is the Gaussian. Thus, it is natural that the Gaussian distribution emerges as the limit in the standard central limit theorem. However, at the same time, Theorem 5.6 suggests that the Gaussian distribution is, in some sense, a corner case since all other limiting distributions of additive processes are heavy-tailed with infinite variance.

The proof of Theorem 5.6 is quite technical, but it is possible to give a proof of a restricted version of the result using elementary techniques. This proof is a good illustration of the use of a Tauberian theorem. Recall that we introduced two Tauberian theorems in Chapter 2: Theorem 2.15 for Laplace–Stieltjes transforms and Theorem 2.17 for characteristic functions. Here we make use of the Tauberian theorem in Theorem 2.17 for characteristic functions, which is due to Pitman [176] (see also page 336 of [31]). Recall that this theorem uses only the real component of the characteristic function, $U_X(t)$, that is,

$$U_X(t) := \text{Re}(\phi_X(t)) = \int_{-\infty}^{\infty} \cos(tx)dF(x).$$

We restate the result here for the convenience of the reader.

Theorem 5.7 (Pitman's Tauberian theorem)   *For slowly varying $L(x)$, and $\alpha \in (0, 2)$, the following are equivalent:*

$$\Pr\left(|X| > x\right) \sim x^{-\alpha}L(x) \text{ as } x \to \infty,$$

$$1 - U_X(t) \sim \frac{\pi}{2\Gamma(\alpha)\sin(\pi\alpha/2)}t^{\alpha}L(1/t) \text{ as } t \downarrow 0.$$

The power of this result is that it connects the tail of the distribution to the behavior of the characteristic function around zero. However, this particular Tauberian theorem applies only to the tail of $|X|$, not the tail of X. Thus, it cannot be used to distinguish the behavior of the right and left tails of the distribution. Rather, it provides information about the *sum* of the two tails. But, we choose to use this particular Tauberian theorem here because it deals only with the real part of the characteristic function, which makes it much simpler to work with analytically, which allows the proof to be more instructive. Naturally, using the Tauberian theorem in Theorem 5.7, we cannot hope to prove the entirety of Theorem 5.6; instead we prove the following restricted version.

Theorem 5.8 (Restricted version of Theorem 5.6) *If X is α-stable for $\alpha \in (0,2)$, then $|X| \in \mathcal{RV}(-\alpha)$.*

Though not as precise as Theorem 5.6, this restricted result already highlights the most important aspect of Theorem 5.6: that α-stable distributions with $\alpha \in (0,2)$ are heavy-tailed. Note the key difference between the proof of the restricted version and the proof of the full result is the use of a more detailed Tauberian theorem than Theorem 5.7, and the added analytic complexity that comes along with the need to work with the complex-valued characteristic function $\phi_X(t)$ instead of just the real-valued component $U_X(t)$.

Proof of Theorem 5.8 Let Z denote an α-stable random variable. Thus, the characteristic function of Z is given by the representation theorem as (5.2).

For simplicity, we treat only the case of $\alpha \in (0,2)$, $\alpha \neq 1$, and leave the case of $\alpha = 1$ as an exercise for the reader. In this case, the representation theorem gives, for $t > 0$, $\phi_Z(t) = \exp\{-t^\alpha(1 - i\beta \tan\left(\frac{\pi\alpha}{2}\right))\}$. This means that

$$U_Z(t) = \exp\{-t^\alpha\}\cos(\delta t^\alpha),$$

where $\delta = \beta \tan\left(\frac{\pi\alpha}{2}\right)$.

To prove the theorem, it is sufficient to show that

$$1 - U_Z(t) \sim t^\alpha \quad (t \downarrow 0). \tag{5.3}$$

The theorem follows from (5.3) because we can apply the Tauberian theorem in Theorem 5.7 to conclude that

$$\Pr(|Z| > x) = \Pr(Z > x) + \Pr(Z < -x) \sim \frac{2\Gamma(\alpha)\sin(\pi\alpha/2)}{\pi} x^{-\alpha}.$$

To show (5.3), we first consider the case when $\delta = 0$. In this case,

$$U_Z(t) = \exp\{-t^\alpha\} = 1 - t^\alpha + o(t^\alpha),$$

which implies (5.3).

All that remains is the case when $\delta \neq 0$. In this case,

$$U_Z(t) = [1 - t^\alpha + o(t^\alpha)]\cos(\delta t^\alpha).$$

Therefore,

$$\begin{aligned}1 - U_Z(t) &= (1 - \cos(\delta t^\alpha)) + \cos(\delta t^\alpha)t^\alpha - \cos(\delta t^\alpha)o(t^\alpha)\\ &= 2\sin^2(\delta t^\alpha/2) + \cos(\delta t^\alpha)t^\alpha - \cos(\delta t^\alpha)o(t^\alpha).\end{aligned}$$

Using $\sin(\delta t^\alpha/2) \sim \delta t^\alpha/2$ as $t \downarrow 0$, we note that $\sin^2(\delta t^\alpha/2) = o(t^\alpha)$ as $t \downarrow 0$, which is enough to guarantee that (5.3) holds and thus complete the proof. $\square$

5.4 The Generalized Central Limit Theorem

The characterization of stable distributions in the previous section gives us the context we need in order to move to the generalized central limit theorem. We already know that the limiting distributions of additive processes are stable (Theorem 5.4). Combining this with our characterization of stable distributions in Theorem 5.5 allows us to understand when different stable distributions emerge as the limiting distribution.

Theorem 5.9 (Generalized central limit theorem) *Consider an infinite sequence of i.i.d. random variables $X_1, X_2, \ldots$ with distribution F. There exist deterministic sequences $\{a_n\}, \{b_n\}$ $(a_n > 0)$ such that*

$$\frac{(X_1 + X_2 + \cdots + X_n) - b_n}{a_n} \xrightarrow{d} Z,$$

if and only if Z is α-stable for some $\alpha \in (0, 2]$. Further,

(i) Z is Gaussian, that is, $\alpha = 2$, if and only if $\int_{-x}^{x} y^2 dF(y)$ is slowly varying as $x \to \infty$;
(ii) $|Z| \in \mathcal{RV}(\alpha)$, that is, $\alpha \in (0, 2)$, if and only if

$$\bar{F}(x) = (p + o(1))x^{-\alpha}L(x), \quad F(-x) = (q + o(1))x^{-\alpha}L(x)$$

as $x \to \infty$, where $L(x)$ is slowly varying, and $p, q \geq 0$, $p + q > 0$.

The most striking aspect of this result comes from contrasting it with the standard central limit theorem. Here, the emergence of the Gaussian distribution is just one of many possible options, and one that is, in some sense, a corner case since it corresponds only to $\alpha = 2$. Another interesting contrast to the standard central limit theorem is that the Gaussian distribution occurs as the limiting distribution even in some cases where the variance is infinite.[5] In these cases, the normalizing constant a_n must be chosen to be larger than $\sqrt{n}$ and includes a slowly varying function to counter the growth of $\int_{-x}^{x} y^2 dF(y)$. For a discussion of exactly how to determine a_n in such cases, see [82, Chapter 17].

The most novel aspect of the generalized central limit theorem compared to the classical version is that it highlights the possibility of heavy-tailed stable distributions emerging as the limit of additive processes. In particular, if one starts with finite variance distributions, the Gaussian emerges, but if one starts with regularly varying distributions with infinite variance, then heavy-tailed distributions can emerge. Further, the statement shows that the emerging distribution can even have an infinite mean, and so the tail can be extremely heavy. Figure 5.3 illustrates the contrast with the behavior in the case when the X_is have finite variance. Note that there are big jumps in the process in the case of infinite variance that are not present in the classical case illustrated in Figure 5.2(a) and, as a result, the limiting distribution looks dramatically different from the Gaussian. This is a visceral illustration of the catastrophe principle discussed in Chapter 3.

Not only does the generalized central limit theorem specifically highlight which limiting distributions may occur, it also characterizes exactly when specific stable distributions

---

[5] Indeed, note that the condition that $\int_{-x}^{x} y^2 dF(y)$ is slowly varying is satisfied by all finite variance distributions, as well as some distributions with infinite variance (see Exercise 4).

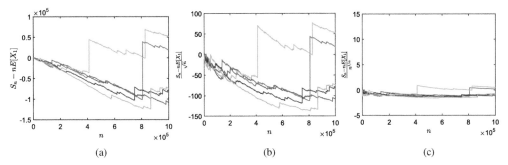

Figure 5.3 Illustration of the error of the first-order approximation of S_n provided by the law of large numbers for Pareto distributed X_1 with $\alpha = 1.2$ and mean 1 (note that X_1 has an infinite variance). The plots illustrate five realizations of $S_n - n\mathbb{E}[X_1]$, each showing a different scaling: (a) shows the unscaled deviation around the mean; (b) shows the classical central limit theorem scaling, which show a similar pattern and therefore does not lead to a limiting distribution; and (c) shows the generalized central limit theorem scaling, which leads to a non-Gaussian limiting distribution.

emerge in the limit, that is, the "domain of attraction" for limiting distributions. Formally, the domain of attraction of a limiting distribution Z is the set of distributions for X_i that have Z as the limiting distribution. For example, the domain of attraction of the Gaussian distribution is the set of distributions where $\int_{-x}^{x} y^2 dF(y)$ is slowly varying as $x \to \infty$. The domain of attraction is also tightly coupled with the sequences of $\{a_n\}$, $\{b_n\}$ that are used for scaling the process. Though the statement of the generalized central limit theorem we have given does not explicitly give $\{a_n\}$, $\{b_n\}$, they can be precisely specified. For example, when $\alpha \in (0, 2)$, the scaling constants a_n must be chosen to satisfy

$$\lim_{n\to\infty} \frac{nL(a_n)}{a_n^\alpha} = c \in (0, \infty)$$

and the translation constants may be chosen to satisfy

$$b_n = \begin{cases} 0 & \text{if } 0 < \alpha < 1, \\ na_n \int_{-\infty}^{\infty} \sin(x/a_n)dF(x) & \text{if } \alpha = 1, \\ n\mathbb{E}[X_1] & \text{if } 1 < \alpha < 2. \end{cases}$$

The case of $\alpha = 2$ is more involved, and so we refer the reader to [75, Section 2.2]. Note that the latter two sequences are both regularly varying, which again emphasizes the fundamental importance of the concept of regular variation.

Finally, it is important to discuss the proof of the generalized central limit theorem. Unfortunately, the full proof is quite technical and not appropriate for inclusion here. The interested reader can find the full details in [82, Chapter 17]. However, to give a sense for the generalized central limit theorem, we prove a restricted form using an approach that mimics the way the standard central limit theorem is typically proven. This makes clear that the generalized central limit theorem is no more mysterious than the standard central limit theorem; it simply requires a bit more technical machinery.

In particular, we prove a version of the generalized central limit theorem restricted to the case of symmetric random variables. The key reason that this restriction is simpler to prove is that symmetry means that the imaginary part of the characteristic function disappears, which allows the use of the Tauberian theorem in Theorem 5.7. Also, because of symmetry, the translation constants $\{b_n\} = 0$, and so they do not clutter the statement or the proof.

Theorem 5.10 (Generalized central limit theorem for symmetric random variables) *Suppose that $\{X_i\}_{i\geq 1}$ are i.i.d. symmetric random variables with distribution F, where $\Pr(|X| > x) \sim cx^{-\alpha}$, with $\alpha \in (0, 2)$. Then*

$$\frac{\sum_{i=1}^n X_i}{n^{1/\alpha}} \xrightarrow{d} G,$$

where G is a symmetric α-stable distribution.

Proof Since the X_i are symmetric, the imaginary part of their characteristic function equals zero. Thus, denoting the characteristic function of the i.i.d. X_i by ϕ_X, we have that $\phi_X = U_X(t)$. Moreover, since U_X is a symmetric function, that is, $U_X(-t) = U_X(t)$, so is ϕ_X.

The proof of the standard central limit theorem begins by considering the Taylor expansion of the characteristic function of $Y_i = X_i - \mathbb{E}[X]$. Of course, since X_i are symmetric around zero here, we do not need to consider such a shift. But, a more fundamental difference is that, since the $\mathrm{Var}[X_i] = \infty$, we cannot write such a Taylor expansion of ϕ_X. Instead, we invoke the Tauberian theorem in Theorem 5.7 to obtain a similar representation of ϕ_X. In particular, we have

$$\phi_X(t) = 1 - b|t|^\alpha(1 + o(1)) \text{ as } t \downarrow 0,$$

where $b = \frac{c\pi}{2\Gamma(\alpha)\sin\left(\frac{\pi\alpha}{2}\right)}$ and $\alpha \in (0, 2)$. Recall that the corresponding representation in the proof of the standard central limit theorem has the form $1 - \frac{\sigma^2 t^2}{2} + o(t^2)$, which is parallel to the form above if one were to use $\alpha = 2$.

Given this representation of ϕ_X, the proof now mimics that for the standard central limit theorem. In particular, denote $Z_n = \frac{\sum_{i=1}^n X_i}{n^{1/\alpha}}$. Then the characteristic function of Z_n satisfies the following

$$\phi_{Z_n}(s) = \left(\phi_X(s/n^{1/\alpha})\right)^n = \left(1 - b(1 + o(1))\frac{|t|^\alpha}{n}\right)^n \to e^{-b|t|^\alpha} \text{ as } n \to \infty.$$

The limit is simply a scaled version of the canonical characteristic function of the symmetric ($\beta = 0$) α-stable distribution, as desired. Again, applying the Lévy continuity theorem then completes the proof. □

5.5 A Variation: The Emergence of Heavy Tails in Random Walks

To this point in the chapter, we have considered a general, but abstract version of an additive process: $S_n = X_1 + \cdots + X_n$. Within this context, we focused entirely on understanding the behavior of S_n as $n \to \infty$, and we showed that if the X_i have finite variance, then the Gaussian distribution emerges; but that if the X_i have infinite variance, then heavy-tailed distributions can emerge.

Of course, additive processes come in many guises, and, depending on the setting, one may ask a variety of other questions about S_n besides simply how it behaves as $n \to \infty$. In this section, we show that, when different aspects of additive processes are studied, heavy-tailed distributions may emerge even in contexts where the random process is made up of entirely light-tailed components.

In particular, in this chapter we consider a classical example of an additive process, *random walks*. Random walks are an example of a very simple process that has found application in a surprising number of disciplines. In fact, few mathematical models have found applications in as diverse a range of areas, including finance, computer science, physics, biology, and more.

The most simple example of a random walk is one in which a walker is equally likely to take a single step in one direction or the other at each time step. Formally, the walker starts at 0 and takes a sequence of independent steps $X_1, \ldots, X_n$, where

$$X_i = \begin{cases} 1, & \text{with probability } 1/2; \\ -1, & \text{with probability } 1/2. \end{cases}$$

Thus, the position of the walker at time n is simply the additive process

$$S_n = X_1 + X_2 + \cdots + X_n.$$

This is referred to as a simple symmetric random walk in one dimension, where simple refers to the fact that the step size is one, symmetric refers to the fact that steps up and down are equally likely, and one dimension refers to the fact that the random walk is on a line. You have almost certainly come across this version of a random walk in your introductory probability course. It is often described under the guise of a drunken person leaving a bar and wandering aimlessly up and down the street trying to get home.

Of course, there are many more complicated versions of random walks too. In general, the random walk may be asymmetric or biased, for example, the probability of taking a positive step may be $p \neq 1/2$; the random walk may be in more than one dimension or over a general graph; or the random walk might allow step sizes other than 1 (e.g., the step size could be random). But, for this section, we stick to the simple, symmetric, one-dimensional case because this case is already sufficient to highlight the emergence of heavy-tailed distributions. We study other forms of random walks in Chapters 3 and 7.

One of the most natural questions to ask about a random walk is where the walker is likely to be after a certain (large) number of steps, that is, what is the behavior of S_n as n grows large. Of course, this is the same question that we have addressed throughout this chapter for more general additive processes. In particular, the law of large numbers and the central limit theorem are enough to give us an answer. In fact, in the case of the simple one-dimensional random walk we are studying, the X_i are light-tailed, and so the classical version of the central limit theorem applies, which means that $S_n/\sqrt{n}$ converges to a Gaussian distribution. However, it is not even necessary to apply the central limit theorem, since the distribution of the position of the random walk can be easily seen to be a Binomial distribution. Note that if the random walk is no longer "simple" and has random step sizes, this question becomes more complex. We discuss that case in Section 3.4, where we highlight the impact of heavy-tailed step sizes on the behavior of the walker.

Though the position of the random walk at time n can be understood quite easily, this is only one of many questions one may ask about a random walk. For example, two other important questions that are often asked are "when will the walker first return to its starting point?" and "what is the maximum position of the walker after n steps?" It is the first of these that we focus on here. We discuss the second question in Section 7.4.

More specifically, in this section we study the "return time" of a random walk, which is denoted by T and defined as the first time when the walker returns to its starting point. The return time of a random walk is often of crucial importance.

The return time is of particular interest in the context of this chapter because it is an example where a heavy-tailed distribution emerges from an additive process. Further, it is a jarring example of the emergence of heavy-tailed distributions because a heavy-tailed distribution emerges even though the underlying process is defined entirely using bounded (and thus extremely light-tailed) distributions. Specifically, the tail of the distribution of the return time of a simple, symmetric, one-dimensional random walk can be characterized as follows.

Theorem 5.11   *Consider a simple, symmetric, one-dimensional random walk. The distribution of the return time T satisfies* $\Pr\left(T > x\right) \sim \dfrac{\sqrt{2/\pi}}{\sqrt{x}}.$

This result emphasizes that not only is the return time heavy-tailed, it is extremely heavy-tailed. In particular, it is regularly varying with index $1/2$, which implies that both the mean and variance are infinite.

There are many methods for proving Theorem 5.11. In what follows, we use the proof as an opportunity to illustrate another application of Tauberian theorems. In this case, we make use of a generalized version of Karamata's Tauberian theorem introduced in Chapter 2 (i.e., Theorem 2.16). Thus, the bulk of the proof is simply to derive the form of the Laplace–Stieltjes transform of the return time and then we apply Karamata's Tauberian theorem (Theorem 2.15) in order to infer the tail behavior of the return time from the behavior of the Laplace–Stieltjes transform around zero.

Proof   For $n = 0, 1, 2, \ldots,$ let $u(n)$ denote the probability that the random walk hits zero at time n, that is, $u(n) = \Pr\left(S_n = 0\right).$ [6] Clearly, $u(n) = 0$ for all odd n, since the random walk can only return to zero in an even number of steps. Also, since the random walk starts at zero, $u(0) = 1$.

We analyze the distribution of T by relating it to the sequence $u(\cdot)$, which, as we will see, can be computed explicitly. Let $f(n) = \Pr\left(T = n\right).$ As before, note that $f(n) = 0$ for all odd n. Also, note that since $T > 0$, $f(0) = 0$.

We first relate $u(\cdot)$ and $f(\cdot)$ as follows. For $n \geq 1$, if the random walk hits zero at time $2n$, then the time T of *first* return to zero is necessarily $\leq 2n$. We can therefore represent the probability $u(2n)$ as

$$u(2n) = \Pr\left(S_{2n} = 0\right) = \sum_{j=1}^{n} \Pr\left(T = 2j\right) \Pr\left(S_{2n} = 0 \mid T = 2j\right).$$

---

[6] It is important to note that when we consider the event $S_n = 0$, the random walk could have hit zero several times before time n. $u(n)$ captures the probability of all such paths.

Moreover, note that $\Pr\left(S_{2n}=0 \mid T=2j\right)$ is simply the probability that the random walk hits zero after starting at zero after $2n - 2j$ steps, that is, $\Pr\left(S_{2n}=0 \mid T=2j\right) = u(2n - 2j)$. Therefore, we obtain the following recursive relation for $n \geq 1$.

$$u(2n) = \sum_{j=1}^{n} f(2j)u(2n - 2j) \tag{5.4}$$

We now use the above recursion to relate the Laplace–Stieltjes transform of the random variable T to that of the function $u(\cdot)$. Specifically, define

$$\psi_T(s) := \mathbb{E}\left[e^{-sT}\right] = \sum_{n=1}^{\infty} f(2n)e^{-2ns},$$

$$\psi_u(s) := \sum_{m=0}^{\infty} u(m)e^{-ms} = \sum_{n=0}^{\infty} u(2n)e^{-2ns}.$$

Now, using (5.4), we may express ψ_u as follows.

$$\psi_u(s) = \sum_{n=0}^{\infty} u(2n)e^{-2ns}$$

$$= 1 + \sum_{n=1}^{\infty}\sum_{j=1}^{n} f(2j)u(2n - 2j)e^{-2ns}.$$

Interchanging the order of the summations above, we obtain

$$\psi_u(s) = 1 + \sum_{j=1}^{\infty} e^{-2js} f(2j) \sum_{n=j}^{\infty} e^{-s(2n-2j)} u(2n - 2j)$$

$$= 1 + \psi_T(s)\psi_u(s),$$

which gives us

$$\psi_T(s) = 1 - \frac{1}{\psi_u(s)}. \tag{5.5}$$

Note that this interchange can be justified via Fubini's theorem.

Now that we have related the Laplace–Stieltjes transform of T to that of the sequence $u(\cdot)$, we move on to computing $u(\cdot)$ and its transform explicitly.

For $n \geq 1$, let us consider the event $S_{2n} = 0$. This event simply means that, after starting at zero, the random walk made n positive steps and n negative steps. Therefore, the total number of paths of the random walk that correspond to the event $S_{2n} = 0$ equals $\binom{2n}{n}$. Since the total number of possible paths after $2n$ steps equals 2^{2n}, and each of these is equally likely, we conclude that

$$u(2n) = \frac{\binom{2n}{n}}{2^{2n}}.$$

Therefore, the transform ψ_u is given by

$$\psi_u(s) = 1 + \sum_{n=1}^{\infty} \frac{\binom{2n}{n}}{2^{2n}} e^{-2ns}.$$

Using the binomial expansion

$$\frac{1}{\sqrt{1-x}} = 1 + \sum_{n=1}^{\infty} \frac{\binom{2n}{n}}{2^{2n}} x^n,$$

we may express ψ_u as

$$\psi_u(s) = \frac{1}{\sqrt{1 - e^{-2s}}}.$$

Therefore, using the relation (5.5), we obtain

$$\psi_T(s) = 1 - \sqrt{1 - e^{-2s}}.$$

With an explicit expression for the Laplace transform in hand, we can now apply Theorem 2.16 to deduce the tail behavior of T from the behavior of $\psi_T(s)$ near the origin. Specifically, as $s \downarrow 0$, since $e^{-2s} = 1 - 2s(1 + o(1))$,

$$\psi_T(s) = 1 - \sqrt{1 - (1 - 2s(1 + o(1)))}$$
$$= 1 - \sqrt{2s(1 + o(1))}$$
$$= 1 - \sqrt{2s}(1 + o(1)).$$

Therefore, from Theorem 2.16,

$$\Pr(T > x) \sim \frac{\sqrt{2}}{\Gamma(1/2)} \frac{1}{\sqrt{x}}.$$

Noting that $\Gamma(1/2) = \sqrt{\pi}$, we have our desired tail characterization.   □

5.6 Additional Notes

The central limit theorems presented in this chapter are part of a rich literature with a long history that originated in the study of coin tossing-type problems by De Moivre, Gauss, and Laplace in the eighteenth and nineteenth centuries. Within this literature, the (partial) universality of the Gaussian distribution emerged around 1900 and was the result of contributions from a number of renowned mathematicians including Markov, Lyapunov, Lindeberg, and Kolmogorov, culminating in the classical central limit theorem. A detailed history can be found in [84].

Our treatment of the central limit theorem focuses on its role as a second-order approximation of additive processes. For more discussion of this viewpoint, see [217]. Given this viewpoint, it is natural to consider higher-order expansions. It is indeed possible to derive such higher-order expansions, and an overview of these can be found in [29, 172, 173].

The process of moving beyond the classical central limit theorem toward the generalized central limit theorem was initiated by Lévy, who was the first to show that stable distributions beyond the Gaussian can occur when the summands of the random walk have infinite variance [140]. This led to a number of variations of "generalized" central limit theorems for stable distributions such as the one presented in this chapter. Detailed mathematical treatments of these results can be found in [15, 31, 82, 100, 118, 143, 228]; see [75, Section 2.2] for a discussion. A comprehensive history of the development of generalized central limit theorems can be found in [84].

Motivated by the importance of stable distributions for generalized versions of the central limit theorem, stable distributions have become a topic worthy of study in their own right. Lévy was among the first to develop properties of stable distributions, and now there is a rich literature focused on characterizing stable distributions and stable processes, which generalize the Gaussian processes. A comprehensive overview can be found in [190].

We ended the chapter by studying the return times of random walks in Section 5.5. This is a classical topic, and a rich theory exists. In particular, we have illustrated the heavy-tailed nature of return times using a simple random walk, but the phenomenon is more general. Power laws persist when the step size distribution is generally distributed, as long as the mean is zero; in this case one typically considers the event of crossing, rather than hitting the zero boundary. More complicated boundary-crossing probabilities (particularly time-dependent boundaries) have been considered as well. In all of these cases, heavy-tailed behavior persists. An excellent survey of classical results in this context is [66].

5.7 Exercises

1. To ground the ideas in this chapter it is important to get a feel of the classical and generalized central limit theorems using data.

   (a) *Classical central limit theorem:* Simulate a sequence of i.i.d. Weibull random variables $\{X_i\}$. For some (large) fixed n, generate a large number of samples of $\frac{\sum_{i=1}^n X_i - n\mathbb{E}[X_1]}{\sqrt{n\mathrm{Var}(X_1)}}$. Plot a histogram of the data you have generated. Does your data "look" Gaussian?

   (b) *Generalized central limit theorem:* Simulate a sequence of i.i.d. Pareto random variables $\{X_i\}$ with shape parameter $\alpha \in (1, 2)$. For some (large) fixed n, generate a large number of samples of $\frac{\sum_{i=1}^n X_i - b_n}{a_n}$, where the scaling constants $\{a_n\}$ and translation constants $\{b_n\}$ are as obtained in Exercise 6.

      i. Plot a histogram of the data you have generated. Does it look qualitatively different from the histogram from Exercise 1? What are the issues with visualizing your data using a histogram?
      ii. Next, plot the empirical c.c.d.f. of the data, that is, plot the fraction of data points exceeding x as a function of x. Does the visualization get better?
      iii. Finally, re-plot the above empirical c.c.d.f. using a logarithmic scale for both axes. (You will have to restrict yourself to positive x for this.) Explain the picture you see using the generalized central limit theorem.

2. In this chapter, we gave a number of examples of stable distributions but did not verify them formally. In this problem you will verify that they are indeed stable. Specifically, prove that the following distributions are stable by checking that they satisfy the condition in Definition 5.3. See Chapter 1 for definitions of these distributions.

   (a) The standard Gaussian.
   (b) The Cauchy distribution.
   (c) The Lévy distribution.

3. The goal of this exercise is to come up with an alternative, equivalent definition of stable distributions that can sometimes be easier to work with. Specifically, prove that a distribution F is stable if and only if, for i.i.d. random variables X_1, X_2 with distribution F, and any constants $a_1, a_2 > 0$, there exist constants $a > 0$ and $b \in \mathbb{R}$ such that

$$a_1 X_1 + a_2 X_2 \stackrel{d}{=} a X_1 + b.$$

4. As mentioned in the chapter, the Gaussian distribution is the limiting distribution for additive processes beyond the setting of the classical central limit theorem. Your task in this problem is to give an example of this phenomenon. Specifically, construct a distribution having infinite variance that belongs to the domain of attraction of the Gaussian distribution.

5. To get a better understanding of the limiting distributions for the generalized central limit theorem, in this problem your task is to construct specific distributions that lead to α-stable limiting distributions. In particular, for each $\alpha \in (0, 2)$, construct an explicit distribution over the nonnegative reals that lies in the domain of attraction of an α-stable distribution.

6. To practice applying the generalized central limit theorem, let us consider the case of Pareto random variables. Consider the running sum $S_n = \sum_{i=1}^{n} X_i$ of i.i.d. Pareto random variables $\{X_i\}_{i \geq 1}$ with shape parameter $\alpha \in (0, 2)$. Identify the scaling and translation constants required for the generalized central limit theorem to apply to the sequence $\{S_n\}_{n \geq 1}$.

7. Suppose that $X_1, X_2, \ldots, X_n$ are i.i.d., standard Cauchy random variables. Prove that $\frac{1}{n} \sum_{i=1}^{n} X_i$ is a standard Cauchy random variable.

Hint: You might want to exploit the characteristic function of the standard Cauchy.

6

Multiplicative Processes

The widely held belief that heavy-tailed phenomena are rare curiosities stems, in large part, from the view that the emergence of the Gaussian distribution is almost a "law of nature." As we have discussed, this view stems from the prominence of the central limit theorem, and more generally the view that additive processes frequently govern growth in the world around us.

Though it is certainly true that additive processes are prominent in the world around us, they are far from the only way for things to evolve. In fact, they are probably not even the most common. One could argue, with foundation, that it is as or even more common for things to evolve according to some form of *multiplicative process*, where growth happens proportionally to the current size. For example, investments and incomes tend to grow multiplicatively since interest payments and raises tend to be expressed in terms of percentages (e.g., a 5% interest rate or a 3% raise). Another example is the growth of populations: the number of kids that families have on average, which varies generation to generation, provides a multiplicative growth rate for a population (e.g., if families tend to have 2.5 kids, then the population will grow multiplicatively at a rate of 2.5/2 = 1.25 per generation).

Multiplicative processes are relevant not only for growth processes like the simple examples just given, they are also relevant for studying situations involving fragmentation. For example, consider a stick or rock breaking into two pieces, and then the resulting pieces further breaking into two pieces, and so on. In each case, the resulting pieces represent some fraction of the original item, and so the fragmentation follows a multiplicative process. Such processes have been suggested as models for the evolution of the sizes of meteors (and consequently craters), which tend to break up as they collide with each other, for example.

These examples show that multiplicative processes tend to be more representative models than additive processes for many parts of our physical and economic world. The same is true for the digital world. For example, when someone watches a video online and likes it, she is likely to share it with friends, and thus the number of views will grow proportionally with the number of people who have watched the video already. Similarly, if someone has a lot of Twitter followers, this may lead to a lot of retweets, which then leads to more followers, and so on.

The prevailing theme of these examples is a sense that the "rich get richer," that is, once something becomes large or popular it tends to grow more quickly, often in proportion to its increased size. If a song has been downloaded 100,000 times, then it is likely that it will quickly get many more downloads. If one considers the number of links pointing to a website, then as the website is linked to more often, more people are likely to find out about it, and thus the number of links pointing to it will grow even more quickly. This concept of "rich get richer" is fundamentally tied to multiplicative processes and heavy tails.

In fact, the examples of multiplicative processes that we have discussed should sound very familiar, as we have used them often in this book as examples of heavy-tailed distributions. This is not an accident or coincidence. Multiplicative processes are fundamentally tied to the emergence of heavy-tailed distributions. We illustrate and explain this link in this chapter by studying a number of variations of multiplicative processes, including multiplicative processes with barriers, with noise, and the celebrated preferential attachment, aka Yule process, model. Our study of these variations reveals that different limiting distributions may emerge depending on the specific form of the multiplicative process, but that the emerging distribution from a multiplicative process is nearly always heavy-tailed. This is quite different from the behavior of additive processes. Though heavy-tailed distributions can emerge from additive processes, they emerge only when the additive process is made up of infinite variance heavy-tailed distributions. Thus, additive processes do not *create* heavy-tailed distributions from light-tailed distributions, they only *maintain* heavy tails. In contrast, multiplicative processes can *create* heavy-tailed distributions from light-tailed distributions, and so they provide a much more satisfying explanation for the prevalence of heavy-tailed phenomena.

6.1 The Multiplicative Central Limit Theorem

Let us begin our discussion of multiplicative processes with a simple, generic example of a process that grows via the product of random events. In particular, we consider the following multiplicative process:

$$P_n = Y_1 \cdot Y_2 \cdots Y_n = P_{n-1} \cdot Y_n,$$

where $P_0 = 1$ and Y_i are i.i.d. and strictly positive. See Figure 6.1(a) for an illustration of such a process.

As in the case of additive processes, our goal is to study the behavior of the multiplicative process P_n as $n \to \infty$. However, to derive intuition for why multiplicative processes are tied to heavy-tailed distributions it is useful to start with some simple examples of small, finite n. In particular, let us first look at the case where the Y_i are exponentially distributed with rate λ. In this case, though the Y_i are light-tailed, P_2 already turns out to be heavy-tailed. To see this, we can use the following simple bound on P_2:

$$\Pr\left(P_2 > t\right) \geq \Pr\left(Y_1 > \sqrt{t}\right) \Pr\left(Y_2 > \sqrt{t}\right) = e^{-2\lambda\sqrt{t}}.$$

This is already enough to show that P_2 is heavy-tailed.

More generally, the same phenomenon happens with distributions that have lighter than exponential tails, but as the tail gets lighter an increasing number of terms is necessary in order to generate a heavy tail. For example, consider the case where Y_i follows a Weibull distribution with scale parameter β and shape parameter $\alpha > 1$. Then for all $k > \alpha$,

$$\Pr\left(P_k > t\right) \geq \left(\Pr\left(Y_i > t^{1/k}\right)\right)^k = e^{-k\beta t^{\alpha/k}},$$

which guarantees that P_k is heavy-tailed.

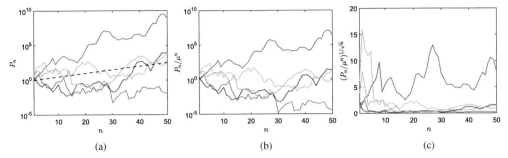

(a)   (b)   (c)

Figure 6.1 Illustration of the multiplicative process P_n. The plots illustrate five realizations of P_n with $Y_i \sim \mathrm{Exp}(1/2)$. Figure (a) shows the (unscaled) process itself. Note that the scale on the y-axis is logarithmic, indicating an exponential growth in the process P_n. The dotted line is the first-order approximation μ^n, where $\mu = e^{\mathbb{E}[\log(Y_1)]}$. Figure (b) shows the same process normalized with respect to the first-order approximation. Note that the y-axis is still on a logarithmic scale. Finally, Figure (c) shows the process scaled as per the multiplicative central limit theorem. Note that the y-axis is now linear. Under this scaling, Theorem 6.1 shows that the process converges in distribution to the LogNormal distribution.

The two examples we have looked at so far already show that multiplicative processes behave quite differently from additive processes, and are much more strongly tied to heavy-tailed distributions. In fact, given these two examples, it is quite natural to expect P_n as $n \to \infty$ to yield a heavy-tailed distribution. This is indeed true and, in fact, it turns out that it is possible to prove this fact via a connection between multiplicative processes and additive processes. In particular, a simple translation of P_n allows its behavior to be described using the law of large numbers and the central limit theorem. The key observation that provides this connection is that the logarithm of a multiplicative process yields an additive process. Specifically, defining $X_i = \log Y_i$ yields

$$\log P_n = \log Y_1 + \log Y_2 + \cdots + \log Y_n$$
$$= X_1 + X_2 + \cdots + X_n.$$

A consequence of this translation is that the emergence of heavy-tailed distributions under multiplicative processes is quite natural. It is essentially a consequence of the central limit theorem: the logarithm of a multiplicative process is an additive process and so, instead of the Gaussian distribution, the LogNormal distribution emerges. To show this more concretely, let us walk through the application of the law of large numbers and the central limit theorem. Assume that the X_i have finite mean $\mathbb{E}[X_i]$ and finite variance σ^2. Starting with the law of large numbers, we have

$$\log P_n \overset{a.s.}{=} \mathbb{E}[X_i]\, n + o(n),$$

which yields

$$P_n \overset{a.s.}{=} e^{\mathbb{E}[X_i]n+o(n)} \overset{a.s.}{=} \left(e^{\mathbb{E}[\log Y_i]}\right)^n e^{o(n)}. \tag{6.1}$$

Just as we saw in the case of additive processes, the law of large numbers provides a first-order approximation of the process P_n. However, there are interesting contrasts in the structure of the first-order approximation between the two cases. First, for the multiplicative process P_n, the first-order approximation is an exponential function $\left(e^{\mathbb{E}[\log Y_i]}\right)^n$ as opposed to the linear approximation of additive processes provided by the law of large numbers. Second, note that the error of this first-order approximation is much larger for multiplicative processes than for additive processes. Expressed multiplicatively, the error is $e^{o(n)}$ in the case of multiplicative processes; whereas there is only an additive $o(n)$ error in the case of additive processes. Figure 6.1(b) provides an illustration of the error of the first-order approximation of the multiplicative process P_n that emphasizes these contrasts.

That the first-order approximation of a multiplicative process is an exponential should not in itself be surprising. However, what might be surprising is that the base of this exponential approximation is $e^{\mathbb{E}[\log Y_i]}$. Indeed, one might guess from the fact that we are applying the law of large numbers that the dominant growth of P_n should be $\mathbb{E}[Y_i]^n$, since the Y_i are independent and thus $\mathbb{E}[P_n] = \prod_{i=1}^{n} \mathbb{E}[Y_i]$. However, instead, the dominant growth is $\left(e^{\mathbb{E}[\log Y_i]}\right)^n$, which can be seen to be smaller than $\mathbb{E}[P_n]$ as a consequence of Jensen's inequality.[1] Thus, the mean is not actually a good predictor of the behavior of P_n.

A simple example makes this point clear. Consider Y_i that takes on either $1/2$ or $3/2$ with equal probability. Thus, $\mathbb{E}[Y_i] = 1$ and so $\mathbb{E}[P_n] = 1$ too. However, the median can be far from the mean. To see this, note that the number of Y_i that are equal to $1/2$ follows a Binomial distribution, and thus the median of P_n has $n/2$ samples as $1/2$ and $3/2$. So, the median of P_n is

$$\left(\frac{1}{2}\right)^{n/2} \left(\frac{3}{2}\right)^{n/2} = \left(\frac{3}{4}\right)^{n/2} < 1 = \mathbb{E}[P_n].$$

In fact, not only is the median of P_n smaller than the mean, the gap between them grows exponentially with n. Thus, the mean is not a good predictor of the behavior of the distribution, which is typical for heavy-tailed distributions.

Of course, we can improve the characterization of P_n provided by the law of large numbers by additionally applying the central limit theorem. In particular, this gives the following approximation for $\log P_n$:

$$\log P_n \stackrel{d}{=} \mathbb{E}[X]n + Z\sqrt{n} + o(\sqrt{n}), \text{ where } Z \sim \text{Gaussian}(0, \sigma^2).$$

Again, we can then recover P_n by exponentiating:

$$P_n \stackrel{d}{=} e^{\mathbb{E}[X_i]n + Z\sqrt{n} + o(\sqrt{n})} \stackrel{d}{=} \left(e^{\mathbb{E}[X_i]}\right)^n \left(e^Z\right)^{\sqrt{n}} e^{o(\sqrt{n})}, \text{ where } Z \sim \text{Gaussian}(0, \sigma^2).$$

This shows that, similarly to the case of additive processes, the combination of the law of large numbers and the central limit theorem gives a second-order approximation for the growth of a multiplicative process. But, the form of this approximation illustrates that it is exponentially less accurate for multiplicative processes than the parallel approximation is for additive processes, that is, the error term is of the order $e^{o(\sqrt{n})}$ instead of $o(\sqrt{n})$. However, this increased error does not affect the use of the approximation to study the limiting behavior of P_n as $n \to \infty$. In particular, by looking at the centered and scaled version of

---

[1] Jensen's inequality (see [34]) states that given a random variable X and a function f that is concave over the support of X, $\mathbb{E}[f(X)] \leq f(\mathbb{E}[X])$, provided that the expectations exist. Taking $f(x) = \log(x)$, this implies that $\mathbb{E}[\log(Y_i)] \leq \log(\mathbb{E}[Y_i])$, which implies that $e^{\mathbb{E}[\log(Y_i)]} \leq \mathbb{E}[Y_i]$.

P_n we can obtain the following characterization of the limiting distribution of multiplicative processes:

$$\left(\frac{P_n}{e^{\mathbb{E}[X_i]n}}\right)^{1/\sqrt{n}} \stackrel{d}{=} e^{Z+o(\sqrt{n})/\sqrt{n}} \stackrel{d}{\to} e^Z, \text{ as } n \to \infty, \text{ where } Z \sim \text{Gaussian}(0, \sigma^2).$$

This calculation shows that, when properly normalized, P_n converges to a LogNormal distribution in the limit, and thus the behavior of multiplicative processes is tightly tied to heavy-tailed distributions. See Figure 6.1(c) for an illustration of P_n under this normalization.

Though the preceding sketch is not completely rigorous, it can easily be made so, which gives the following statement of the multiplicative central limit theorem. This is sometimes referred to as Gibrat's law, named after Robert Gibrat, who studied multiplicative growth in the context of the growth of firms in the 1930s [98].

Theorem 6.1 (The multiplicative central limit theorem) *Suppose $\{Y_i\}_{i\geq 1}$ is an i.i.d. sequence of strictly positive random variables satisfying* $\text{Var}\left[\log Y_i\right] = \sigma^2 < \infty$ *and define* $\mu = e^{\mathbb{E}[\log Y_i]}$. *Then*

$$\left(\frac{Y_1 \cdots Y_n}{\mu^n}\right)^{\frac{1}{\sqrt{n}}} \stackrel{d}{\to} H, \text{ where } H \sim \text{LogNormal}(0, \sigma^2).$$

The form of the multiplicative central limit theorem essentially parallels that of the central limit theorem except that the centering and scaling of the process happens in a different manner due to the fact that the process grows multiplicatively instead of additively. Note that we have stated a version of the multiplicative central limit theorem that uses only the standard central limit theorem, not the generalized central limit theorem. However, because the condition of finite variance applies only to $\log Y_i$ rather than to Y_i itself, this case already includes all common distributions, even many distributions with infinite mean. For example, recall that if Y_i is Pareto distributed, then $\log Y_i$ is Exponentially distributed, and thus all of its moments are finite. More generally, a sufficient condition to ensure that $\text{Var}\left[\log Y_i\right] = \sigma^2 < \infty$ is $\bar{F}_Y(x) = o\left(\frac{1}{\log^2(x)}\right)$, which shows that all regularly varying distributions with nonzero index satisfy the condition – and thus only extremely heavy-tailed distributions are excluded. Of course, if one desires to state a "generalized" multiplicative central limit theorem for that case, the derivation can be adjusted accordingly.

The contrast between the multiplicative central limit theorem and the central limit theorem for additive processes is striking. While the emergence of heavy-tailed limiting distributions under the central limit theorem requires starting with heavy-tailed distributions with infinite variance, the emergence of heavy-tailed limiting distributions under the multiplicative process is *guaranteed* – even when the Y_i are extremely light-tailed. So, one truly should *expect* heavy-tailed distributions whenever multiplicative growth governs a process.

6.2 Variations on Multiplicative Processes

The simple form of the multiplicative central limit theorem that we have described so far already highlights the strong connection between multiplicative processes and heavy-tailed distributions, and specifically LogNormal distributions. More broadly, the connection

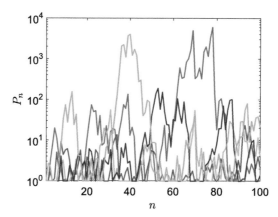

Figure 6.2 Illustration of a multiplicative process with a lower barrier. The plots
show 5 realizations of P_n with $Y_i \sim \mathrm{Exp}(3/4)$ and $l = 1$. Note that the process gets
occasionally 'reset' to the lower barrier, but can take large values between
successive resets.

between heavy-tailed distributions and multiplicative processes is extremely robust, but one
should not always expect the LogNormal distribution to emerge.

To illustrate the robustness of the connection between multiplicative processes and heavy-
tailed distributions, as well as the brittleness of the emergence of the LogNormal distribution,
we study two variants of multiplicative processes in this section. The first variation is a
multiplicative process that has a strictly positive lower barrier which cannot be crossed, and
the second variation is a multiplicative process that includes additive noise at each step. Each
of these variations leads to the emergence of heavy-tailed distributions; however, in both
cases, though the variations are small, the specific heavy-tailed distributions that emerge
are no longer LogNormal distributions. Instead, distributions that are approximately power-
law emerge. This illustrates that, even though the emergence of the LogNormal distribution
under multiplicative processes is brittle, the connection between multiplicative processes and
heavy-tailed distributions is robust.

6.2.1 A Multiplicative Process with a Lower Barrier

An important variation of the classical multiplicative process that we have considered so far
is to add a lower barrier on the process that ensures that the process never drops below a
certain level, $l > 0$. Specifically,

$$P_n = \max(P_{n-1}Y_n, l) = \max(\cdots \max(\max(Y_1, l) \cdot Y_2, l) \cdots Y_n, l),$$

where the Y_n are strictly positive i.i.d. random variables and $P_0 = 1$ is the starting point of
the process. See Figure 6.2 for a visualization of this process.

This variation of multiplicative processes captures the idea that it is often unreasonable to
have the process take on arbitrarily small values. A frequently cited example where this is the
case is incomes: if there is a minimum wage, then there is a point beneath which the income
can never drop. Similarly, minimum levels also are natural if one considers the population of
a city, the number of links to a website, and many other common examples of multiplicative
processes.

Initially, it is natural to think that the addition of a lower bound should have little impact on the behavior of the multiplicative process, especially since the unbounded process is already guaranteed to be strictly positive, and so has a lower bound already. Indeed, in many cases it is true that the lower bound plays no role. For example, if the process grows quickly then, no matter what the lower bound, the process quickly grows away from it and is then not influenced by it for large n. Thus, for the lower bound to play a role, it is necessary to impose that the process drifts downward, that is, that $\mu := e^{\mathbb{E}[\log(Y_i)]} < 1$. Alternatively, there are also some situations where the lower bound plays too strong a role. For example, if the multiplicative process never grows, then the process simply converges to the lower bound. To avoid this trivial case it is enough to ensure that the process grows with some probability, that is, that $\Pr(Y_i > 1) > 0$.

It turns out that the two conditions we have just described are enough to ensure that the lower bound plays a role, but does not lead to trivial limiting behavior. In particular, they ensure that the multiplicative process P_n neither drifts upward to infinity nor downward to the lower bound. As a consequence, the process no longer needs to be centered or scaled as in the multiplicative central limit theorem. This leads to a slightly simpler form for the multiplicative central limit theorem in this case.

Theorem 6.2 *Consider the multiplicative process with a lower barrier* $P_n = \max (P_{n-1}Y_n, l)$, *where* $\{Y_n\}_{n\geq 1}$ *is an i.i.d. sequence of strictly positive random variables,* $l > 0$ *is a lower barrier enforced on the process, and* $P_0 = 1$. *Suppose that the distribution of* Y_1 *satisfies the following conditions:*

(a) $\mu := e^{\mathbb{E}[\log(Y_1)]} < 1$.
(b) $\Pr(Y_1 > 1) > 0$.
(c) $\mathbb{E}[Y_1^s] < \infty$ *for some* $s > 0$.

Then, $P_n \xrightarrow{d} H$, *where* H *is heavy-tailed and has a distribution* F *that satisfies* $\lim_{x\to\infty} -\frac{\log \bar{F}(x)}{\log(x)} = s^*$, *with* $s^* = \sup\{s \geq 0 \mid \mathbb{E}[Y_1^s] \leq 1\}$.

Interestingly, Theorem 6.2 shows that, though it is in some sense a mild variation of the classical multiplicative process, the addition of a lower barrier leads to a qualitatively different limiting behavior. In particular, the LogNormal distribution is no longer the emergent distribution; instead a distribution that is approximately power-law, and thus has a much heavier tail, emerges. The limiting distribution is asymptotically linear on a log-log scale, that is, $\log \bar{F}(x)/\log x$ converges to a constant. This condition is not satisfied by the LogNormal distribution, but is satisfied by regularly varying distributions (e.g., the Pareto distribution). More specifically, this property nearly corresponds to the class of regularly varying distributions; however, it is more general. Recall that Lemma 2.13 states that a regularly varying distribution with index $-\alpha$, has $\lim x \to \infty \log \bar{F}(x)/\log(x) = -\alpha$, but the converse does not hold.

The conditions imposed by Theorem 6.2 appear technical at first glance, but are all quite natural. We have already explained the motivation for conditions (a) and (b). Condition (c) is also natural since it is parallel to the condition that $\mathrm{Var}[\log Y_i] < \infty$ in the multiplicative central limit theorem. That is, condition (c) ensures the distribution of Y_i is not "too heavy-tailed" for standard analytic techniques to be used, that is, that the tail of Y_1 is bounded above by a power law, $\Pr(Y_1 > x) = o(x^{-\alpha})$ for some $\alpha > 0$. Of course, condition (c) is not particularly restrictive since it already includes distributions that are quite heavy-tailed, for example, regularly varying distributions with an infinite mean. In particular, the condition corresponds to imposing that $\log Y_i$ is light-tailed.

Proof of Theorem 6.2   As in the case of the multiplicative central limit theorem, the key step in the proof of Theorem 6.2 is to apply a logarithmic transformation on our multiplicative process to convert it into an additive process. This allows us to deduce the limiting behavior of our multiplicative process from analysis of the behavior of random walks.

   For simplicity, we prove the result only in the case of $l = 1$. However, the case of general l can be reduced to this case without too much additional effort.

   Assuming that $l = 1$, we may inductively express P_n as follows:

$$\begin{aligned}
P_n &= \max(\max(P_{n-1}Y_{n-1}, 1)Y_n, 1) \\
&= \max(P_{n-2}Y_{n-1}Y_n, Y_n, 1) \\
&= \max(\max(P_{n-3}Y_{n-2}, 1)Y_{n-1}Y_n, Y_n, 1) \\
&= \max(P_{n-3}Y_{n-2}Y_{n-1}Y_n, Y_{n-2}Y_{n-1}Y_n, Y_{n-1}Y_n, Y_n, 1).
\end{aligned}$$

Proceeding in this manner, and noting that $P_0 = 1$, it is easy to see that

$$P_n = \max(Y_1Y_2, Y_3 \cdots Y_{n-1}Y_n,\ Y_2, Y_3 \cdots Y_{n-1}Y_n,\ \ldots,\ Y_{n-1}Y_n,\ Y_n,\ 1).$$

Now, since Y_i are i.i.d., we may write

$$P_n \overset{d}{=} \max(1,\ Y_1,\ Y_1Y_2,\ Y_1Y_2Y_3,\ \ldots,\ Y_1Y_2 \cdots Y_{n-1},\ Y_1Y_2 \cdots Y_{n-1}Y_n).$$

At this point, we make the logarithmic transformation $X_i = \log(Y_i)$ for $i \geq 1$, and $Z_n = \log(P_n)$ to obtain

$$Z_n \overset{d}{=} \max(S_0, S_1, \ldots, S_n), \tag{6.2}$$

where $S_0 = 0$, and $S_i = \sum_{j=1}^{i} X_j$ for $j \geq 1$. We have thus expressed Z_n as the maximum over n time steps of a random walk starting at zero, with i.i.d. increments X_i. Moreover, our assumptions on the distribution of Y_1 imply that

(a) $\mathbb{E}[X_1] < 0$, that is, the random walk has a negative drift,
(b) $\Pr(X_1 > 0) > 0$, that is, the random walk makes positive increments with some probability,
(c) $\mathbb{E}[e^{sX_1}] < \infty$ for some $s > 0$, that is, the increments are light-tailed to the right.

Now, it is clear that as $n \to \infty$, $\max(S_0, S_1, \ldots, S_n) \to S_{\max} = \max_{j \geq 0}\{S_j\}$, the all-time maximum of the random walk, whenever $S_{\max}$ is finite. We study $S_{\max}$ in detail in Section 7.4, and we can now apply one of the results we prove there. In particular, under the preceding assumptions on X_1, we show that $S_{\max}$ is finite with probability 1, and

$$\lim_{x \to \infty} -\frac{\log \Pr\left(S_{\max} > x\right)}{x} = s^*, \tag{6.3}$$

where $s^* = \sup\{s \geq 0 \mid \mathbb{E}\left[e^{sX_1}\right] \leq 1\}$ (see Theorem 7.8).

From (6.2), it now follows that as $n \to \infty$,

$$Z_n \xrightarrow{d} S_{\max},$$

implying that

$$P_n = e^{Z_n} \xrightarrow{d} e^{S_{\max}}.$$

Finally, we note that the distribution F of $e^{S_{\max}}$ satisfies

$$\lim_{x \to \infty} -\frac{\log \bar{F}(x)}{\log(x)} = \lim_{x \to \infty} -\frac{\log \Pr\left(e^{S_{\max}} > x\right)}{\log(x)}$$
$$= \lim_{x \to \infty} -\frac{\log \Pr\left(S_{\max} > \log(x)\right)}{\log(x)} = s^*,$$

where $s^* = \sup\{s \geq 0 \mid \mathbb{E}\left[e^{sX_1}\right] \leq 1\} = \sup\{s \geq 0 \mid \mathbb{E}\left[Y_1^s\right] \leq 1\}$. □

6.2.2 A Noisy Multiplicative Process

We have just seen that the phenomenon of multiplicative processes leading to heavy-tailed distributions is robust to a lower barrier. Here, we demonstrate that it is also robust to additive noise in the process. In particular, in this section we discuss a "noisy" multiplicative process where there is a small amount of additive noise between the steps of multiplicative growth. Specifically,

$$P_n = P_{n-1}Y_n + Q_n, \tag{6.4}$$

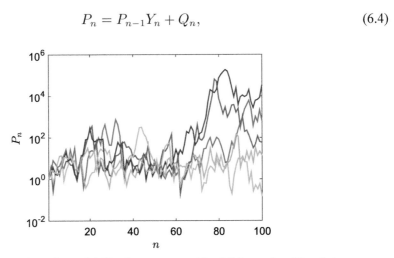

Figure 6.3 Illustration of a multiplicative process with additive noise. The plots show five realizations of P_n with $Y_i \sim \text{Exp}(0.75)$, Q_i being distributed as the absolute value of a standard Gaussian.

where the Y_n are i.i.d. and strictly positive, the Q_n provide additive noise and are i.i.d. nonnegative, and $P_0 = 1$ is the starting point of the process.[2] See Figure 6.3 for a visualization of this process.

This noisy variation of a multiplicative process captures the idea that there is often an opportunity for small additive changes to occur between multiplicative growth stages. For example, in the case of incomes, the effect of a bonus or a promotion is an additive change that happens between yearly raises; and, in the case of populations, the effect of immigration may be viewed as an additive change that combines with the multiplicative effect of birth rate. Alternatively, the noise could be a result of error in observations of the growth of some physical process.

Interestingly, the addition of additive noise to a multiplicative process causes a qualitative change to the limiting distribution that emerges. In fact, the impact is the same as in the previous case where we studied the addition of a lower barrier: the limiting distribution is no longer always LogNormal; instead, it is approximately power law.

Theorem 6.3 *Consider the noisy multiplicative process $P_n = P_{n-1} \cdot Y_n + Q_n$, where $P_0 = 1$ and $\{Y_n, Q_n\}_{n \geq 1}$ is an i.i.d. sequence of random variable pairs, where Y_n is strictly positive and Q_n is nonnegative. Suppose that the distribution of Y_1 satisfies the following conditions.*

(a) $\Pr(Y_1 > 1) > 0.$
(b) $\mu := e^{\mathbb{E}[\log(Y_1)]} < 1.$
(c) $\mathbb{E}[Y_1^s] < \infty$ for some $s > 0$

Additionally, suppose that Q_1 is light-tailed and $\Pr(Q_1 > 0) > 0$. Then, $P_n \xrightarrow{d} F$, where F is heavy-tailed and satisfies $\lim_{x \to \infty} -\frac{\log \bar{F}(x)}{\log(x)} = s^*$, with $s^* = \sup\{s \geq 0 \mid \mathbb{E}[Y_1^s] \leq 1\}.$

Note that Theorem 6.3 essentially parallels Theorem 6.2 for the case of a multiplicative process with a lower barrier. The same conditions are sufficient for a limiting distribution to emerge, and the limiting distribution that emerges satisfies the same logarithmic asymptotics. Further, it is again the case that a LogNormal distribution does not emerge, and instead the emerging distribution is approximately power law (i.e., asymptotically linear on a log-log scale).

Though it may not be apparent initially why Theorem 6.3 and Theorem 6.2 parallel each other, there is some intuition for the similarity. Roughly, the additive noise provides a "soft" lower bound on the process. That is, because the noise is i.i.d., the multiplicative process cannot spend much time below the median of Q_n. Thus, though the noise does not give a strict lower barrier, it does provide a significant boost to the process when it is near zero.

However, the intuitive connection between noisy multiplicative processes and multiplicative processes with lower bounds does not translate directly to a proof. In fact, the addition of noise to the process means that a structurally different proof technique is required to prove

---

[2] We adopt the terminology of "noise" to describe Q_n following Newman [163]. However, because Q_n is non-negative it may be more descriptive to think of Q_n as a "repulsive force" that keeps the process from shrinking to zero, which is the phrasing used in Sornette [200] and Gabaix [93].

Theorem 6.3. The consequence is that the full proof is more technical, and we do not include it here. The interested reader can find a proof in [93, 125, 200]. Here, we give a proof of only the special case where $Q_n = 1$, which can be done cleanly and highlights the connection between the case of additive noise and the case of a lower barrier.

Proof of Theorem 6.3 for $Q_n = 1$ We begin by obtaining a clearer characterization of our process for the special case $Q_n = 1$. It is not hard to show from (6.4) that

$$P_n = 1 + Y_n + Y_n Y_{n-1} + Y_n Y_{n-1} Y_{n-2} + \cdots + Y_n Y_{n-1} \cdots Y_1.$$

Thus,

$$P_n \stackrel{d}{=} 1 + Y_1 + Y_1 Y_2 + Y_1 Y_2 Y_3 + \cdots + Y_1 Y_2 \cdots Y_n.$$

As in the proof of Theorem 6.2, we now make the logarithmic transformation $X_i = \log(Y_i)$ for $i \geq 1$, yielding

$$P_n \stackrel{d}{=} \sum_{k=0}^{n} e^{S_k},$$

where $S_0 = 0$, and $S_i = \sum_{j=1}^{i} X_j$ for $j \geq 1$. It now follows that

$$P_n \stackrel{d}{\to} P = \sum_{k=0}^{\infty} e^{S_k}. \tag{6.5}$$

Having identified the limiting distribution of P_n, it now remains to the characterize the tail behavior of this limiting distribution. Our approach will be to derive asymptotically matching lower and upper bounds on the tail. In both cases there is strong connection to the case of a multiplicative process with a lower barrier. We start with the lower bound.

Lower bound: The lower bound is obtained by noting that (6.5) implies that

$$P \geq P' := \sup_{k \geq 0} e^{S_k}$$

almost surely. This lower bound has a simple interpretation: note that P' is simply the limit of the multiplicative process with a lower barrier defined by $P'_n = \max\{P'_{n-1} Y_n, 1\}$, $P'_0 = 1$. Indeed, it is not hard to see directly that the $P_n \geq P'_n$ for all n. This bound allows us to invoke Theorem 6.2 for multiplicative processes with a lower barrier to conclude that

$$\liminf_{x \to \infty} \frac{\log \Pr(P > x)}{\log x} \geq \lim_{x \to \infty} \frac{\log \Pr(P' > x)}{\log x} = -s^*.$$

Upper bound: The proof of the upper bound is also highly related to the proof of Theorem 6.2. It follows from (6.5) that for $\epsilon > 0$,

$$P \leq e^{\sup_{k \geq 0}[S_k + \epsilon k]} \sum_{k=0}^{\infty} e^{-\epsilon k} = \frac{1}{1 - e^{-\epsilon}} e^{\sup_{k \geq 0}[S_k + \epsilon k]} =: c_\epsilon P'_\epsilon.$$

We now note that $\sup_{k \geq 0}[S_k + \epsilon k]$ is simply the supremum of the random walk with increment process $X_n + \epsilon$. Taking ϵ small enough so tha $\mathbb{E}[X_n + \epsilon] < \infty$, we can conclude from Theorem 6.2 that

$$\lim_{x \to \infty} \frac{\log \Pr(P'_\epsilon > x)}{\log x} = -s^*_\epsilon,$$

where

$$s_\epsilon^* = \max\{s : \mathbb{E}\left[e^{s(X_1+\epsilon)}\right] \leq 1\} = \max\{s : \mathbb{E}\left[Y_1^s\right] \leq e^{-\epsilon s}\}.$$

We thus obtain the upper bound

$$\limsup_{x\to\infty} \frac{\log \Pr\left(P_\epsilon' > x\right)}{\log x} \leq -s_\epsilon^*.$$

To see that the upper bound matches the lower bound, observe that s_ϵ^* is decreasing in ϵ. Furthermore, as $E[Y_1^s]$ is continuous on $(0, s^*)$, it follows that $s_\epsilon^* \uparrow s^*$ as $\epsilon \downarrow 0$, removing the gap between the lower and upper bound. $\qquad\qquad\square$

6.3 An Example: Preferential Attachment and Yule Processes

So far in this chapter we have focused on generic multiplicative processes without delving into any particular example in depth. To end the chapter, we focus in some depth on one specific, celebrated example of a multiplicative process that comes from the area of *network science*.

Network science has provided some of the most popular, and controversial, examples of heavy-tailed phenomena. There was an enormous amount of excitement surrounding the discovery of heavy-tailed phenomena in complex networks during the late 1990s and early 2000s as large datasets about social, biological, and communication networks began to emerge. This period was particularly exciting due to the diversity of settings where large datasets were becoming available and the apparent "universality" of the features observed in these networks. One of the particularly striking features that was observed repeatedly in complex networks from a variety of settings was that the degree distribution tended to be heavy-tailed, often power law, and thus scale invariant. Such observations of scale-invariant behavior emerged in the context of social networks, citation networks, biological networks, communication networks, power networks, and beyond.[3]

The emergence of power law degree distributions in diverse complex networks, aka the emergence of scale-free networks, motivated a deep scientific question that was (and still is) particularly seductive for the network science research community:

> *Is there some sort of unifying explanation, or fundamental law, governing the emergence of heavy-tailed, scale-invariant behavior in complex networks?*

The seductive nature of this question has led to an enormous body of literature that seeks to develop and analyze models for the evolution of networks and, at this point, there is an enormous array of models that lead to heavy-tailed degree distributions in networks. For example, [8, 24, 42, 135, 147, 168]; excellent surveys in this context can be found in [23,

---

[3] As we discuss in Chapter 8, in some cases the initial claims of power-law distributions have since been refuted in favor of either other heavy-tailed distributions (e.g., LogNormal distributions or Weibull distributions) or, in some cases, light-tailed distributions. Examples of such refutations include the internet graph [48, 222] and protein–protein interaction networks [206]. Debate about which networks exhibit power-law degree distributions continues to this day (e.g., [37, 212]).

163, 209]. These networks use a variety of growth mechanisms; however, many of them rely on some form of multiplicative process at their core. To highlight this fact, in this section we describe a particularly celebrated example of such a mechanism: *preferential attachment*.

The Preferential Attachment Mechanism

The preferential attachment mechanism has a long history and has been reinvented under many guises over the years. In particular, it is essentially equivalent to the growth of the so-called Yule processes [224], which are named after the statistician who first studied them in 1925, Udny Yule. Initially, Yule processes were introduced as a simplified model for studying the sizes of biological taxa; however, it did not take long for them to be adopted in other disciplines as well. For example, Herbert Simon [196] applied Yule processes to study the distribution of wealth and, in the process, gave them their name. It did not take long for Yule processes to find application in complex networks. In 1968 Derek Price [179] adopted the Yule process to study the evolution of citation networks, renaming the mechanism "cumulative advantage." More recently, during the excitement surrounding the discoveries of heavy-tailed phenomena in complex networks in the 1990s, Albert-László Barabási and Réka Albert [24] rediscovered Price's mechanism, giving it the name that is commonly used today: preferential attachment. It is this name and context that we adopt in the following; however, we would like to emphasize that there is nothing inherent in preferential attachment that restricts its application to complex networks, and in fact it has many applications outside of this context.

The preferential attachment mechanism essentially formalizes the "rich get richer" phenomenon in the context of a network. In particular, under preferential attachment the network grows in a manner where new edges are "preferentially attached" to vertices that already have high degree (i.e., the "rich" vertices).

More specifically, consider a discrete time process on $t = 0, 1, \ldots, n$ for creating a directed graph. At each time t, one vertex arrives, with a single link to itself, and chooses a preexisting vertex to connect to via a directed edge from itself to the selected vertex. The key to the preferential attachment mechanism is that the vertex to connect with is chosen with probability proportional to its in-degree.

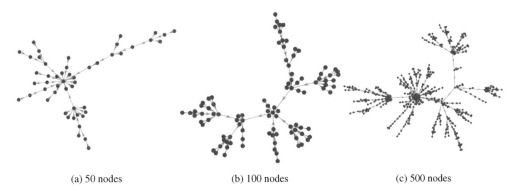

(a) 50 nodes   (b) 100 nodes   (c) 500 nodes

Figure 6.4 Illustrations of graphs generated by the preferential attachment model. For clarity, self-edges are not shown. Note that there are a few "core" high-degree nodes, surrounded by a periphery of many low-degree nodes.

Formally, we study a model where, with probability $1 - \alpha > 0$, the newly arriving vertex uses a preferential attachment mechanism and chooses a preexisting vertex to connect to proportionally to its in-degree; and, with probability α, the newly arriving vertex chooses a preexisting vertex uniformly at random. So, $1 - \alpha$ is the probability of preferential attachment and α is the probability of random choice. Thus, the probability that the vertex arriving at time t chooses a certain vertex with in-degree k to connect to equals

$$\frac{\alpha}{t} + \frac{(1 - \alpha)k}{2t}.$$

Note that there are $2t$ preexisting edges in the network when the new vertex arrives at time t.[4]

Some examples of networks that emerge from the preferential attachment mechanism are shown in Figure 6.4. These examples highlight some of the properties that are commonly associated with preferential attachment. In particular, it is easy to see that there is extreme diversity in the in-degrees of the vertices.[5] A consequence is that high-degree nodes tend to be very "central" to the network and, in particular, the network exhibits a so-called "core-periphery" structure where older, higher-degree nodes form a tightly connected core surrounded by a periphery of lower-degree vertices. Similarly to heavy-tailed degree distributions, this core-periphery structure is another aspect of complex networks that is, in some sense, "universal."

The Degree Distribution under Preferential Attachment

The preferential attachment mechanism is closely tied to the multiplicative processes that we have studied in this chapter. At each time, the probability that a new vertex chooses to connect to a given vertex is proportional to the in-degree of the vertex, and thus the expected increase in the in-degree is proportional to the current degree. As a result, we should *expect* the emergent degree distribution under the preferential attachment mechanism to be heavy-tailed. Indeed this is the case. However, a formal proof does not follow directly from the generic results about multiplicative processes we have discussed to this point in the chapter. It requires a more detailed analysis, which we perform in the following.

More specifically, Theorem 6.4 states that the degree distribution that emerges under the preferential attachment mechanism is not just heavy-tailed, but also power-law and scale invariant. To state the theorem, let F_t denote the distribution of the in-degree when the total number of nodes equals t, that is, $F_t(k)$ equals the probability that a randomly selected node has in-degree $\leq k$, just before the arrival of the node at time t.

Theorem 6.4 *Under the preferential attachment model, $F_t \xrightarrow{d} F$, as $t \to \infty$, where $\bar{F}(x) \sim \beta x^{-\frac{2}{1-\alpha}}$ for some positive constant β.*

Interestingly, this theorem not only shows that the degree distribution under preferential attachment is a power-law, it also gives the dependence of the power-law on the probability

---

[4] To make this model consistent at $t = 0$, let us assume that the first node creates two links to itself, i.e., it starts with an in-degree of 2. Note that each subsequent node starts with in-degree 1.

[5] The degree of a vertex is strongly correlated with its arrival time, i.e., older vertices tend to have larger in-degree.

of preferential attachment $(1 - \alpha)$. In particular, as the probability of preferential attachment increases (i.e., as α decreases), the tail of the limiting degree distribution becomes heavier. Moreover, by varying the parameter α over $[0, 1)$, the model generates power-law degree distributions with the exponent varying over $[2, \infty)$. However, it is important to note that the specific range on the exponent of the power-law is not a fundamental consequence of preferential attachment itself; it is simply a consequence of the particular way we have chosen to model it. There are a wide variety of variations on the preferential attachment model that lead to different expressions for the power-law exponent (see, for example, [24, 67, 131, 179, 196]). Examples of such model variations include allowing the incoming node to attach to more than one preexisting node, allowing the preexisting nodes to add edges to one another in a preferential manner, and so on.

It is also important to point out that the preferential attachment mechanism is only one of many mechanisms that can lead to heavy-tailed phenomena in networks. In particular, though the mechanism provides a simple explanation for the emergence of a heavy-tailed degree distribution, one should be careful to conclude that it is an accurate model for complex networks more broadly. In particular, the simplicity of the preferential attachment mechanism means that many other properties that are associated with complex networks are not captured. And, in fact, if one simply compares the examples of preferential attachment networks with real-world examples of complex networks, it is clear that there are significant differences. As a result, since the introduction of the preferential attachment model, many other examples of generative mechanisms for networks have been proposed (e.g., the configuration model [209], hyperbolic graphs [169], Kronecker graphs [134], the forest-fire model [136], dot-product graphs [223], and many more).

Finally, let us move on to the proof of Theorem 6.4. The proof may appear technical at first glance, but the structure of the proof is actually quite simple.

Proof of Theorem 6.4 The idea of this proof is to first understand the evolution of the average number of nodes with degree k at time t and then use that understanding to study the limiting degree distribution as $t \to \infty$. While the second piece of the proof requires some technical calculation, the first piece is simple and already provides clear intuition as to why preferential attachment is related to multiplicative processes and, thus, why a heavy-tailed degree distribution emerges.

To start, let $m_t(k)$ denote the average number of preexisting nodes with degree k at time t. Let $p_t(k) = \frac{m_t(k)}{t}$. Note that $p_t(k)$ is the probability that a randomly selected (preexisting) node at time t has degree k, that is, $p_t(\cdot)$ is the probability mass function corresponding to the distribution F_t. Our objective is to analyze the behavior of $p_t(\cdot)$ as $t \to \infty$.

Our first step toward this goal is to develop a recursive relation for $m_t(k)$. Consider the newly arriving node at time t. As we have noted, the probability that this node connects to a particular preexisting node with in-degree k equals $\frac{\alpha}{t} + \frac{(1-\alpha)k}{2t}$. Since there are, on average, $m_t(k)$ nodes with in-degree k, the probability that the newly arriving node connects with a node with in-degree k equals $m_t(k)\left(\frac{\alpha}{t} + \frac{(1-\alpha)k}{2t}\right)$. We can therefore express the evolution of the average number of nodes with in-degree $k > 1$ as follows:

$$m_{t+1}(k) = m_t(k) + m_t(k-1)\left(\frac{\alpha}{t} + \frac{(1-\alpha)(k-1)}{2t}\right) - m_t(k)\left(\frac{\alpha}{t} + \frac{(1-\alpha)k}{2t}\right)$$

$$= m_t(k) \left(1 - \frac{2\alpha + (1-\alpha)k}{2t}\right) + m_t(k-1) \left(\frac{2\alpha + (1-\alpha)(k-1)}{2t}\right).$$

$$(6.6)$$

Note that this evolution of the average number of nodes with in-degree k accounts for an increase coming from the possibility that a node with in-degree $k-1$ at the previous time step gained an incoming edge, and a decrease coming from the possibility that a node with in-degree k at the previous time step gained an incoming edge. The evolution for the case $k = 1$ is different. Since the increase in the number of nodes with degree 1 comes only from the newly arrived node,

$$m_{t+1}(1) = m_t(1) + 1 - m_t(1) \left(\frac{\alpha}{t} + \frac{(1-\alpha)}{2t}\right)$$

$$= m_t(1) \left(1 - \frac{1+\alpha}{2t}\right) + 1. \tag{6.7}$$

Already, the form of the evolution of $m_t(k)$ suggests that a heavy-tailed degree distribution is likely. Note that the existence of terms $m_t(k-1)\cdot(k-1)$ highlights the multiplicative nature of the growth process and, given the results in this chapter, that already indicates that a heavy-tailed degree distribution is likely.

Proceeding with the proof, we now use the recursions for $m_t(k)$ to analyze the asymptotic behavior of $p_t(\cdot)$ as $t \to \infty$. We denote, for $k \geq 1$,

$$p(k) := \lim_{t\to\infty} p_t(k) = \lim_{t\to\infty} \frac{m_t(k)}{t}.$$

We prove that the limit exists later in the proof. Note that $p(k)$ may be interpreted as long-term probability of picking a node with degree k. In other words, $p(\cdot)$ is the probability mass function corresponding to our limiting distribution F.

Our analysis of the behavior of $p_t(\cdot)$ as $t \to \infty$ is based on the technical result, the proof of which is left to the reader as Exercise 7. Consider the recursion

$$a_{t+1} = a_t \left(1 - \frac{b_t}{t}\right) + c_t.$$

If $\lim_{t\to\infty} b_t = b \in (0, \infty)$ and $\lim_{t\to\infty} c_t = c \in \mathbb{R}$, then $\frac{a_t}{t}$ converges as $t \to \infty$, and

$$\lim_{t\to\infty} \frac{a_t}{t} = \frac{c}{1+b}.$$

Applying this result to the recursion (6.7) (taking $a_t = m_t(1)$, $b_t = \frac{1+\alpha}{2}$, and $c_t = 1$), we conclude that $\lim_{t\to\infty} \frac{m_t(1)}{t}$ exists, and

$$p(1) = \lim_{t\to\infty} \frac{m_t(1)}{t} = \frac{1}{1 + \frac{1+\alpha}{2}} = \frac{2}{3+\alpha}.$$

Next, we study the asymptotic behavior of $p(k)$ for $k \geq 2$ by applying the same result to the recursion (6.6). To show that $\frac{m_t(k)}{t}$ converges as $t \to \infty$, we can proceed inductively. Suppose that $\frac{m_t(k-1)}{t}$ converges as $t \to \infty$. We can now apply this result to recursion (6.6), setting $a_t = m_t(k)$, $b_t = \frac{2\alpha+(1-\alpha)k}{2}$, and $c_t = m_t(k-1)\left(\frac{2\alpha+(1-\alpha)(k-1)}{2t}\right)$. This gives that $\frac{m_t(k)}{t}$ converges as $t \to \infty$, and

$$p(k) = \lim_{t\to\infty} \frac{m_t(k)}{t} = \frac{c}{1 + \frac{2\alpha+(1-\alpha)k}{2}},$$

where

$$c = \lim_{t\to\infty} m_t(k-1)\left(\frac{2\alpha+(1-\alpha)(k-1)}{2t}\right) = p(k-1)\left(\frac{2\alpha+(1-\alpha)(k-1)}{2}\right).$$

Therefore, we have

$$p(k) = p(k-1)\left[\frac{2\alpha+(1-\alpha)(k-1)}{2\left(1+\frac{2\alpha+(1-\alpha)k}{2}\right)}\right]$$

$$= p(k-1)\left[\frac{\frac{2\alpha}{1-\alpha}+k-1}{\frac{2+2\alpha}{1-\alpha}+k}\right].$$

This relation can be iterated to obtain

$$p(k) = p(1)\left[\frac{\left(\frac{2\alpha}{1-\alpha}+1\right)\left(\frac{2\alpha}{1-\alpha}+2\right)\cdots\left(\frac{2\alpha}{1-\alpha}+k-1\right)}{\left(\frac{2+2\alpha}{1-\alpha}+2\right)\left(\frac{2+2\alpha}{1-\alpha}+3\right)\cdots\left(\frac{2+2\alpha}{1-\alpha}+k\right)}\right].$$

The remainder of the argument consists of manipulating the preceding expression into a more intuitive form. To this end, we can represent $p(k)$ in terms of the gamma function,[6] using the property $\Gamma(a) = (a-1)\Gamma(a-1)$ as follows:

$$p(k) = p(1)\left[\frac{\Gamma\left(\frac{2\alpha}{1-\alpha}+k\right)}{\Gamma\left(\frac{2\alpha}{1-\alpha}+1\right)} \frac{\Gamma\left(\frac{2+2\alpha}{1-\alpha}+2\right)}{\Gamma\left(\frac{2+2\alpha}{1-\alpha}+k+1\right)}\right].$$

Collecting the factors that do not depend on k into a single constant β', we obtain

$$p(k) = \beta'\frac{\Gamma\left(\frac{2\alpha}{1-\alpha}+k\right)}{\Gamma\left(\frac{2+2\alpha}{1-\alpha}+k+1\right)}.$$

We can now deduce the asymptotic behavior of $p(k)$, using the following property of the gamma function: for fixed y,[7]

$$\frac{\Gamma(x)}{\Gamma(x+y)} \sim x^{-y}.$$

---

[6] The gamma function, for nonnegative z, is defined as $\Gamma(z) = \int_0^\infty x^{z-1}e^{-x}dx$. It is easy to see that $\Gamma(0) = 1$, and $\Gamma(z+1) = z\Gamma(z)$. It thus follows that for $n \in \mathbb{N}$, $\Gamma(n+1) = n!$. Thus, the gamma function may be interpreted as an extension of the factorial function over the real line.

[7] See Exercise 6.

Therefore, we conclude that

$$p(k) \sim \beta' \left(k + \frac{2\alpha}{1-\alpha} \right)^{-\left(1 + \frac{2}{1-\alpha} \right)} \sim \beta' k^{-\left(1 + \frac{2}{1-\alpha} \right)}.$$

This form shows that the probability mass function of F decays as a power-law. It now follows easily (see Exercise 8) that

$$\bar{F}(k) = \sum_{j>k} p(j) \sim \frac{\beta'(1-\alpha)}{2} k^{-\frac{2}{1-\alpha}},$$

which completes the proof. $\square$

6.4 Additional Notes

Multiplicative processes are tightly tied to many generative models for heavy-tailed phenomena. The observation that the simple multiplicative process we consider has a LogNormal limit is quite classical; the book by Aitchison and Brown [7] is an early reference. This idea has been used to claim the emergence of the LogNormal distribution in several domains, including finance (see, for example, [50, 117]), basic sciences like ecology, geology, and atmospheric science (see the edited volume [53]), and more recently, communication networks (see, for example, [6, 116]).

The observation that multiplicative mechanisms with a lower barrier result in a power-law limiting distribution was first made by Champernowne [44]. The additive noise variant was first studied by Kesten [125], who analyzed a multidimensional version of the processes we introduce in Section 6.2.2. Kesten's result has been extended and simplified during the past several decades (e.g., see [39, 128, 154]). A recent book devoted to this subject is [40]. These mechanisms have been used to argue the emergence of power laws in distributions of incomes [44], city sizes [93], and stock market returns [139], to name a few.

More generally, all of the multiplicative processes we consider in this chapter are special cases of a general recursion of the form

$$P_{n+1} = \Phi_n(P_n), n \geq 0, \tag{6.8}$$

with $\Phi_n, n \geq 0$ a sequence of i.i.d. random functions satisfying a linear growth condition, this is, $\Phi_n(x) = A_n x + o(x)$ as $x \to \infty$. Under certain conditions (especially on the distribution of the slope A_n), it has been shown that power laws appear for processes of this form. A seminal paper on this topic is due to Goldie, [102]. These results have further been extended to settings with branching mechanisms in [122].

To illustrate the application of multiplicative processes in a specific application setting, we discussed the preferential attachment generative model for complex networks. This model has a long history and has been reinvented under many guises. The model is usually credited to Yule, who used the model to study biological taxa in 1925 [224]. However, Pólya actually used the same model a few years earlier in the study of his celebrated urn model [74]. Later, Simon applied them to the modeling of the populations of cities and the distribution of wealth in the 1950s [196]. In the 1960s, Price applied the model to citation networks [179] and, most recently, Barabási and Albert suggested that the model could be used more generally to study complex networks such as the internet graph and the world wide web [24].

Our focus on preferential attachment should not suggest that it is the only (or even the most realistic) generative model for complex networks that exhibits a heavy-tailed degree distribution. Heavy-tailed degrees are only one of many salient features that have been observed in complex networks, and the preferential attachment mechanism does not generate networks that have other important properties, such as (i) searchability, the ability for agents to find short paths quickly [126]; (ii) densification, the fact that complex networks often become more dense as they grow [137]; and (iii) shrinking diameters, the fact that the diameter of complex networks often shrinks as the network grows [137].

There is a wide variety of other probabilistic generative models (e.g., the configuration model [209], hyperbolic graphs [169], Kronecker graphs [134], the forest-fire model [136], dot-product graphs [223], and many more). Nearly all of these can be tied to some form of multiplicative process, as was the case for the preferential attachment model.

There are also many other classes of generative models that are used to provide explanations for heavy-tailed phenomena in complex networks. One particularly important class of models is based on optimization rather than randomness. This direction was pioneered by Mandelbrot, who argued that power-laws form a sort of information-theoretic optimal for language [147]. In particular, Mandelbrot argued that power-law frequency of word lengths optimizes the average amount of information per letter in a language such as English. This idea was later adopted and extended to a variety of other settings (e.g., Carson and Doyle suggest models for file sizes and forest fires [42], and Fabrikant et al. suggest an application to modeling the internet graph [78]).

Interestingly, Mandelbrot's optimization-based model and Simon's preferential attachment model were introduced around the same time, and there was a heated debate between the two about the contrasting assumptions and explanations of heavy-tailed behavior; see [148–150,197–199]. This was mirrored decades later, as there was again a heated debate between the preferential attachment model and optimization-based approaches in the early 2000s; see [141, 222].

6.5 Exercises

1. The goal of this exercise is to get a feel for the multiplicative mechanisms described in this chapter using data. Let $\{Y_i\}_{i\geq1}$ be a sequence of i.i.d. exponential random variables with mean $1/\lambda$.

   (a) *Simple multiplicative process:* Set λ such that $\mu = e^{\mathbb{E}[\log(Y_1)]} > 1$. For some (large) fixed N, generate a large number of samples of $\left(\frac{P_N}{\mu^N}\right)^{1/\sqrt{N}}$, where $P_0 = 1$, $P_n = P_{n-1} \cdot Y_n$ for $n \geq 1$. Does your data look LogNormal, as predicted by the multiplicative central limit theorem? You might want to make a logarithmic transformation on your data to see this.
   Repeat this exercise by setting λ such that $\mu < 1$.

   (b) *Multiplicative process with a lower barrier:* Set λ such that $\mu = e^{\mathbb{E}[\log(Y_1)]} < 1$. For some (large) fixed N, generate a large number of samples of P_N, where $P_0 = 1$, $P_n = \max(P_{n-1} \cdot Y_n, 1)$ for $n \geq 1$.

Plot the empirical c.c.d.f. of the data (i.e., the fraction of data points exceeding x as a function of x), using a logarithmic scale on both axes. Is your visualization consistent with the conclusion of Theorem 6.2?

(c) *Multiplicative process process with additive noise:* Set λ such that $\mu = e^{\mathbb{E}[\log(Y_1)]} < 1$. For some (large) fixed N, generate a large number of samples of P_N, where $P_0 = 1$, $P_n = P_{n-1} \cdot Y_n + Q_n$ for $n \geq 1$, where $\{Q_n\}_{n\geq 1}$ is a sequence of i.i.d. uniform random variables taking values in $[0, 1]$. Once again, plot the empirical c.c.d.f. of the data using a logarithmic scale on both axes. Is your visualization consistent with the conclusion of Theorem 6.3?

2. Show that the LogNormal distribution is *stable under multiplication*. Specifically, for $n \geq 2$, if $Y_1, Y_2, \ldots, Y_n$ are i.i.d. LogNormal random variables, show that there exist positive constants $c, d > 0$ such that

$$\prod_{i=1}^{n} Y_i \stackrel{d}{=} cY_1^d.$$

3. Consider the simple multiplicative process defined in Section 6.1. Prove that
   (a) If $\mu > 1$, then $P_n \to \infty$ almost surely as $n \to \infty$.
   (b) If $\mu < 1$, then $P_n \to 0$ almost surely as $n \to \infty$.
   Now, construct an example where, as $n \to \infty$, $P_n \to 0$ almost surely, but $\mathbb{E}[P_n] \to \infty$. How can you explain this phenomenon?

4. The goal of this exercise is to gain some experience in computing the power-law tail index s^* that emerges in the limit of a multiplicatve process with a lower barrier (Theorem 6.2) or with additive noise (Theorem 6.3). Specifically, compute $s^* := \{s \geq 0 \mid \mathbb{E}[Y^s] \leq 1\}$ for the following examples (you should also verify that these examples are consistent with assumptions of Theorems 6.2 and 6.3).
   - For $p > 1/2$,

$$Y = \begin{cases} 1/2 & \text{w.p. } p, \\ 2 & \text{w.p. } 1 - p. \end{cases}$$

   - $Y \sim \text{LogNormal}(\mu, \sigma^2)$, where $\mu < 0$.

5. The goal of this exercise is to get a feel for the preferential attachment model. For $\alpha \in [0, 1]$, write a program for generating a random graph with n nodes using the preferential attachment model presented in Section 6.3.
   (a) Plot the graphs you obtain for $n = 10, 100, 1000$. Do you see the "rich get richer" phenomenon?
   (b) For large enough n, plot the empirical c.c.d.f. of the in-degree distribution on a log-log scale. Is your visualization consistent with the statement of Theorem 6.4?

6. For $y \in \mathbb{N}$, show that as $x \to \infty$,

$$\frac{\Gamma(x)}{\Gamma(x + y)} \sim x^{-y}.$$

Note: This statement also holds for all positive real y, though this is harder to prove.

7. The goal of this exercise is to prove the following result, which was used in the proof of Theorem 6.4.

Consider a sequence $\{a_t\}$ that satisfies the recursion

$$a_{t+1} = a_t \left(1 - \frac{b_t}{t} \right) + c_t,$$

where $\lim_{t \to \infty} b_t = b \in (0, \infty)$ and $\lim_{t \to \infty} c_t = c \in \mathbb{R}$. Prove that

$$\lim_{t \to \infty} \frac{a_t}{t} = \frac{c}{1+b}.$$

8. The goal of this exercise is to prove the following result, which was used in the proof of Theorem 6.4. It shows that if a probability mass function is asymptotically power law, then the corresponding c.c.d.f. is also asymptotically power law. Note that this may be viewed as a highly simplified Karamata's theorem.

   Let p_X denote the probability mass function of a random variable X taking integer values. As $k \to \infty$, if $p_X(k) \sim ck^{-(1+\gamma)}$, where $c, \gamma > 0$, show that $\bar{F}_X(k) \sim \frac{c}{\gamma} k^{-\gamma}$.

   Hint: You can bound $\sum_{m>k} m^{-(1+\gamma)}$ *as follows.*

   $$\int_{k+1}^{\infty} x^{-(1+\gamma)} dx \leq \sum_{m>k} m^{-(1+\gamma)} \leq \int_{k}^{\infty} x^{-(1+\gamma)} dx.$$

7

Extremal Processes

Extreme events, the good and the bad, shape our lives dramatically. Catastrophic events such as large earthquakes, hurricanes, pandemic viruses, floods, and stock market crashes have huge societal tolls, while exhilarating events such as world record-breaking times at the Olympics leave a sense of amazement. Though such events are almost always unexpected, as a society we must plan for them carefully – dams and dikes must be built in anticipation of possible flooding, buildings must be built to withstand the largest earthquakes, and insurance companies must be prepared for any number of catastrophic events. These situations motivate the importance of understanding the behavior of *extremal processes*, which grow as the maximum (or minimum) of a sequence of events. Further, as we illustrate in Part III of this book, extremal processes are crucial tools for statistical analysis of heavy-tailed distributions. In particular, they are key to the analysis of procedures for fitting the tail of heavy-tailed distributions.

Though extremal processes are more exotic than additive and multiplicative processes, we all have strong intuition about their behavior based on our life experiences. In particular, most typically we think of extremal processes as being analogous to the evolution of world record times. For example, consider the evolution of the world record for the half marathon (Figures 7.1(a) and 7.1(b)). Initially, the world record dropped quickly, sometimes by over a minute every few years, but recently the world record has tended to drop only a minute per *decade*. A similar slowing of the progression of world record times is evident for other events too (e.g., the 100 m dash (Figures 7.1(c) and 7.1(d))). We see related phenomena in many other aspects of our lives as well. Consider the progression of the size of the largest snowfall you have experienced or the height of the tallest mountain you have climbed. Thus, it is natural to assume that extremal processes have a decreasing rate of change and may even eventually asymptote to some limiting value – thus they seem to be light-tailed.

This intuition encourages the idea that heavy-tailed distributions are rare in the context of extremal processes. Of course, at this point in the book we have already seen how heavy-tailed distributions can emerge in the context of both additive and multiplicative processes, and so it should not be surprising that they are more than just curiosities here too. In particular, there are many contexts in which extremes yield heavy-tailed distributions, for example, the progression of size of the largest stock market crashes, earthquakes, and floods. As we show in Figure 7.2, these examples can exhibit very different behavior than the progression of world records.

The goal of this chapter is to explain why the emergence of heavy-tailed phenomena in the context of extremal processes is not surprising. Interestingly, the behavior of extremal processes closely parallels the behavior of additive processes, which we study in Chapter 5.

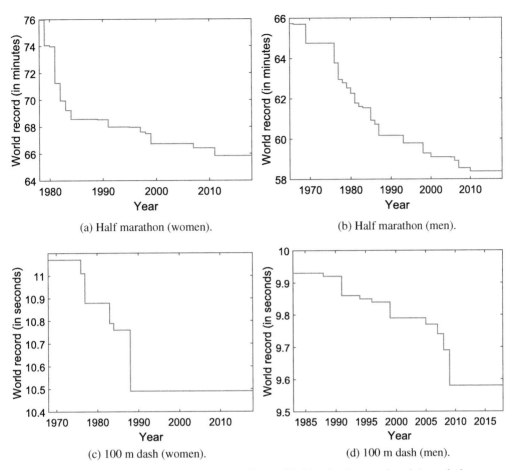

(a) Half marathon (women).   (b) Half marathon (men).

(c) 100 m dash (women).   (d) 100 m dash (men).

Figure 7.1 Progression of world record as ratified by the International Association of Athletics Federations (IAAF).

Specifically, as we saw for additive processes, extremal processes can yield either light-tailed limiting distributions or heavy-tailed limiting distributions. In the following we introduce the "extremal central limit theorem," which characterizes when heavy-tailed distributions can emerge from extremal processes similarly to the characterization of additive processes provided by the generalized central limit theorem. Further, analogously to the class of stable distributions that we discussed in the context of additive processes, we highlight a class of max-stable distributions that precisely characterizes which distributions can emerge as the limit of an extremal process.

Though there are many parallels between additive and extremal processes, one important distinction between the two relates to the emergence of heavy-tailed distributions. Heavy-tailed distributions can emerge from additive processes, but only when one starts with infinite-variance heavy-tailed distributions. In contrast, under extremal processes, heavy-tailed distributions can emerge when starting from distributions with finite variance. In fact, heavy-tailed distributions can emerge even when starting from bounded, light-tailed distributions. Thus, like multiplicative processes, extremal processes can *create* heavy-tailed distributions. However, extremal processes are not as closely connected with heavy-tailed

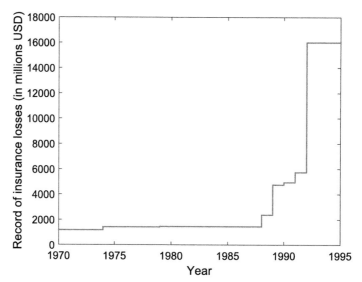

Figure 7.2 Record of insurance loss due to catastrophies worldwide in millions of US dollars (at 1992 prices) between 1970–1995 [3].

distributions as multiplicative processes, and there are many situations where (extremely) light-tailed distributions can also emerge from extremal processes.

7.1 A Limit Theorem for Maxima

The study of extremal processes is typically referred to as "extreme value theory" and has a long history. Our focus in this chapter is on only a small piece of this theory. In particular, we focus on a generic one-dimensional extremal process of the following form:

$$M_n = \max(X_1, X_2, \ldots, X_n), \text{ where } X_i \text{ are i.i.d. with distribution } F.$$

This simple process corresponds to the evolution of the sample maxima; however, it is straightforward to translate our discussion into results for the evolution of the sample minima via the relation

$$\min(X_1, X_2, \ldots, X_n) = -\max(-X_1, -X_2, \ldots, -X_n).$$

Our goal in this chapter is to characterize the behavior of M_n as $n \to \infty$ and to use this simple process to develop intuition about the connection between heavy tails and extremal processes. In some sense, studying extremal processes is easier than studying additive or multiplicative processes. Unlike the cases of additive and multiplicative processes, it is easy to explicitly write down the distribution of M_n in a compact, closed form:

$$\Pr(M_n \leq t) = \Pr(X_1 \leq t, \ldots, X_n \leq t) = F(t)^n.$$

This highlights a simple but important observation about the limiting distribution of M_n: extremes happen only at the upper end of the distribution. In particular, define x_F as the upper end of the support of F, that is, $x_F = \sup\{x : F(x) < 1\}$ and consider a fixed t.

Then

$$\lim_{n\to\infty} \Pr\left(M_n \leq t\right) = \lim_{n\to\infty} F(t)^n = \left\{ \begin{array}{ll} 0, & \text{if } t < x_F; \\ 1, & \text{if } t = x_F. \end{array} \right.$$

This calculation illustrates that, if the distribution has a finite upper bound (e.g., the Uniform distribution), then M_n converges with probability 1 to this upper bound. Similarly, if the distribution has no finite bound, the preceding calculation illustrates that for any finite value t, M_n is eventually larger than t with probability 1.

Of course, one does not learn much about the behavior of the maxima from statements like this one. In fact, those statements are no more informative than simply stating that the additive process $S_n = X_1 + \cdots + X_n$ approaches $+/-\infty$ as $n \to \infty$ depending on whether X_i has positive/negative mean. Just as the law of large numbers and the central limit theorem tell us *how S_n scales* to $+/-\infty$, here we seek a similar understanding of the behavior of M_n. Specifically, the goal is to obtain a more fine-grained understanding of the behavior of the M_n, for example, by understanding the rate at which M_n grows with n and the typical deviations of M_n around this growth rate. Thus, as we did for additive and multiplicative processes, we are interested in studying the centered and normalized maxima, deriving convergence results of the form

$$\frac{M_n - b_n}{a_n} \xrightarrow{d} Z, \text{ where } a_n > 0, b_n \in \mathbb{R}. \tag{7.1}$$

Clearly, this form parallels the form of the cental limit theorems for additive and multiplicative processes we have discussed in the previous two chapters. As in those cases, one can think of this form as providing a second-order approximation of M_n as

$$M_n \overset{d}{=} b_n + a_n Z + o(a_n).$$

However, unlike in those cases, it is not easy to anticipate the proper choices of a_n and b_n nor the limiting distribution of Z.

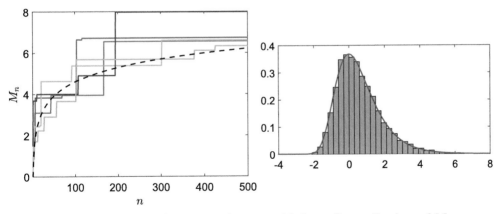

Figure 7.3 Illustration of an extremal process. (a) shows five realizations of M_n with $X_i \sim \text{Exp}(1)$. The dotted line is the first-order approximation $\log(n)$. (b) shows a histogram of the error $M_n - \log(n)$ of the first-order approximation for $n = 500$ over 5000 realizations. The orange curve is the probability density function of the Gumbel distribution.

To obtain intuition for the behavior of the centered and normalized version of M_n it is useful to consider a simple example. For this, let us look at the Exponential distribution.

Example: Exponential distribution

For the case of the Exponential distribution we have a sequence X_i with $F(t) = 1 - e^{-t}$, as illustrated in Figure 7.3. In this case we can immediately write down the limit we are interested in. This gives

$$\Pr\left((M_n - b_n)/a_n \leq t\right) = F(a_n t + b_n)^n = (1 - e^{-a_n t - b_n})^n.$$

It is not immediately clear how to choose a_n and b_n to obtain a meaningful limit, but with some foresight one can see that $a_n = 1$ and $b_n = \log n$ is a natural choice to take advantage of the fact that $(1 - 1/n)^{-n} \to 1/e$. In particular, with these choices, we obtain

$$\Pr\left(M_n - B_n \leq t\right) = F(t + \log n)^n = (1 - e^{-t - \log n})^n = \left(1 - \frac{e^{-t}}{n}\right)^n \to e^{-e^{-t}}$$

$$\text{as } n \to \infty.$$

This shows that a limiting distribution for M_n does indeed emerge as $n \to \infty$ and, though you may not recognize this immediately, the limiting distribution is the Gumbel distribution Γ (see Figure 7.3). The c.d.f. of the Gumbel distribution, typically denoted by Λ, is defined by

$$\Lambda(x) = e^{-e^{-x}} \quad (x \in \mathbb{R}).$$

Note that to obtain this limiting distribution we chose $b_n = \log n$, and so $M_n = \log(n) + Z + o(1)$. This form nicely aligns with the intuition we described at the start of the chapter, which suggests that extremes tend to change quickly initially before eventually settling down. The connection with this intuition is strengthened further by the fact that the Gumbel distribution is extremely light-tailed. In fact, the left tail of the Gumbel decays as a double exponential, and the right tail decays exponentially: $\bar{\Lambda}(x) = e^{-x} + o(e^{-x})$ as $x \to \infty$.

The emergence of the Gumbel distribution in this example provides a candidate for the limiting distribution of extremal processes more broadly than just the case of the Exponential distribution. To investigate this further, let us now repeat the previous calculation starting with the Gumbel distribution instead of the Exponential distribution.

$$\Pr\left(M_n - \log n \leq t\right) = F(t + \log n)^n = \left(e^{-e^{-t - \log n}}\right)^n = e^{-n e^{-t - \log n}} = e^{-e^{-t}}.$$

Note that we again obtain the Gumbel distribution, and this time the equality holds for all finite n, not just in the limit. This indicates that the Gumbel distribution is "stable" with respect to maxima, similarly to the way that the Gaussian distribution is "stable" with respect to sums. Thus, it is natural to define a class of "max-stable" distributions in a parallel manner to the class of stable distributions that we introduced for additive processes in Chapter 5.

Definition 7.1 A distribution F is said to be max-stable if, for any $n \geq 2$ i.i.d. random variables $X_1, X_2, \ldots, X_n$ with distribution F, there exist constants $c_n > 0$ and $d_n \in \mathbb{R}$ such that

$$\max(X_1, X_2, \ldots, X_n) \overset{d}{=} c_n X_1 + d_n.$$

A random variable is said to be max-stable if its distribution function is max-stable.

Like the class of stable distributions, the class of max-stable distributions is difficult to understand from the definition alone. We have seen that the Gumbel distribution is one example, but it is not immediately clear which other distributions satisfy the definition. However, it is clear that the class of max-stable distributions is tightly coupled to the form of the limit theorems for extremal processes that we are looking for (i.e., to (7.1)). In particular, it is easy to see that a max-stable distribution can occur as the limiting distribution in (7.1), and it turns out that max-stable distributions are the *only* nondegenerate distributions that can appear in such a limit. More specifically, just like the class of stable distributions characterizes precisely those distributions that can serve as the limiting distribution of additive processes, max-stable distributions characterize precisely those distributions that can serve as the limiting distribution of extremal processes.

Theorem 7.2 *A random variable Z is max-stable if and only if there exists an infinite sequence of i.i.d. random variables $X_1, X_2, \ldots$, and deterministic sequences $\{a_n\}$, $\{b_n\}$ $(a_n > 0)$, such that*

$$\frac{\max(X_1, X_2, \ldots, X_n) - b_n}{a_n} \overset{d}{\to} Z.$$

In a sense, Theorem 7.2 gives a central limit theorem for extremal processes since it characterizes exactly which limiting distributions can emerge. However, alone, it is quite vague and unsatisfying since it does not provide insight into the properties of the limiting distributions or when different limiting distributions may emerge. In this way, Theorem 7.2 for max-stable distributions exactly parallels Theorem 5.4 for stable distributions. In fact, the parallels between the two theorems are more than just superficial – the proofs of the two results parallel one another as well.

Proof sketch of Theorem 7.2 To prove Theorem 7.2, we first show that if F is a max-stable distribution, then it is the limit, in distribution, of a centered, normalized extremal process. To do this, let $\{X_i\}_{i \geq 1}$ denote an i.i.d. sequence of random variables with distribution F. By Definition 7.1, for any $n \geq 2$, there exist constants $c_n > 0$, $d_n \in \mathbb{R}$ such that

$$\max(X_1, X_2, \ldots, X_n) \overset{d}{=} c_n X_1 + d_n.$$

In other words, for any $n \geq 2$,

$$\frac{\max(X_1, X_2, \ldots, X_n) - d_n}{c_n} \overset{d}{=} X_1.$$

It therefore follows trivially that as $n \to \infty$,

$$\frac{\max(X_1, X_2, \ldots, X_n) - d_n}{c_n} \xrightarrow{d} F.$$

Next, we show that if the distribution F is the limit in distribution of a centered, normalized extremal process, then F is max-stable. Accordingly, suppose that

$$\frac{\max(X_1, X_2, \ldots, X_n) - b_n}{a_n} \xrightarrow{d} F,$$

where $\{X_i\}_{i \geq 1}$ is an i.i.d. sequence of random variables, and $\{a_n\}$, $\{b_n\}$ are deterministic sequences satisfying $a_n > 0$. Now, fix integer $k \geq 2$, and define, for $m = jk$, $j \in \mathbb{N}$,

$$Y_m = \frac{\max(X_1, X_2, \ldots, X_m) - b_m}{a_m},$$

$$Z_m = \frac{\max(X_1, X_2, \ldots, X_m) - b_j}{a_j}.$$

Consider the limit as $m \to \infty$ by taking $j \to \infty$. Clearly, $Y_m \xrightarrow{d} F$. On the other hand, note that Z_m is the maximum of k i.i.d. random variables, each distributed as $\frac{\max(X_1, X_2, \ldots, X_j) - b_j}{a_j}$. Therefore, $Z_m \xrightarrow{d} F^k$. Moreover, since Y_m and Z_m differ only via translation and scaling parameters, it can be shown that their limiting distributions also only differ via translation and scaling parameters. In other words, F^k and F differ only via translation and scaling parameters. Since this is true for all $k \geq 2$, it then follows from Definition 7.1 that F is max-stable.

To formalize the preceding argument, one has to show that F is necessarily a continuous function, and invoke a technical result relating the limits of sequences of random variables that are themselves related via translation and scaling (see [75, Appendix A1.5]). The interested reader is referred to [75, Chapter 2] for the details. □

While we have so far seen parallels between limits of additive and extremal processes, there are also some important differences. For additive processes, the generalized central limit theorem (Theorem 5.9) states that a nondegenerate limiting distribution exists for most commonly encountered distributions of the summands X_i (with suitable normalization coefficients). In contrast, there are several commonly encountered distributions of X_i for which no nondegenerate limiting distribution is possible for the extremal process $\{M_n\}$, for any choice of normalization coefficients. For example, if X_i has an atom at its right endpoint, or is geometric, or Poisson; see Exercises 3–5. In this sense, limits of extremal processes are more fragile than those of additive processes.

7.2 Understanding Max-Stable Distributions

To this point, we have characterized the limiting distributions for extremal processes via the class of max-stable distributions. But, beyond knowing that the Gumbel distribution is an example of a max-stable distribution, we do not yet understand much about max-stable distributions or about when different max-stable distributions may emerge from extremal processes. So, in order to move toward a general statement of the extremal central limit

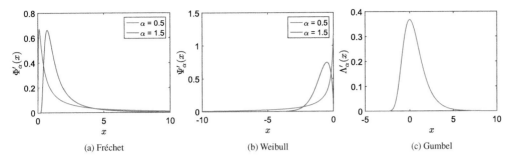

Figure 7.4 Probability density functions corresponding to the standard max-stable distributions.

theorem, we need to first develop a more detailed understanding of the class of max-stable distributions.[1]

In Chapter 5 we were able to characterize the class of stable distributions explicitly via their characteristic functions using the representation in Theorem 5.5. The fact that the representation was in terms of the characteristic function was natural, given that stable distributions emerge from an additive process and the sums of independent random variables can be cleanly represented as the product of their characteristic functions. In the case of extremal processes, the maximum of independent random variables has a clean representation in terms of the product of their distribution functions. Thus, it is natural for a representation theorem to be in terms of the distribution function. Of course, this is appealing since it is much more direct to infer distributional properties from the distribution function than from the characteristic function. In particular, the class of max-stable distributions has a very concrete representation, given by the following theorem.

Theorem 7.3 *A nondegenerate random variable X is max-stable if and only if $X \overset{d}{=} aZ+b$ where $a > 0$, $b \in \mathbb{R}$, and the distribution function of Z is one of the following:*

(i)   Fréchet: $\Phi_\alpha(x) = \begin{cases} 0 & \text{for } x \leq 0 \\ \exp\{-x^{-\alpha}\} & \text{for } x > 0 \end{cases} \quad (\alpha > 0).$

(ii) Weibull: $\Psi_\alpha(x) = \begin{cases} \exp\{-(-x)^\alpha\} & \text{for } x \leq 0 \\ 1 & \text{for } x > 0 \end{cases} \quad (\alpha > 0).$

(iii) Gumbel: $\Lambda(x) = \exp\{-e^{-x}\}.$

This representation of max-stable distributions leads the Fréchet, Weibull, and Gumbel distributions to be termed the "extreme value distributions." Interestingly, these three distributions have quite different properties, as is illustrated in Figure 7.4, which shows the p.d.f. of each. For example, while the Fréchet is a heavy-tailed distribution, the Gumbel is a light-tailed distribution, and the Weibull can be either heavy-tailed or light-tailed. Similarly, the supports of the three distributions also differ.

---

[1] Degenerate distributions are trivially max-stable, but they are not particularly interesting to consider, and so we focus on nondegenerate distributions throughout this chapter.

More specifically, the Fréchet distribution, denoted by Φ_α, is parameterized by $\alpha > 0$, has support $[0, \infty)$, and has a tail that satisfies

$$\bar{\Phi}_\alpha(x) = 1 - (1 - x^{-\alpha} + o(x^{-\alpha})) = x^{-\alpha} + o(x^{-\alpha}).$$

So, $\bar{\Phi}_\alpha(x) \sim x^{-\alpha}$ as $x \to \infty$, and thus is a regularly varying distribution with index $-\alpha$.

In contrast, the Weibull distribution, denoted by Ψ_α is parameterized by $\alpha > 0$, and has support $(-\infty, 0]$. Note that Ψ_α is actually the mirror image of the Weibull that we have typically discussed in this book. That is, we have so far defined the Weibull distribution as having support $[0, \infty)$, and defined by $F_\alpha(x) = 1 - e^{-x^\alpha}$ for $x \geq 0$. However, all the properties that we associate with the Weibull still hold true. Most importantly, Ψ_α is heavy-tailed (to the left) when $0 < \alpha < 1$, and light-tailed (to the left) when $\alpha \geq 1$.

Finally, the Gumbel distribution, $\Lambda(x)$, has support over the full real line and, as we have already discussed, is light-tailed both to the right and to the left. Interestingly, the left tail decays doubly-exponentially fast, that is, $\Lambda(-x) = e^{-e^x}$, while the right tail decays exponentially, that is, $\bar{\Lambda}(x) = 1 - (1 - e^{-x} + o(e^{-x})) = e^{-x} + o(e^{-x})$.

Though the three extreme value distributions that make up the max-stable class behave very differently, it turns out that they are actually related. In particular, the following lemma shows that the Weibull and Gumbel distributions can be represented as simple functions of the Fréchet distribution. These relationships play a crucial role in the proof of the extremal central limit theorem, which we sketch in the next section.

Lemma 7.4 *Consider a random variable X. The following statements are equivalent.*

1. X has the Fréchet distribution Φ_α.
2. $-1/X$ has the Weibull distribution Ψ_α.
3. $\log(X^\alpha)$ has the Gumbel distribution Λ.

We leave the proof of Lemma 7.4 as an exercise for the reader; see Exercise 6.

Finally, it is important to observe that all three extreme value distributions can be represented in one parametric form via the following equation:

$$H(x) = \exp\left\{-\left(1 + \xi\frac{x - \mu}{\psi}\right)^{-1/\xi}\right\}. \tag{7.2}$$

This is often referred to as the *generalized extreme value distribution*. If $\xi > 0$, this corresponds to the Fréchet distribution (i.e., the regularly varying case, with $\xi = 1/\alpha$), while $\xi < 0$ corresponds to the Weibull distribution. Using the fact that $(1 + \xi y)^{-1/\xi} \to e^{-y}$ as $\xi \to 0$, we see that $\xi = 0$ corresponds to the Gumbel distribution. This representation is used heavily in Chapter 9 when applying extremal techniques to estimate the tail behavior of a distribution from data.

7.3 The Extremal Central Limit Theorem

At this point we have seen that the class of max-stable distributions precisely corresponds to the limiting distributions of extremal processes, and we have seen that the class of max-stable distributions is made up of three families of distributions: the Fréchet, the Weibull,

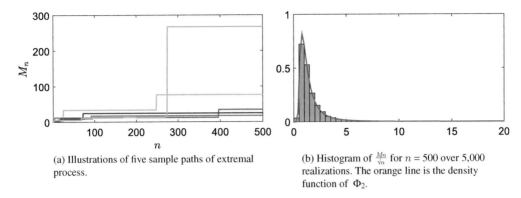

(a) Illustrations of five sample paths of extremal process.

(b) Histogram of $\frac{M_n}{\sqrt{n}}$ for $n = 500$ over 5,000 realizations. The orange line is the density function of Φ_2.

Figure 7.5 $X_i \sim \text{Pareto}\,(x_m = 1, \alpha = 2)$, which lies in the MDA of the Fréchet distribution.

and the Gumbel. However, we have not yet understood when each of these limiting distributions can emerge from extremal processes. Formally, just as the generalized central limit theorem specifies the domain of attraction for stable distributions in the context of additive processes, we would like to specify the *maximum domain of attraction* (MDA) for each of the max-stable distributions. These results are provided by the following extremal central limit theorem.

Theorem 7.5 (Extremal central limit theorem) *Consider an infinite sequence of i.i.d. random variables* $X_1, X_2, \ldots$ *with distribution* F. *Let* $x_F := \sup\{x : \bar{F}(x) > 0\}$ *denote the right endpoint of* F. *There exist deterministic sequences* $\{a_n\}, \{b_n\}$ $(a_n > 0)$ *such that*

$$\frac{\max(X_1, X_2, \ldots, X_n) - b_n}{a_n} \xrightarrow{d} G$$

if and only if G *is max-stable. Further,*

(i) G *follows the Fréchet distribution* Φ_α *if and only if* $\bar{F}(x) = x^{-\alpha}L(x)$, *where* L *is slowly varying;*

(ii) G *follows the Weibull distribution* Ψ_α *if and only if* $x_F < \infty$, *and* $\bar{F}(x_F - 1/x) = x^{-\alpha}L(x)$, *where* L *is slowly varying;*

(iii) G *follows the Gumbel distribution* Λ *if and only if there exists* $z < x_F$, *such that* F *satisfies* $\bar{F}(x) = c(x)\exp\left\{-\int_z^x \frac{\beta(t)}{g(t)}dt\right\}$ *for* $x \in (z, x_F)$, *where* $\lim_{x \uparrow x_F} c(x) = c \in (0, \infty)$, $\lim_{x \uparrow x_F} \beta(x) = 1$, *and* g *is a positive, absolutely continuous function satisfying* $\lim_{x \uparrow x_F} g'(x) = 0$.

See Figures 7.5, 7.6, and 7.7 for an illustration of the extremal central limit theorem under the three regimes described in Theorem 7.5.

To some extent, the extremal central limit theorem takes on an expected form: for the heavy-tailed Fréchet distribution to emerge, it must be that the X_i are already heavy-tailed, specifically regularly varying. This is similar to the statement of the generalized central limit theorem, except that for additive processes heavy-tailed distributions can emerge only from

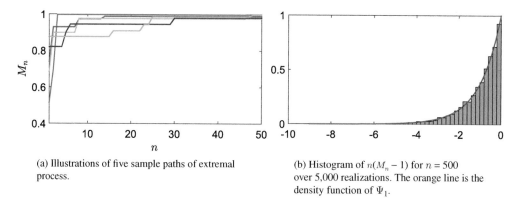

(a) Illustrations of five sample paths of extremal process.

(b) Histogram of $n(M_n - 1)$ for $n = 500$ over 5,000 realizations. The orange line is the density function of Ψ_1.

Figure 7.6 $X_i \sim \text{Uniform}([0, 1])$, which lies in the MDA of the Weibull distribution.

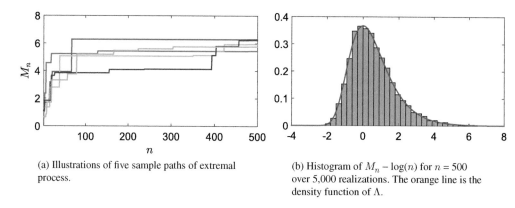

(a) Illustrations of five sample paths of extremal process.

(b) Histogram of $M_n - \log(n)$ for $n = 500$ over 5,000 realizations. The orange line is the density function of Λ.

Figure 7.7 $X_i \sim \text{Exp}(1)$, which lies in the MDA of the Gumbel distribution.

distributions that already have infinite variance, while whether the variance is finite or infinite is irrelevant for extremal processes.

Another difference is that heavy-tailed distributions can emerge from extremal processes even in some situations where the X_i are light-tailed. In particular, the MDA for the Weibull distribution shows that a heavy-tailed distribution (Ψ_α with $\alpha < 1$) can emerge even when the X_i have bounded support. Thus, heavy-tailed distributions are, in some sense, more likely to emerge under extremal processes than under additive processes.

However, light-tailed distributions can also emerge under extremal processes. In fact, the MDA of the Gumbel distribution shows that a light-tailed distribution can emerge even when the X_i are heavy-tailed, as long as the tail decays faster than regularly varying tails. To see this, let us contrast the representation given in case (iii) for $x_F = \infty$ with the representation theorem for regularly varying functions (Theorem 2.12). The two have the same form except that in the representation for $\mathcal{RV}(\rho)$, $\beta(t) \to \rho$ and $g(t) = t$. But, since $g(t) = o(t)$ for distributions in the MDA of the Gumbel distribution, it follows that these distributions have much lighter tails than regularly varying distributions; see Exercise 7.

To keep the statement of the extremal central limit theorem from becoming bloated, we have not described explicitly how to define the scaling and centering sequences, $\{a_n\}$ and

$\{b_n\}$, for each case. However, these can also be defined explicitly. To state them, we need to use the generalized inverse, $F^{\leftarrow}$, which is defined as $F^{\leftarrow}(t) = \inf\{x \in \mathbb{R} \mid F(x) \geq t\}$ for $t \in (0,1)$. Given this definition, we have that if F belongs to the MDA of the Fréchet distribution, Φ_α, then

$$\frac{M_n}{F^{\leftarrow}(1 - 1/n)} \xrightarrow{d} \Phi_\alpha.$$

In contrast, if F belongs to the MDA of the Weibull distribution, Ψ_α, then

$$\frac{M_n - x_F}{x_F - F^{\leftarrow}(1 - 1/n)} \xrightarrow{d} \Psi_\alpha.$$

Finally, if F belongs to the MDA of the Gumbel distribution, Λ, then

$$\frac{M_n - F^{\leftarrow}(1 - 1/n)}{g(F^{\leftarrow}(1 - 1/n))} \xrightarrow{d} \Lambda,$$

where g is defined as in Theorem 7.5, case (iii).

While the MDA of the Fréchet distribution is easy to understand, it is instructive to further explore MDA of the Weibull and the Gumbel distribution.

The MDA of the Weibull distribution is intimately tied to that of the Fréchet distribution. Indeed, the distribution F with right endpoint $x_F < \infty$ belongs to the MDA of the Ψ_α if and only if the distribution G defined by $\bar{G}(x) = \bar{F}(x_F - 1/x)$ belongs to the MDA of Φ_α. This connection is consistent with the relationship between Ψ_α and Φ_α given in Lemma 7.4; see Exercise 8. The simplest example of a distribution that belongs to the MDA of the Weibull distribution is the uniform distribution. If $F \sim \text{Uniform}([a,b])$, then $\bar{F}(x) = \frac{b-x}{b-a}$ for $x \in [a,b]$, which implies that for large enough x,

$$\bar{F}(b - 1/x) = \frac{1}{x(b-a)}.$$

It follows that F lies in the MDA of Ψ_1. Another common distribution in the MDA of the Weibull is the beta distribution (see Exercise 9).

The MDA of the Gumbel distribution includes distributions with a finite right endpoint, as well as distributions with $x_F = \infty$ that have lighter tails than regularly varying distributions. While we have already contrasted the representation in case (iii) with the representation theorem for regularly varying distributions, it is also instructive to contrast it with the the the representation of the complementary c.c.d.f. in terms of the hazard rate (see Equation (4.6) in Chapter 4). Setting $c(\cdot) \equiv c$, $\beta(\cdot) \equiv 1$, we see that $a(x)$ becomes the reciprocal of the hazard rate function $q(x)$. It follows that a sufficient condition for F to be in the MDA of the Gumbel distribution is

$$\lim_{x \uparrow x_F} \frac{d}{dx}\left(\frac{1}{q(x)}\right) = 0. \tag{7.3}$$

For example, consider the (nonnegative) Weibull distribution, characterized by the c.c.d.f. $\bar{F}(x) = e^{-\beta t^\alpha}$, for $\beta, \alpha > 0$. In this case, it is not hard to see that $q(x) = \alpha \beta t^{\alpha-1}$ satisfies (7.3), which means that the class of (nonnegative) Weibull distributions belongs to the MDA of the Gumbel distribution. The condition (7.3) is actually quite powerful – it can be used to show that a large class of distributions, including the Erlang, the hyperexponential, the

Gaussian, and the LogNormal, belong to the MDA of the Gumbel distribution (see Exercises 10–13). Note that this MDA includes heavy-tailed as well as light-tailed distributions. For an example of a distribution with a finite right endpoint that lies in the MDA of the Gumbel, consider the distribution F defined as:

$$\bar{F}(x) = e^{-\frac{\alpha}{x_F - x}}, \qquad x \in (-\infty, x_F),$$

where $\alpha > 0$. An elementary application of the condition (7.3) confirms that F lies in the MDA of the Gumbel. For an in-depth understanding of the relationship between (7.3) and the MDA of the Gumbel, we refer the reader to [75, Chapter 3].

It is important not to get confused by the appearance of heavy-tailed distributions (like the LogNormal) in the MDA of the Gumbel distribution, which is itself light-tailed. For example, taking X_i in our definition of an extremal process to be LogNormal, it follows from Lemma 4.13 that for any n, $M_n = \max(X_1, X_2, \ldots, X_n)$ is long-tailed.

We see on the one hand that the limiting distribution of M_n (with suitable centering and scaling) is light-tailed (specifically, Gumbel)! However, note that the former statement holds for any fixed n and characterizes the asymptotic behavior of $\Pr(M_n > t)$ as $t \to \infty$. On the other hand, the latter statement involves the limit as $n \to \infty$ for fixed t. In particular, for finite (but large) n, the Gumbel approximation would apply to the *body* of M_n, whereas Lemma 4.13 provides an approximation for its *tail*.

Finally, let us move to the proof of the extremal central limit theorem. As in the case of the generalized central limit theorem, the full proof is too technical for inclusion here. However, a proof of a restricted version of the theorem is possible using elementary techniques.

Sketch of the proof of Theorem 7.5   In what follows, we show that distributions that satisfy the conditions of case (i) of Theorem 7.5 belong to the MDA of Φ_α. Using similar aguments, it can be shown that distributions that satisfy the conditions of case (ii) of Theorem 7.5 belong to the MDA of Ψ_α (see Exercise 8). It is harder to prove the converse implications of the preceding statements, and to tackle the MDA of the Gumbel distribution. The interested reader is referred to [75, 186] for the details.

Suppose that F satisfies the condition $\bar{F}(x) = x^{-\alpha} L(x)$, with L slowly varying. To prove that F belongs to the maximum domain of attraction of the Fréchet distribution Φ_α, it suffices to prove that

$$\frac{M_n}{a_n} \xrightarrow{d} \Phi_\alpha, \tag{7.4}$$

where $a_n := F^\leftarrow(1 - 1/n)$. Note that $a_n = \inf\{x \in \mathbb{R} \mid \bar{F}(x) \le n^{-1}\}$, which implies that $a_n \to \infty$ as $n \to \infty$. Moreover, it can be shown that $\bar{F}(a_n) \sim n^{-1}$ (this is trivial if F is continuous; proving this statement for general F is the goal of Exercise 14).

For $x > 0$,

$$\Pr\left(\frac{M_n}{a_n} \le x\right) = F(a_n x)^n = (1 - \bar{F}(a_n x))^n.$$

Since F is regularly varying with index $-\alpha$, $\bar{F}(a_n x) \sim x^{-\alpha} \bar{F}(a_n) \sim x^{-\alpha} n^{-1}$. It then follows that

$$\lim_{n \to \infty} \Pr\left(\frac{M_n}{a_n} \le x\right) = e^{-x^{-\alpha}}.$$

For $x \leq 0$,

$$\Pr\left(\frac{M_n}{a_n} \leq x\right) = F(a_n x)^n \leq F(0)^n.$$

Since $F(0) < 1$, it follows that

$$\lim_{n \to \infty} \Pr\left(\frac{M_n}{a_n} \leq x\right) = 0,$$

which completes the proof of (7.4).

$\square$

7.4 An Example: Extremes of Random Walks

Until this point in the chapter we have considered a generic extremal process, and the results we have presented demonstrate that the behavior of extremal processes parallels the behavior of additive processes in the sense that both light-tailed and heavy-tailed limiting distributions may emerge depending on the distribution of the X_i. However, unlike the case of an additive process, the results illustrate that heavy-tailed distributions may emerge from extremal processes even when the X_i are bounded, light-tailed distributions.

To end the chapter, we move from a generic extremal process to a specific example of an extremal process that is important to a wide variety of disciplines. In particular, we return to the setting of random walks and study the behavior of *extremes* of random walks.

Random walks are one of our favorite examples due to their generic nature and wide applicability, and we have discussed them already in Section 3.4, where we analyzed the probability that a random walk exhibits a large deviation from its expected value, and in Section 5.5, where we focused on the return time of a symmetric random walk where the walker was equally likely to take a unit step up or down at each time.

The random walks we have considered have been simple so far, but random walks come in a variety of forms and, in general, random walks can be quite complex (e.g., they can be asymmetric; they can allow steps to be of different sizes; and they can happen in more than one dimension). In this section, we consider a random walk that is asymmetric and allows arbitrary step sizes, but is still one-dimensional. In particular, we consider a random walk in which a walker starts at 0 and takes a sequence of i.i.d. steps $X_1, \ldots, X_n$. The position of the walker at time n is thus simply the additive process

$$S_n = X_1 + X_2 + \cdots + X_n, \text{ for } n > 0 \text{ and } S_0 = 0.$$

The first natural question to ask about such a random walk is "where will the walker be after n steps?", that is, "what is the behavior of S_n?" Since S_n is simply an additive process, our discussion in Chapter 5 provides the tools to address this question. However, this is only one of many questions one may ask about a random walk. Two other important questions that are often asked are "when will the walker first return to its starting point?" and "what is the maximum position the walker visited?" Since the focus of this chapter is on extremal processes, it is the second of these that we focus on here. Note that we have discussed the first of these already in Section 5.5.

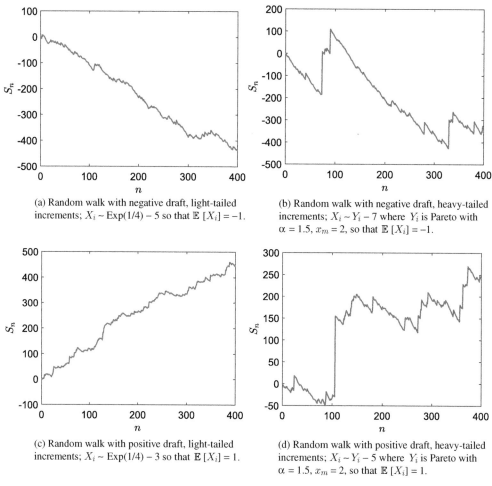

(a) Random walk with negative draft, light-tailed increments; $X_i \sim \text{Exp}(1/4) - 5$ so that $\mathbb{E}[X_i] = -1$.

(b) Random walk with negative draft, heavy-tailed increments; $X_i \sim Y_i - 7$ where Y_i is Pareto with $\alpha = 1.5$, $x_m = 2$, so that $\mathbb{E}[X_i] = -1$.

(c) Random walk with positive draft, light-tailed increments; $X_i \sim \text{Exp}(1/4) - 3$ so that $\mathbb{E}[X_i] = 1$.

(d) Random walk with positive draft, heavy-tailed increments; $X_i \sim Y_i - 5$ where Y_i is Pareto with $\alpha = 1.5$, $x_m = 2$, so that $\mathbb{E}[X_i] = 1$.

Figure 7.8 Illustrations contrasting the behavior of random walks with negative/positive drifts and light-tailed/heavy-tailed increments.

It is important to begin by formalizing the question we are asking. To do this, denote the all-time maximum of the random walk by

$$S_{max} = \max_{n \geq 0}\{S_n\}.$$

Our goal will be to characterize the distribution of S_{max}. Note that while our earlier analyses in this chapter focused on the maximum of i.i.d. random variables, S_{max} is the maximum of the positions S_n of the random walk, which are neither independent nor identically distributed. Because of this, a precise characterization of S_{max} is beyond our reach, and we focus on understanding the behavior of the tail of S_{max}, that is, $\Pr(S_{max} > x)$ as $x \to \infty$.

As a first step toward characterizing the maximum position of a random walk, consider the examples shown in Figure 7.8. These suggest that understanding the maximum is simple if the random walk has positive drift (i.e., if $\mathbb{E}[X_i] > 0$). In such cases, the random walk grows unboundedly with probability 1, and so there is no finite maximum as we let $n \to \infty$. Thus, it makes sense to focus on random walks with negative drift (i.e., with $\mathbb{E}[X_i] < 0$).

In this case, the random walk approaches $-\infty$ as $n \to \infty$ with probability 1, which means that S_{max} is well defined and finite with probability 1.

Another observation from the examples in Figure 7.8 is that there seems to be a distinction between the behavior of S_{max} under light-tailed X_i and under heavy-tailed X_i. This is not too surprising, given the results we have described already in the chapter. We have already seen that light-tailed or heavy-tailed limiting distributions can emerge from extremal processes depending on the distribution of X_i. Indeed a similar phenomenon happens here, and the behavior of S_{max} differs, depending on the weight of the tail of X_i. So, in the following we treat these two cases separately.

Heavy-Tailed Step Sizes

We start with the case where X_i is heavy-tailed, specifically regularly varying. In this case, based on the extremal central limit theorem, it is natural to expect the maximum of the random walk to also be heavy-tailed. So, the more interesting component of the result that follows is the characterization of the tail itself. In particular, it is interesting to see how the tail of S_{max} relates to the tail of X_i.

Theorem 7.6 *Consider a one-dimensional random walk S_n with heavy-tailed i.i.d. step sizes X_i such that $\mathbb{E}[X_i] = -a < 0$ and $X_i \in \mathcal{RV}(-\alpha)$, where $\alpha > 1$. Then, the all-time maximum $S_{max} = \max_{n \geq 0}\{S_n\}$ satisfies*

$$\Pr(S_{max} > x) \sim \frac{1}{a(\alpha - 1)} x \Pr(X_i > x).$$

The key point made in Theorem 7.6 is that the tail of S_{max} is one degree heavier than the tail of X_i. Thus, for example, if X_i has infinite variance and finite mean, then S_{max} has infinite mean. This means that the maximum of the random walk is significantly more variable than the step size. It is interesting to observe the contrast between this result and the extremal central limit theorem. Specifically, in the case where $X_i \in \mathcal{RV}(-\alpha)$, the extremal central limit theorem gives that the scaled M_n converges to a Fréchet distribution, Φ_α, which is regularly varying with the same index $-\alpha$. Thus, the dependency inherent in S_n is leading S_{max} to have a heavier tail than predicted by the extremal central limit theorem.

Theorem 7.6 is proved by establishing asymptotically matching upper and lower bounds on $\Pr(S_{max} > x)$. In what follows, we provide a proof of the lower bound, which confirms the insight that a large all-time maximum is most likely caused by a single big jump in the random walk. The proof of the matching upper bound is more involved. Here, we only prove a weaker upper bound that is of the same order as the lower bound. In other words, our upper bound is asymptotically within a constant factor of the lower bound. For a proof of the precise upper bound, we refer the reader to [226].

Our upper bound is based on the following maximal inequality (see [173]).

Theorem 7.7 *Consider an independent sequence of random variables $\{Y_i\}_{i \geq 1}$, and the associated random walk $\{W_n\}_{n \geq 1}$, where $W_n := \sum_{i=1}^{n} Y_i$. For a given $m \in \mathbb{N}$, if there exist constants $C, q > 0$ such that*

$$\Pr\left(W_m - W_k > -C\right) \geq q \quad \forall\, k \in \{1, 2, \ldots, m-1\},$$

then

$$\Pr\left(\max_{1\leq k\leq m} W_k > x\right) \leq \frac{1}{q}\Pr\left(W_m > x - C\right) \quad \forall\, x.$$

Sketch of the proof of Theorem 7.6 We break the proof into two parts, first showing the lower bound and then moving to the upper bound.

Lower Bound: The idea behind the lower bound is to make use of Karamata's theorem (Theorem 2.10) to bound $\Pr\left(S_{max} > x\right)$. A key step in accomplishing this is the following result, which is a consequence of the weak law of large numbers: given $\epsilon, \delta > 0$, there exists $L > 0$ such that

$$\Pr\left(S_n > -L - n(a + \epsilon)\right) \geq 1 - \delta \quad \forall\, n \geq 0.$$

We leave the proof of this statement as an exercise for the reader (see Exercise 15).

Using this result, we can now bound $\Pr\left(S_{max} > x\right)$ in terms of the running maximum of the random walk, $S_{max}^{(n)} := \max_{0\leq k\leq n} S_k$, as follows:

$$\Pr\left(S_{max} > x\right) = \sum_{n=0}^{\infty} \Pr\left(S_{max}^{(n)} \leq x,\, S_{n+1} > x\right)$$

$$\geq \sum_{n=0}^{\infty} \Pr\left(S_{max}^{(n)} \leq x,\, S_n > -L - n(a+\epsilon),\, X_{n+1} > x + L + n(a+\epsilon)\right)$$

$$= \sum_{n=0}^{\infty} \Pr\left(S_{max}^{(n)} \leq x,\, S_n > -L - n(a+\epsilon)\right) \bar{F}(x + L + n(a+\epsilon))$$

$$\geq \sum_{n=0}^{\infty} \left(\Pr\left(S_{max}^{(n)} \leq x\right) - \delta\right) \bar{F}(x + L + n(a+\epsilon))$$

$$\geq \left(\Pr\left(S_{max} \leq x\right) - \delta\right) \sum_{n=0}^{\infty} \bar{F}(x + L + n(a+\epsilon))$$

$$\geq \frac{\Pr\left(S_{max} \leq x\right) - \delta}{a + \epsilon} \int_{x+L}^{\infty} \bar{F}(s)\,ds.$$

Noting that Karamata's theorem implies that $\int_x^{\infty} \bar{F}(s)\,ds \sim x\bar{F}(x)$, we obtain the lower bound:

$$\liminf_{x\to\infty} \frac{\Pr\left(S_{max} > x\right)}{x\bar{F}(x)} \geq \frac{1 - \delta}{a + \epsilon}.$$

The desired bound now follows, letting $\epsilon, \delta \downarrow 0$.

Order-matching upper bound: The proof of a tight upper bound is quite technical, and so we prove a weaker order-matching bound here. Additionally, we assume the increments X_i have finite second moment in order to keep the argument simple and informative. The idea exhibited in what follows appeared before in [153].

The key idea of the proof is to use the union bound to upper bound the probability that S_{max} exceeds a threshold x by a sum of probabilities of the random walk exceeding x within suitably chosen finite intervals. We then invoke the maximal inequality of Theorem 7.7 to further upper bound the probability that the random walk exceeds x over finite intervals.

We start with the following union bound.

$$\Pr\left(S_{max} > x\right) \le \sum_{k=0}^{\infty} \Pr\left(\max_{2^k \le n \le 2^{k+1}} S_n > x\right)$$

$$= \sum_{k=0}^{\infty} \Pr\left(\max_{2^k \le n \le 2^{k+1}} (S_n + na) > x + na\right)$$

$$\le \sum_{k=0}^{\infty} \Pr\left(\max_{2^k \le n \le 2^{k+1}} (S_n + na) > x + a2^k\right)$$

$$\le \sum_{k=0}^{\infty} \Pr\left(\max_{1 \le n \le 2^{k+1}} (S_n + na) > x + a2^k\right).$$

Next, we rewrite the bound using a change of variables. Define $Y_i = X_i + a$, $W_n = \sum_{i=1}^{n} Y_i = S_n + na$. Thus,

$$\Pr\left(S_{max} > x\right) \le \sum_{k=0}^{\infty} \Pr\left(\max_{1 \le n \le 2^{k+1}} W_n > x + a2^k\right).$$

Now, since Y_i are zero mean and i.i.d., it follows from the central limit theorem that

$$\lim_{n \to \infty} \Pr\left(\sum_{i=1}^{n} Y_i > -\sigma\sqrt{n}\right) \ge \frac{1}{2}.$$

It now follows that there exists large enough $D > 0$ such that

$$\Pr\left(\sum_{i=1}^{n} Y_i > -D\sqrt{n}\right) \ge \frac{1}{2} \quad \forall n.$$

Thus, we see that the conditions of Theorem 7.7 are satisfied, with $C = D\sqrt{m}$, and $q = 1/2$. It therefore follows that

$$\Pr\left(\max_{1 \le k \le m} W_k > y\right) \le 2\Pr\left(W_m > y - D\sqrt{m}\right).$$

Taking $m = 2^{k+1}$ and $y = x + a2^k$, we obtain

$$\Pr\left(\max_{1 \le n \le 2^{k+1}} W_n > x + a2^k\right) \le 2\Pr\left(W_{2^{k+1}} > x + a2^k - D2^{(k+1)/2}\right).$$

For x large enough, one can find $\epsilon \in (0, 1)$ such that for all k,

$$\Pr\left(W_{2^{k+1}} > x + a2^k - D2^{(k+1)/2}\right) \le \Pr\left(W_{2^{k+1}} > (x + a2^k)(1 - \epsilon)\right).$$

Consequently,

$$\Pr\left(S_{max} > x\right) \le \sum_{k=0}^{\infty} \Pr\left(\max_{1 \le n \le 2^{k+1}} W_n > x + a2^k\right)$$

$$\le 2 \sum_{k=0}^{\infty} \Pr\left(W_{2^{k+1}} > (x + a2^k)(1 - \epsilon)\right).$$

We now invoke (3.24) to further bound the sum in this inequality. Specifically, it follows from (3.24) that, given $\delta > 0$, there exists $n_0 \in \mathbb{N}$ such that for $n > n_0$,

$$\sup_{t:\, t > C\sqrt{n \log(n)}} \left| \frac{\Pr(W_n > t)}{n \Pr(Y_1 > t)} - 1 \right| < \delta.$$

Since, for large enough k,

$$a 2^k (1 - \epsilon) > C \sqrt{2^{k+1} \log(2^{k+1})},$$

it follows that there exists $k_0 \in \mathbb{N}$ such that for $k > k_0$,

$$\Pr\left(W_{2^{k+1}} > (x + a 2^k)(1 - \epsilon)\right) \leq (1 + \delta) 2^{k+1} \Pr\left(Y_1 > (x + a 2^k)(1 - \epsilon)\right).$$

For $k \leq k_0$, it follows from Exercise 2.7 that for large enough x,

$$\Pr\left(W_{2^{k+1}} > (x + a 2^k)(1 - \epsilon)\right) \leq (1 + \delta) 2^{k+1} \Pr\left(Y_1 > (x + a 2^k)(1 - \epsilon)\right).$$

Finally, combining the bounds, for large enough x, we have

$$\Pr(S_{max} > x) \leq 2(1 + \delta) \sum_{k \geq 0} 2^{k+1} \Pr\left(Y_1 > (x + a 2^k)(1 - \epsilon)\right).$$

$$\leq \gamma \int_{x(1-\epsilon)}^{\infty} \Pr(Y_i > t)\, dt$$

$$\overset{(a)}{\sim} \hat{\gamma}\, x \Pr(Y_1 > x)$$

$$\overset{(b)}{\sim} \hat{\gamma}\, x \Pr(X_1 > x),$$

for positive constants γ, $\hat{\gamma}$. Note that step (a) follows from Karamata's theorem (Theorem 2.10), while (b) is a consequence of the long-tail property. □

Light-Tailed Step Sizes

We now move from the case where X_i is heavy-tailed to the case where X_i is light-tailed. In this setting it is natural to expect the maximum position of the random walk to also be light-tailed, and the following result shows that this is indeed the case. Further, the result provides a precise description of the logarithmic asymptotics of the tail of the maximum of the random walk.

Theorem 7.8 *Consider a one-dimensional random walk S_n with light-tailed i.i.d. step sizes X_i such that $\mathbb{E}[X_i] < 0$. Then, the all-time maximum $S_{max} = \max_{n \geq 0}\{S_n\}$ satisfies*

$$\lim_{x \to \infty} -\frac{\log \Pr(S_{max} > x)}{x} = s^*,$$

where $s^* = \sup\{s \geq 0 \mid \mathbb{E}[e^{s X_i}] \leq 1\}$.

The characterization of the tail of S_{max} in the preceding result is less precise than the characterization in Theorem 7.6. In this case, only the logarithmic asymptotics of the tail are characterized, that is, the asymptotics of $\log \Pr\left(S_{max} > x\right)$. However, this is already enough to provide some interesting information. In particular, we see that the tail decays approximately exponentially. This is similar to what the extremal central limit theorem gives in the case of i.i.d. processes. In particular, the Gumbel distribution would be the emergent distribution in this setting if the steps were independent, and the right tail of the Gumbel distribution decays exponentially. However, the specific decay rate of S_{max} is more mysterious, though s^* may appear familiar. This familiarity is because we already applied this result in Chapter 6 in order to characterize the behavior of multiplicative processes with a lower bound, and so the same s^* appeared in Theorem 6.2.

Proof For simplicity, we prove the result under the assumption that there exists $\bar{s} > 0$ such that $\mathbb{E}\left[e^{\bar{s}X_i}\right] > 1$. (For a proof of the result without this assumption, we refer the reader to [166].) Under this assumption, $\mathbb{E}\left[e^{sX_i}\right] < \infty$ for all $s \in (0, \bar{s})$ and moreover, there exists a unique $s^* \in (0, \bar{s})$ such that $\mathbb{E}\left[e^{s^*X_i}\right] = 1$, $\mathbb{E}\left[e^{sX_i}\right] < 1$ for $s \in (0, s^*)$, and $\mathbb{E}\left[e^{sX_i}\right] > 1$ for $s \in (s^*, \bar{s})$.

The key idea behind the proof is that for any $m \in \mathbb{N}$,

$$\Pr\left(S_m > x\right) \le \Pr\left(S_{max} > x\right) \le \sum_{n \ge 0} \Pr\left(S_n > x\right). \tag{7.5}$$

While the second inequality in (7.5) yields the desired upper bound on the tail of $\log \Pr\left(S_{max} > x\right)$, the matching lower bound is obtained by utilizing the first inequality of (7.5) with a carefully chosen value of m. Indeed, our proof highlights that the all-time maximum of the random walk is caused due to a "conspiracy" between a large number of "larger than usual" increments. Moreover, the proof reveals the most likely time when the maximum occurs.

We first utilize the second inequality in (7.5) to obtain the asymptotic upper bound on $\log \Pr\left(S_{max} > x\right)$. Let $s^\delta = s^* - \delta$ and write

$$\Pr\left(\max_{n \ge 0}\{S_n\} > x\right) \le \sum_{m \ge 0} \Pr\left(S_m > x\right)$$

$$= \sum_{m \ge 0} \Pr\left(e^{s^\delta S_m} > e^{s^\delta x}\right)$$

$$\le e^{-s^\delta x} \sum_{m \ge 0} \mathbb{E}\left[e^{s^\delta X_1}\right]^m$$

$$= e^{-s^\delta x} \frac{1}{1 - \mathbb{E}\left[e^{s^\delta X_1}\right]},$$

which is finite for any $\delta > 0$. Consequently,

$$\liminf_{x \to \infty} \frac{-\log \Pr\left(\max_{n \ge 0}\{S_n\} > x\right)}{x} \ge s_\delta. \tag{7.6}$$

Letting $\delta \downarrow 0$, we obtain

$$\liminf_{x \to \infty} \frac{-\log \Pr\left(\max_{n \ge 0}\{S_n\} > x\right)}{x} \ge s_0. \tag{7.7}$$

Next, we utilize the first inequality in (7.5) to obtain an asymptotically matching lower bound on $\log \Pr\left(S_{max} > x\right)$. Let $a = \mathbb{E}\left[X_1 e^{s^* X_1}\right]$ and set $m = x/a$. Invoking Cramér's theorem (see Theorem 3.12), we obtain

$$\lim_{x \to \infty} \frac{-\log \Pr\left(S_{x/a} > x\right)}{x} = \lim_{y \to \infty} \frac{-\log \Pr\left(S_y > ya\right)}{ya}$$

$$= \frac{1}{a} \sup_{s > 0}[as - \log \mathbb{E}\left[e^{sX_1}\right]]$$

$$= s^*,$$

since s^* solves this optimization problem. Consequently,

$$\limsup_{x \to \infty} \frac{-\log \Pr\left(\max_{n \geq 0}\{S_n\} > x\right)}{x} \leq \lim_{x \to \infty} \frac{-\log \Pr\left(S_{x/a} > x\right)}{x} = s^*.$$

$\square$

7.5 A Variation: The Time between Record-Breaking Events

We started this chapter with a discussion of the progression of world records in the half marathon and the 100 m dash (Figure 7.1). Given that these are such canonical examples of extremal processes, it is natural to end by coming full circle back to these examples. As we already highlighted, at first glance these world record progressions show a leveling off, and thus suggest a light-tailed limiting distribution. But, on a second look, you may start to see that the time between improvements looks like it could be heavy-tailed. Basically, there are lots of periods where the record changes frequently, but there are also others where the records go unchanged for long periods. In fact, this has been observed and tested empirically in a variety of contexts, for example, records for rainfall, earthquakes, and other extreme events (see, for example, [152, 157]). Very commonly, the times between "records" seem to exhibit heavy-tailed behavior.

With this observation in mind, it is natural to ask "Why?" and, unfortunately, the results we have discussed so far in the chapter do not provide an explanation. So, providing an explanation is the goal of this section.

To formalize the setting, let us consider the following. Suppose we observe a sequence $\{X_i\}_{i \geq 1}$ of i.i.d. random variables, with distribution F. As before, define $M_n := \max(X_1, X_2, \ldots, X_n)$. Let L_k denote the instant of the kth record, that is, $L_1 = 1$ and $L_{k+1} = \min\{i > L_k \mid X_i > M_{i-1}\}$ for $k \geq 1$. For $k \geq 1$, let $T_k := L_{k+1} - L_k$ denote the time between the kth record and the $k + 1$st record.

Then, we can prove the following theorem, which shows that the time between records is indeed heavy-tailed, specifically regularly varying.

Theorem 7.9   *Suppose that F is continuous. Then for any $k \geq 1$, T_k is heavy-tailed, with*

$$\Pr\left(T_k > n\right) \sim \frac{2^{k-1}}{n}.$$

Note that this is a somewhat delicate situation that we are studying since T_k is not stationary with respect to k. Indeed, you expect that, as the record gets bigger, the time it takes to break it gets larger. Also, an interesting observation about the theorem is that it is not required that F has infinite support for T_k to be heavy-tailed. So, looking at records can create heavy tails from things that are extremely light-tailed.

Proof   To begin the proof, we start by showing that we may assume that F is exponentially distributed with no loss of generality, since the distribution of T_k does not depend on the distribution F. To do this, we focus on the function $Q(x) = -\log \bar{F}(x)$, which you should recall is the cumulative hazard function corresponding to the distribution F (see Chapter 4). The key step in this first part of the proof is to show that the random variable $Q(X_1)$ is exponentially distributed with mean 1. This follows easily from the fact that $U := \bar{F}(X_1)$ is a uniform random variable over $[0, 1]$:

$$\Pr\left(Q(X_1) > x\right) = \Pr\left(-\log(U) > x\right) = \Pr\left(U < e^{-x}\right) = e^{-x}.$$

Let $E_i = Q(X_i)$. Clearly, $\{E_i\}$ is an i.i.d. sequence of exponential random variables. Moreover, since Q is nondecreasing, records of the sequence $\{X_i\}$ coincide with those of the sequence $\{E_i\}$. Thus, for the purpose of studying the time between records, we may assume without loss of generality that F is an exponential distribution with mean 1. We make this assumption in the remainder of this proof.

Next, we study the distribution of the records. Specifically, we show that the kth record $M_{(k)} := X_{L_k}$ has an Erlang distribution with shape parameter k and rate parameter 1 (i.e., $M_{(k)}$ is distributed as the sum of k i.i.d. exponential random variables, each having mean 1). We proceed inductively as follows. Clearly, the claim is true for $k = 1$, since $M_{(1)} = X_1$. Assume that the claim is true for some $k \in \mathbb{N}$. Note that for $x > 0$,

$$\Pr\left(M_{(k+1)} > M_{(k)} + x\right) = \Pr\left(E > M_{(k)} + x \mid E > M_{(k)}\right),$$

where E is exponentially distributed with mean 1, and independent of $M_{(k)}$. From the memoryless property of the exponential distribution, it now follows that $\Pr\left(M_{(k+1)} > M_{(k)} + x\right) = e^{-x}$, which implies that $M_{(k+1)} \stackrel{d}{=} M_{(k)} + E$. This proves our claim that $M_{(k+1)}$ has an Erlang distribution.

We are now ready to analyze the tail of T_k. Once again, we proceed inductively, and first consider the case $k = 1$. Note that conditioned on the value of X_1, T_1 is geometrically distributed with

$$\Pr\left(T_1 > n \mid X_1 = x\right) = (1 - e^{-x})^n.$$

Therefore, unconditioning with respect to X_1,

$$\Pr\left(T_1 > n\right) = \int_0^\infty (1 - e^{-x})^n e^{-x} dx.$$

Making the substitution $y = e^{-x}$, we get

$$\Pr\left(T_1 > n\right) = \int_0^1 (1 - y)^n dy = \frac{1}{n+1}.$$

It follows that $\Pr\left(T_1 > n\right) \sim \frac{1}{n}$.

Next, we assume that, for some $k \in \mathbb{N}$, $\Pr\left(T_k > n\right) \sim \frac{2^{k-1}}{n}$ and analyze the tail of T_{k+1}. Recall that $M_{(k+1)} \stackrel{d}{=} M_{(k)} + E$, where E is exponentially distributed with mean 1, and independent of $M_{(k)}$. Therefore, we can think of T_{k+1} as the time until a new sample exceeds $M_{(k)} + E$. Note that the time until a new sample exceeds $M_{(k)}$ is distributed as T_k. Moreover, conditioned on a new sample X_i exceeding $M_{(k)}$, the probability that it exceeds $M_{(k)} + E$ equals

$$\Pr\left(X_i > M_{(k)} + E \mid X_i > M_{(k)}\right) = \Pr\left(X_i > E\right) = 1/2.$$

This calculation exploits the memoryless property of the exponential distribution, and the fact that X_i and E are i.i.d. Thus, when a new sample exceeds $M_{(k)}$, it also exceeds $M_{(k)} + E$ (and thus sets a new record) with probability 1/2. Therefore, T_{k+1} is simply distributed as a geometric random sum of i.i.d. random variables, each distributed as T_k, that is,

$$T_{k+1} \overset{d}{=} \sum_{i=1}^{N} Y_k(i),$$

where $\{Y_k(i)\}$ is an i.i.d. sequence of random variables with the same distribution as T_k, and N is a geometric random variable independent of $\{Y_k(i)\}$ with success probability 1/2.

Finally, since T_k is assumed to be regularly varying (and therefore subexponential), we may invoke Theorem 3.9 to obtain the tail behavior of T_{k+1}. We therefore have

$$\Pr\left(T_{k+1} > n\right) \sim \mathbb{E}\left[N\right] \Pr\left(T_k > n\right) = 2\Pr\left(T_k > n\right),$$

which proves our desired induction step. $\qquad\qquad\square$

7.6 Additional Notes

The focus in this chapter has been on a few examples of extremal processes, and there is much more material that interested readers can consult in continuing their study of this important area. In particular, our discussion of max-stable distributions and the extremal central limit theorem are only a brief introduction into the area of extreme value theory, which is one of the main subdisciplines within probability theory and statistics, and is still a very active area of research. There exist many papers and books with which the interested reader can continue their study. Fisher and Tippett [85] is a classic in the field, as are the works of Gnedenko [99], Gumbel [107], and the PhD thesis of De Haan [56]. Excellent textbooks include [9], focusing on the phenomenology of large exceedances of dependent variables appearing in clumps, [133], covering maxima of dependent Gaussian sequences, and books by Resnick [185, 186] exhibiting the deep connections with regular variation and point processes. Embrechts covers many links with financial and insurance models, and also the theory of order statistics in [75].

An important note about the presentation here is that all of the processes we have considered have a discrete time parameter. In the mathematical literature, it is common for a continuous-time process $M(t), t \geq 0$, to be called an extremal process, if, for any nondecreasing sequence $t_1, \ldots, t_n$, the following holds: there exist independent random variables $U_1, \ldots, U_n$ such that $(M(t_1), \ldots, M(t_n))$ has the same joint distribution as $\{U_1, U_1 + U_2, \ldots, U_1 + \cdots + U_n\}$ and

$$\Pr\left(U_i \leq u\right) = \Pr\left(M(t_i - t_{i-1}) \leq u\right).$$

Beyond the simple discrete extremal process that we consider in the chapter, we also focus on examples related to all-time maxima of random walks. This material is classical and can be found in many textbooks, such as [16] and [17]. An important distinctive feature of the heavy-tailed and light-tailed examples in this chapter is perhaps not so much whether the ruin probability has an exponential tail, or a power tail, but the behavior of *the time until ruin, given ruin occurs*. In particular, set $\tau(x) = \inf\{n : S_n > x\}$. We see that $\sup_{n \geq 0} S_n > x$ if and only if $\tau(x) < \infty$. We are interested in the time until ruin, given that ruin occurs, that is, in the behavior of $\tau(x)$, given $\tau(x) < \infty$. It now turns out that, if the claim sizes are light-tailed, $\tau(x)/x$, conditional on $\tau(x) < \infty$ converges to a deterministic constant. This is in sharp contrast with the case of power-law claim sizes, where $\tau(x)/x$, conditional on $\tau(x) < \infty$, converges to a random variable, which is heavy-tailed in itself. This provides an interesting explanation of the unpredictability of so-called "black swans" [204], and a rigorous proof can be found in [18].

A final note about this chapter is that we have not touched upon connections with statistical applications at all. This will be the subject of Part III of this book, where several of the results appearing in this chapter will find application.

7.7 Exercises

For all the exercises in this chapter we will use the following notation: $\{X_n\}_{n \geq 1}$ is an i.i.d. sequence with distribution F, and $M_n = \max(X_1, X_2, \ldots, X_n)$ for $n \geq 1$. Also, we use x_F to denote the right endpoint of the distribution F (i.e., $x_F = \sup\{x : \bar{F}(x) > 0\}$).

1. Prove that for a real sequence $\{x_n\}_{n \geq 1}$ and $\tau \in [0, \infty]$,

$$\lim_{n \to \infty} n\bar{F}(x_n) = \tau \iff \lim_{n \to \infty} \Pr(M_n \leq x_n) = e^{-\tau}.$$

2. For deterministic sequences $\{a_n\}$ and $\{b_n\}$, where $a_n > 0$, and a max-stable distribution G, prove that

$$\frac{M_n - b_n}{a_n} \xrightarrow{d} G \iff \lim_{n \to \infty} n\bar{F}(a_n x + b_n) = -\log G(x) \; \forall \, x \in \mathbb{R}.$$

Here $-\log G(x)$ should be interpreted as ∞ when $G(x) = 0$.
Hint: Use the result of Exercise 1.

3. The goal of this exercise is to show that if F has a jump at its right endpoint, then the corresponding extremal process cannot have a nondegenerate limit for any choice of translation and scaling parameters.
Specifically, show that if $x_F < \infty$ and $\Pr(X_1 = x_F) > 0$ (or equivalently, F has a jump discontinuity at x_F), then $\lim_{n \to \infty} \Pr(M_n \leq x_n) = \theta$ implies that θ is either 0 or 1.
Hint: Use the result of Exercise 1.

4. In this exercise you will show that if F is geometrically distributed, then the corresponding extremal process cannot have a nondegenerate limit for any choice of translation and scaling parameters.
Specifically, if F is geometric with parameter $p \in (0, 1)$, show that $\lim_{n \to \infty} \Pr(M_n \leq x_n) = \theta$ implies that θ is either 0 or 1.

Hint: You have to use the result of Exercise 1. Prove that if $n\bar{F}(x_n) \to \tau \in (0, \infty)$, then that would imply that $x_n \to \infty$ and $\bar{F}(x_n)/\bar{F}(x_{n-1}) \to 1$, which is not possible given that F is geometric.

5. As in the previous exercises, in this problem you will show that if F is Poisson, then the corresponding extremal process cannot have a nondegenerate limit for any choice of translation and scaling parameters. Specifically, if F is Poisson with parameter $\lambda > 0$, show that $\lim_{n \to \infty} \Pr(M_n \le x_n) = \theta$ implies that θ is either 0 or 1.

6. Prove Lemma 7.4.

7. The goal of this exercise is to show that distributions in the maximum domain of attraction of the Gumbel distribution have lighter tails than regularly varying distributions.

   Consider a distribution F with $x_F = \infty$ in the maximum domain of attraction of the Gumbel distribution. Prove that if the distribution G is regularly varying with index $-\rho$, where $\rho > 0$, then

   $$\lim_{x \to \infty} \frac{\bar{F}(x)}{\bar{G}(x)} = 0.$$

   Hint: Compare the representation of F given by Theorem 7.5 with that for G by Karamata's representation theorem (Theorem 2.12). Exploit the fact that the function g in the former representation satisfies $\frac{g(t)}{t} \to 0$ as $t \to \infty$.

8. Suppose that the distribution F satisfies the conditions of the case (ii) of Theorem 7.5. Prove that

   $$\frac{M_n - x_F}{x_F - F^{\leftarrow}(1 - 1/n)} \xrightarrow{d} \Psi_\alpha.$$

   Hint: From the result of Exercise 2, it suffices to show that $n\bar{F}(a_n x + b_n) \to -\log \Psi_\alpha(x)$ for $x < 0$, where a_n and b_n are the given normalization constants. Define G such that $\bar{G}(y) = \bar{F}(x_F - 1/y)$ and $\tilde{a}_n := G^{\leftarrow}(1 - 1/n)$. Prove that $\tilde{a}_n = 1/a_n$. Finally, show that for $x < 0$,

   $$n\bar{F}(a_n x + b_n) = n\bar{G}(a_n/x) \to \Phi_\alpha(1/x) = \Psi_\alpha(x).$$

9. The beta distribution has support $[0, 1]$ and is characterized by the density function

   $$f(x) = \frac{\Gamma(a + b)}{\Gamma(a)\Gamma(b)} x^{a-1}(1 - x)^{b-1} \quad (x \in [0, 1]),$$

   where $a, b > 0$ and $\Gamma(\cdot)$ denotes the gamma function. Prove that the beta distribution lies in the MDA of Ψ_b.
   Hint: Use Karamata's theorem (Theorem 2.10) to show that $\bar{F}(b - 1/x)$ is regularly varying with index $-b$.

10. Recall that the Erlang distribution with parameters (k, μ), where $k \in \mathbb{N}$ and $\mu > 0$, is associated with the c.c.d.f.

    $$\bar{F}(x) = \begin{cases} e^{-\mu x} \sum_{i=0}^{k-1} \frac{(\mu x)^i}{i!} & \text{for } x \ge 0, \\ 1 & \text{for } x < 0. \end{cases}$$

    Prove that the Erlang distribution belongs to the MDA of the Gumbel distribution.

11. Recall that the hyperexponential distribution is the mixture of independent exponentials. Specifically, consider the hyperexponential distribution F defined by the c.c.d.f.

$$\bar{F}(x) = \sum_{i=1}^{n} p_i e^{-\mu_i x} \qquad (x \geq 0).$$

Here, $\mu_i > 0$ and $p_i > 0$ for $1 \leq i \leq n$, with $\sum_{i=1}^{n} p_i = 1$. Prove that the hyperexponential distribution belongs to the MDA of the Gumbel distribution.

12. Prove that the standard Gaussian belongs to the MDA of the Gumbel distribution. *You may want to use the fact that $q_N(x) \sim x$ as $x \to \infty$, where q_N denotes the hazard rate of the standard Gaussian. This can be proved easily using L'Hospital's rule.*

13. Prove that the LogNormal distribution belongs to the MDA of the Gumbel distribution. *Hint: Use the result of Exercise 12.*

14. Suppose that the distribution F is regularly varying. Define

$$a_n := \inf\{x \in \mathbb{R} \mid \bar{F}(x) \leq n^{-1}\}.$$

Prove that $\bar{F}(a_n) \sim n^{-1}$.
Hint: Use Karamata's representation theorem (Theorem 2.12) to show that there exists a continuous distribution G such that $\bar{F}(x) \sim \bar{G}(x)$ as $x \to \infty$.

15. Suppose that $\mathbb{E}[X_i] = \mu$. Let $S_0 = 0$, and $S_n = \sum_{i=1}^{n} X_i$ for $n \geq 1$. Prove that given $\epsilon, \delta > 0$, there exists $L > 0$ such that

$$\Pr(S_n > n(\mu - \epsilon) - L) \geq 1 - \delta \quad \forall n \geq 0.$$

Part III

Estimation

Given the mystique and excitement that surrounds the discovery of heavy-tailed phenomena, the detection and estimation of heavy tails in data is a task that is often (over)zealously pursued. However, it is important that enthusiasm does not overcome care because the detection and estimation of heavy-tailed distributions is fraught with pitfalls and booby traps. Many simple and intuitive approaches that are widely used can be misleading. In fact, it is not uncommon for the misuse of statistical tools to result in mistaken detections of heavy-tailed phenomena, which in turn leads to the controversy surrounding heavy tails that still exists today.

In Part III of this book we focus on introducing the statistical tools used for the estimation of heavy-tailed phenomena. Unfortunately, there is no perfect recipe for how to "properly" detect and estimate heavy-tailed distributions in data. Thus, our treatment seeks to highlight a handful of important approaches and to provide insight into when each approach is appropriate and when each may be misleading. In particular, in Chapter 8 we focus on classical parametric approaches for estimating power-law distributions, such as linear regression and maximum likelihood estimation. These approaches assume the data come from a precise power-law distribution, and can thus use data from the body of the distribution to estimate the tail. However, often it is not the case that exact power-law *distributions* are present in data; instead only power-law *tails* are present. When only a power-law tail is present, classical parametric approaches can be misleading. Thus, Chapter 9 focuses on semi-parametric statistical tools that are appropriate for detecting power-law tails. These techniques provide a dramatic contrast to the classical approaches; they tend to throw away large amounts of data about the body of the distribution, keeping only data about the tail (i.e., the outliers).

Combined, these chapters highlight a crucial point: *one must proceed carefully when estimating heavy-tailed phenomena in real-world data.* In particular, it is typically naive to seek to estimate exact heavy-tailed distributions in data. Instead, the focus should be on estimating the tail of heavy-tailed phenomena. However, even in doing this, one should not rely on a single method for estimation. Instead, it is necessary to build confidence through the use of multiple, complementary estimation approaches. These are important lessons to take to heart since there are many examples of researchers enthusiastically reaching flawed conclusions about heavy-tailed phenomena as a result of methodological mistakes.

8

Estimating Power-Law Distributions: Listen to the Body

The discovery of heavy-tailed phenomena has a long history. One of the earliest (and most celebrated) instances came in the early 1900s when Vilfredo Pareto observed that the distribution of wealth tends to follow a power-law distribution [171]. Pareto's study of wealth led to the coining of the so-called *Pareto Principle* or *80/20 rule*, which states that, for many phenomena, 80% of the effects come from 20% of the causes. Specifically, Pareto observed that 80% of Italy's land at the time was owned by 20% of the population. Not surprisingly, this discovery prompted curiosity and controversy at the time. In fact, statistics of this form are still often met with surprise and consternation today, for example, during the 2009 economic crisis in the US, observations that 1% of the population earned more than 30% of the total US income became a rallying cry for the Occupy Wall Street movement.

Following Pareto's initial observations about wealth, the observation of heavy-tailed phenomena in many other social and economic contexts came quickly (e.g., the population of cities [92, 163], word frequencies [77, 227], the number of copies of books sold [14, 110]) before eventually spreading to other disciplines across science and engineering more broadly (e.g., degree distribution of the web graph [36, 116], computer file sizes [52, 146], the length of protein sequences in genomes [130, 145], earthquake magnitudes and the size of avalanches [109, 144], the returns of stocks [49, 94], and more). Of course, this is just a short list, and an interested reader can find more examples by referring to [23, 163].

These discoveries have nearly always been met with a mixture of surprise and controversy – even today, over a century since Pareto's initial recognition of heavy-tailed phenomena. Time after time, when a new area first encounters an example of heavy-tailed phenomena, the discovery is viewed with excitement because of the prevailing view of heavy-tailed distributions as strange anomalies. This excitement spurs a desire to seek out other cases where heavy-tailed phenomena are present. However, because researchers in the new field are not trained in the statistical tools relevant for heavy-tailed distributions, this enthusiastic pursuit tends to use simple, intuitive, and sometimes flawed statistical tools. As a result, the tools used can give misleading, or simply wrong conclusions about the existence of heavy-tailed phenomena, which then leads to controversy: *which, if any, of the supposedly heavy-tailed phenomena really are heavy-tailed?*

Of course, heavy-tailed phenomena should be expected to be common (as the mechanisms we study in Part II of this book illustrate) and so, for the most part, these discoveries do in fact correspond to heavy-tailed phenomena. Consequently, the controversies are eventually resolved as more appropriate statistical tools find their way into the community, allowing validation and correction to occur. However, this process can take decades.

A recent, and still ongoing instance of this unfortunate parable is in the field of network science. Heavy-tailed phenomena in networks began to draw attention in 1999 with the emergence of papers finding power-law degree distributions in the WWW graph [36, 116] and the Internet [80]. These papers were met with surprise and excitement, and in the span of just a few years so-called *scale-free* degree distributions had been discovered in networks ranging from the power grid to metabolic networks. While Price had identified power-law degree distributions in the citation network decades earlier [179], the time was right since the large-scale networks that shape our lives today, such as the web and Facebook, were emerging. However, it did not take long before controversy over these results began to bubble up. Researchers began to question both the quality of the data and the statistical tools used in the initial studies identifying scale-free degree distributions. The observation of heavy-tailed phenomena in the Internet graph was one of the initial points of controversy. Papers emerged showing that the way the data was gathered (using a tool called traceroute) could create the appearance of heavy-tailed phenomena when there was none [4]. Beyond that, more careful statistical analysis of that dataset and others seemed to further weaken the evidence for the scale-free conclusion in many cases [48, 222]. It became clear that not all detections of power-law degree distributions were correct and that many discoveries needed to be revisited. This controversy persists, and nearly two decades later, papers are being written with the goal of understanding which networks truly exhibit power-law degrees and which do not [37, 212].

Unfortunately, the controversy surrounding the estimation of heavy-tailed phenomena in the network science community is not an isolated occurrence; it is one of the more prominent recent examples, but other examples can be found in a variety of areas, including computer science [68], biology [119], chemistry [160], ecology [10], and astronomy [216].

However, there is no need for the estimation of heavy-tailed distributions to be surrounded by controversy. Just as Part I of this book illustrates that properties of heavy-tailed distributions need not be mysterious and Part II of this book illustrates that the emergence of heavy-tailed distributions need not be surprising, in the coming chapters we show that the identification of heavy-tailed phenomena need not be controversial. One simply needs to be careful with both the data being used and the application of statistical techniques, building confidence through the use of multiple, complementary approaches and datasets.

The following chapters introduce the statistical tools used for the estimation of heavy-tailed phenomena. We highlight the strengths and weaknesses of the simple, intuitive approaches that are commonly used and, in the process, motivate the need for more sophisticated approaches, as well as the intuition behind them.

In this chapter our focus is on the most classical and well-known approaches for estimating heavy-tailed phenomena. In particular, we discuss approaches for parametric estimation based on regression and maximum likelihood estimation, which are perhaps the most intuitive methods available. These approaches work by "listening" to the *body* of the distribution, hence the tagline of the chapter. In particular, data from the body is less noisy than data from the tail and so, when the measurements come from a parametric distribution, data about the body should be used to estimate the tail.

However, though simple and intuitive, these classical approaches have enormous potential for misuse and abuse. The reason is that such approaches assume that the data comes from an exact, parametric power-law *distribution*; this assumption is nearly always false. Instead,

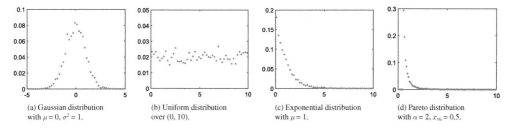

(a) Gaussian distribution with $\mu = 0$, $\sigma^2 = 1$.

(b) Uniform distribution over $(0, 10)$.

(c) Exponential distribution with $\mu = 1$.

(d) Pareto distribution with $\alpha = 2$, $x_m = 0.5$.

Figure 8.1 *Frequency plots aka histograms.*

real-world data typically only has a *tail* that approximately follows a power law. Unfortunately, the classical approaches we discuss in this chapter can be misleading in situations where only the tail of the observations is power-law, since the data from the (non-power-law) body tends to corrupt the estimates of parameters corresponding to the (power-law) tail. Hence, while the classical approaches we study in this chapter are appropriate for the estimation of parametric power-law *distributions*, different approaches must be used for the estimation of power-law *tails*.

We introduce classical approaches in this chapter primarily to show why you typically *should not* use them in practice. We drive this point home in Chapter 9, where we illustrate the misleading conclusions that can result from these techniques when estimating power-law tails. Thus, we want to emphasize that all of the techniques for estimation that we discuss in this chapter are informative in parametric settings, but they can be misleading if not used appropriately.

8.1 Parametric Estimation of Power-Laws Using Linear Regression

Perhaps the most common first step toward understanding the distribution underlying a set of observations is to look at it graphically, and the natural visualization to begin with is the frequency plot, aka histogram. To construct a histogram, the first step is to "bin" the data values by dividing up the range of values into a series of smaller intervals. Then, the number of observations within each interval is counted and displayed on the plot. Formally, throughout this section we consider a simple situation where we are given a set of i.i.d. (scalar) observations $X_1, X_2, \ldots, X_n$ sampled from a Pareto distribution and we would like to estimate the shape parameter α of the distribution. To define the histogram of this data, we consider a fixed number of bins, k, defined by $b = \{[b_0, b_1), [b_1, b_2), \ldots, [b_{k-1}, b_k)\}$. The histogram plots points $(\bar{b}_j, m_j/n)$ where $\bar{b}_j = (b_j + b_{j-1})/2$ and $m_j = \sum_{i=1}^n I(X_i \in [b_{j-1}, b_j))$. Thus, the histogram provides an empirical estimate of the probability of taking values within each bin.

Figure 8.1 shows an example of histograms of samples from Gaussian, Uniform, Exponential, and Pareto distributions. These are generated with 3,000 samples from each distribution using equal-width bin sizes.[1] These examples demonstrate that a simple histogram can

---

[1] Note that the size of the bin used can dramatically impact the histogram, and so it must be chosen carefully. We discuss the use of logarithmic binning later in the chapter as an example of a more sophisticated approach. However, our recommendation is to use procedures that do not require binning the data.

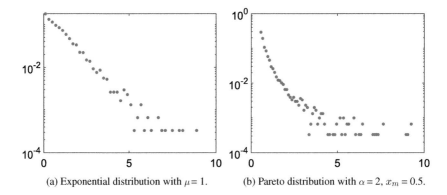

(a) Exponential distribution with $\mu = 1$.   (b) Pareto distribution with $\alpha = 2$, $x_m = 0.5$.

Figure 8.2 *Histograms on a log-linear scale.*

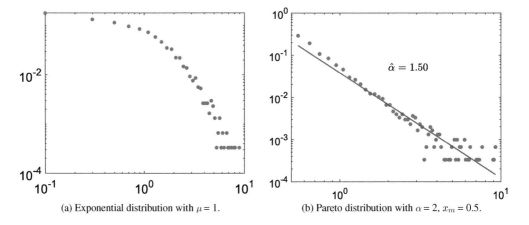

(a) Exponential distribution with $\mu = 1$.   (b) Pareto distribution with $\alpha = 2$, $x_m = 0.5$.

Figure 8.3 *Histograms on a log-log scale.*

already give you a lot of information about the p.d.f. of the underlying distribution the observations are sampled from. For example, when the underlying distribution is Gaussian, you see the distinctive bell curve (as in Figure 8.1(a)), and if the distribution is Uniform, you see a roughly "flat" form (as in Figure 8.1(b)). However, in many cases, the histogram itself provides very little information because the scaling hides much of the behavior. For example, from the histogram alone it is impossible to distinguish which of Figures 8.1(c) and 8.1(d) decays exponentially and which decays polynomially.

To distinguish properties of the tail of the distribution visually, the natural next step is to scale the histogram to get a better look at the underlying behavior. Moving to log-linear and log-log versions of the histogram is typical since these scalings can be used to identify distributions with exponential and power-law tails, respectively. In particular, if the data comes from an Exponential distribution (or a distribution with an exponential tail), then it will look linear on a log-linear plot since

$$f(x) = e^{-\mu x} \quad \Longrightarrow \quad \underbrace{\log f(x)}_{'y'} = -\mu \underbrace{x}_{'x'}. \tag{8.1}$$

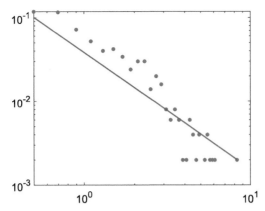

Figure 8.4 *Histogram of 500 samples from an Exponential distribution ($\mu = 1$) plotted on a log-log scale, along with the best-fit regression line.*

Similarly, if the data comes from a Pareto distribution, then it will look linear on a log-log plot since

$$f(x) = \frac{\alpha x_m^\alpha}{x^{\alpha+1}} \quad \Longrightarrow \quad \underbrace{\log f(x)}_{'y'} = \underbrace{\log(\alpha) + \alpha \log(x_m)}_{y\text{-intercept}} + \underbrace{(-\alpha - 1)}_{\text{slope}} \underbrace{\log(x)}_{'x'}. \quad (8.2)$$

Of course, this justification assumes that the bin widths are small, so that the probability of taking values in the bin $[b_{j-1}, b_j)$ is approximately $(b_j - b_{j-1}) f(\bar{b}_j)$.

While it was difficult to distinguish the tail behavior of the Exponential and Pareto distributions by looking at the unscaled histograms in Figure 8.1, Figures 8.2 and 8.3 show how it becomes easy to distinguish these differing tail behaviors when looking at the data on the log-linear and log-log scales. In particular, the linear behavior of the Pareto data on the log-log plot is immediately clear, standing out starkly from the sharp curve in the observations from the Exponential distribution. Similarly, the linear behavior of the Exponential data on the log-linear scale is easily distinguished from the sharp curve in the Pareto data.

Without knowing what underlying distribution the observations are sampled from, looking at the linear form of the data on a log-log scale in Figure 8.3(b) might tempt one to conclude that the data is from a Pareto distribution. One might even be tempted to go further and attempt to estimate the scaling parameter α using linear regression. In particular, (8.2) shows that, if the data is coming from a Pareto distribution, the slope of the line should provide an estimate of $-\alpha - 1$.

Formally, we can define an estimator of α based on linear regression, $\hat{\alpha}_{ls}$ as follows. Let m^* denote the slope of the least squares regression line on the histogram data (on a log-log scale). This yields the estimate $\hat{\alpha}_{ls} = -m^* - 1$. We illustrate this procedure in Figure 8.3(b). In particular, the procedure yields an estimate of $\hat{\alpha}_{ls} = 1.498$, with coefficient of determination $R^2 = 0.92$. Note that even though R^2, which measures the goodness of fit of the regression line, is quite high, the estimate of the tail index we have obtained is not that great – the correct value of α here is 2.

This illustrates that one should not place too much faith in conclusions from this sort of exploratory analysis. While using linear regression in this way is simple and intuitive, it is also quite unreliable. To see this, consider Figure 8.4. Here, we show the same procedure,

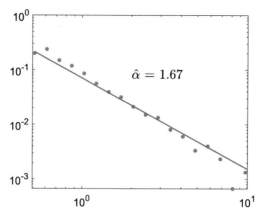

Figure 8.5 *Histogram of 3,000 samples from a Pareto distribution with $x_m = 0.5$ $\alpha = 2$, using logarithmic binning; each bin has width 10 times the width of the previous one, to create 40 bins over the interval $[0.1, 10]$. Bin centers are chosen to be the geometric means of the bin boundaries.*

but with fewer data points. Again, the data appears linear on a log-log scale. Further, when carrying out the regression, we obtain $\hat{\alpha}_{ls} = 0.4$ with $R^2 = 0.84$, which seems to be a strong signal about the quality of the fit. However, in this case we have been fooled. Here, it is not just that our estimate of α is off – the underlying distribution these observations are sampled from is not Pareto. In fact, it is not even heavy-tailed – the observations are sampled from an Exponential distribution!

What Can Be Done to "Fix" This Approach?

When trying to understand why the conclusions from applying linear regression are so unreliable, the natural candidate for blame is the noise in the tail that is visible in all of the histograms we have seen so far. Clearly, noise in the tail can hide the tail behavior that distinguishes the Pareto distribution from light-tailed distributions like the Exponential.

Thus, one might hope that issues with this procedure can be fixed by improving the procedure for binning used when creating the histogram. Optimistically, one might hope that by doing something more sophisticated than equal-width binning it is possible to reduce the noise and jaggedness in the tail of the histogram and thus improve the statistical power of regression in this context. In fact, it is indeed possible to reduce the noise in the tail with such approaches, and there are many sophisticated approaches for binning that have been proposed.

Logarithmic binning. Logarithmic binning is the most common approach for binning in the context of the identification of power-law phenomena. Logarithmic binning works by having the width of bins grow by a fixed multiplicative factor a (e.g., with $a = 2$ the bins are of width $0.1, 0.2, 0.4, 0.8, \dots$). This is appealing in the context of power-law identification because it yields bins that are equally spaced on the log-log scale being considered.

Figure 8.5 shows that logarithmic binning can lead to a significant reduction of the visible noise in the tail of the histogram. As before, linear regression can be used to estimate the tail index, though in this case, the slope of the best-fit line gives an estimate of $-\alpha$ (as opposed to $-(\alpha + 1)$ when using constant bin sizes); the justification for this is left as an exercise

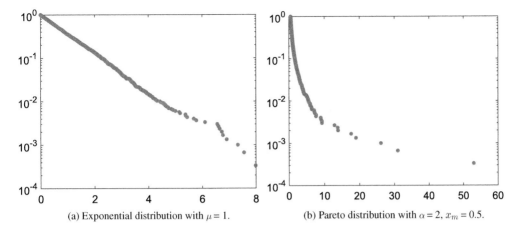

(a) Exponential distribution with $\mu = 1$.   (b) Pareto distribution with $\alpha = 2$, $x_m = 0.5$.

Figure 8.6 *Rank plots on a log-linear scale.*

for the reader (see Exercise 2). In our running example, logarithmic binning leads to a much stronger linear fit and a better estimate of α, $\hat{\alpha} = 1.67$ with $R^2 = 0.96$.

While logarithmic binning is more robust than uniform binning, all histogram-based approaches have certain limitations. First, the estimate depends critically on how the data is binned, and there is no structured procedure to arrive at the best binning scheme. Second, histograms tend to be fairly noisy at estimating the tail end of the underlying distribution, since there are (naturally) fewer data points here. Indeed, bins far into the tail are often empty, making any logarithmic transformation of the histogram problematic!

The rank plot. Our next approach throws out binning all together. The motivation for this is that no matter how binning is performed, it loses significant information from the data. This is because all data within a given bin are lumped together, which loses all information about the specific samples of data with that range. The impact of this is dramatic under logarithmic binning because the tail uses increasingly large bins, which leads to significant loss of information on the relative size of data within each bin.

While binning is necessary when one uses a histogram to visualize the empirical p.d.f. of the distribution, binning can be avoided if one shows the empirical c.c.d.f. of the data instead. Recall the c.c.d.f. is $\bar{F}(x) = \Pr(X > x)$, and so the empirical c.c.d.f. can be written as

$$\bar{F}_n(x) = \frac{1}{n}\sum_{i=1}^{n} I(X_i \geq x). \tag{8.3}$$

Plots of the empirical c.c.d.f. are termed *rank plots* or *frequency plots*, and we show examples in the case of the Exponential and Pareto distributions in Figures 8.6 and 8.7. Compared to the corresponding histograms in Figures 8.2 and 8.3, notice the significant noise reduction in the rank plots.

Rank plots are appealing because they eliminate the need to bin the data, but enable exactly the same procedure for estimating a Pareto distribution as the histogram. In particular, the c.c.d.f. of the Pareto distribution is also linear on a log-log scale, just like the p.d.f. (though the slope of the c.c.d.f. on a log-log scale equals $-\alpha$, whereas the p.d.f. has slope $-(\alpha + 1)$ on a log-log scale).

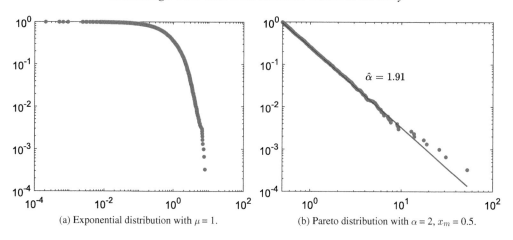

(a) Exponential distribution with $\mu = 1$.   (b) Pareto distribution with $\alpha = 2$, $x_m = 0.5$.

Figure 8.7 *Rank plots on a log-log scale.*

$$\bar{F}(x) = \left(\frac{x_m}{x}\right)^{\alpha} \implies \underbrace{\log \bar{F}(x)}_{'y'} = \underbrace{\alpha \log x_m}_{y\text{-intercept}} + \underbrace{(-\alpha)}_{\text{slope}} \underbrace{\log(x)}_{'x'}. \tag{8.4}$$

Thus, we can again identify a Pareto distribution by looking at the data on a log-log scaled rank plot and estimate α using linear regression.

Figure 8.7(b) shows the impact of using the rank plot instead of the frequency plot for estimation of α. This leads to a stronger linear fit and an improved estimate of α: $\hat{\alpha} = 1.91$ with $R^2 = 0.99$. However, one should still view conclusions made using linear regression with skepticism, as we will now explain.

The Fundamental Problem with Using Linear Regression

The procedure outlined in the previous section is simple and intuitive and, as such, it is frequently used when identifying power-law behavior. However, we cannot emphasize enough that this approach provides very little certainty about its conclusions. In fact, perhaps the most important message in this chapter is that the approach we have outlined so far is fundamentally flawed and any conclusions about the existence of heavy-tailed phenomena reached via naive linear regression should be viewed with skepticism. It is a valuable exploratory tool, but its conclusions should be tested more carefully (using the tools we introduce in Chapter 9) before being accepted.

Perhaps the most fundamental issue underlying the failure of identification via linear regression on a log-log scaled histogram is that, while it is true that one can reject the hypothesis that the data follows a power law distribution by observing that the linear fit on a log-log scale is poor, the opposite is not true. More specifically, a strong linear fit (high R^2) does not provide any guarantee that the underlying distribution is a Pareto. This is because, while it is indeed true that a Pareto histogram looks linear under a log-log scaling, the Pareto distribution is not alone in this fact. In particular, while the p.d.f. of other distributions may not be precisely linear on a log-log scale, many are *nearly* linear. For example, Figure 8.8 shows that histograms generated using samples from other heavy-tailed distributions can also have a high R^2 score for the regression line. However, the problem with using linear regression to estimate power law distributions is more fundamental than this since, given the noise present

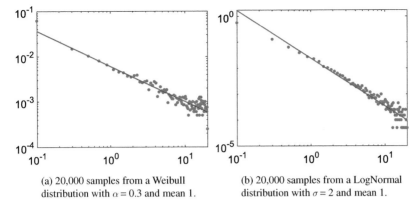

(a) 20,000 samples from a Weibull distribution with $\alpha = 0.3$ and mean 1.

(b) 20,000 samples from a LogNormal distribution with $\sigma = 2$ and mean 1.

Figure 8.8 *Histograms corresponding to samples from a heavy-tailed Weibull (a) and a LogNormal (b) distribution. The least squares regression line has an R^2 of 0.92 in the case of the Weibull, and 0.94 in the case of the LogNormal.*

in samples, it is not uncommon for even light-tailed distributions to yield data that appears to be linear on a log-log scale (as we saw in Figure 8.4). Thus, while a low R^2 is informative, it is rarely observed in practice, and so the value of R^2 has little power to test the hypothesis that the underlying data is Pareto.

While switching from the histogram to the rank plot is a more robust method of detecting power laws, the estimates obtained using linear regression on a log-log scaled rank plot can still be unreliable. Underlying the lack of statistical power of linear regression in this context is that linear regression was not designed for use in this setting and so the ordinary formulas for calculating the best-fit regression line simply do not apply. In particular, the underlying assumption behind least squares linear regression is that the noise in the dependent variable for each value of the independent variable is independent and identically distributed. This certainly is not true in this context of parametric estimation of power laws; the noise in the rank plot is *not* independent, and moreover the tail is far noisier than the body. This leads to large, systematic errors in the estimate of the slope. Indeed, using a specific scheme of *nonuniform* weights for the errors between the (log-log scaled) rank plot and regression line, which accounts for the nonuniformity in the noise on the rank plot, results in considerably more robust estimators of the tail index for Pareto distributions, as we show in Section 8.4.

8.2 Maximum Likelihood Estimation for Power-Law Distributions

Given the issues with using linear regression for the estimation of power-law distributions, it is clear that a different approach is needed. To that end, we now turn to a very different approach: *maximum likelihood estimation (MLE)*. Maximum likelihood estimation is a classical approach in parametric estimation that is applicable well beyond the context of estimating Pareto distributions. The key idea of maximum likelihood estimation is to assume that observations come from a particular distribution, in this case a Pareto, and then find the specific parametrization of the distribution that maximizes the likelihood of the data. While this approach is less visual than using linear regression and thus less suitable for exploratory analysis, it is appealing because it gives much more reliable estimates.

Maximum likelihood estimation is a rich area with many general results, and so we do not attempt to provide a general treatment here; the interested reader is referred to [26, 43, 182]. Instead, we focus only on the context of Pareto distributions. In this context, the application of maximum likelihood estimation is quite straightforward and we are able to provide a derivation of the maximum likelihood estimator of α and prove quality guarantees about this estimator using only elementary tools.

To simplify the exposition, we focus on the case where the lower bound of the Pareto distribution, x_m, is known and our goal is to estimate the shape parameter, α, from the data. We discuss the task of estimating x_m at the end of the chapter in Section 8.5. As in the case of linear regression, we consider a simple situation where we are given a set of i.i.d. (scalar) observations $X_1, X_2, \ldots, X_n$ sampled from a Pareto distribution.

The starting point for maximum likelihood stimulation is, naturally, the likelihood function $\mathcal{L}(\alpha; X_1, \ldots, X_n)$, which characterizes how "likely" the data $X_1, \ldots, X_n$ would have been if the underlying distribution had shape parameter α. Formally, $\mathcal{L}$ is defined as follows:

$$\mathcal{L}(\alpha; X_1, \ldots, X_n) = f(X_1; \alpha) f(X_2; \alpha) \cdots f(X_n; \alpha),$$

where $f(x; \alpha) = \alpha x_m^\alpha / x^{\alpha+1}$ is the density function corresponding to the Pareto distribution.

Given the likelihood function, a maximum likelihood estimator for α, which we denote by $\hat{\alpha}_{\mathrm{mle}}$, is a value of α that maximizes the likelihood function (as the name implies). This can be written as

$$\hat{\alpha}_{\mathrm{mle}} \in \arg\max_{\alpha \in \mathbb{R}} \mathcal{L}(\alpha; X_1, \ldots, X_n). \tag{8.5}$$

From this characterization it is not hard to derive an explicit form for the maximum likelihood estimator by solving the preceding optimization. We do exactly that to obtain the following theorem.

Theorem 8.1 *Suppose that $X_1, X_2, \ldots, X_n$ are i.i.d. samples from a $\mathrm{Pareto}(x_m, \alpha)$ distribution, where x_m is known. The maximum likelihood estimator $\hat{\alpha}_{\mathrm{mle}}$ of α is given by*

$$\hat{\alpha}_{\mathrm{mle}} = \frac{1}{\frac{1}{n} \sum_{i=1}^{n} \log\left(\frac{X_i}{x_m}\right)}.$$

Proof The most common approach for deriving a maximum likelihood estimator is to focus on the logarithm of the likelihood function instead of the likelihood function itself. Of course, optimizing $\log \mathcal{L}$ is equivalent to optimizing $\mathcal{L}$. The reason for the change is that taking the logarithm yields a much simpler optimization by converting the product to a summation. In particular, the log-likelihood function for the $\mathrm{Pareto}(x_m, \alpha)$ distribution has the following form:

$$\log(\mathcal{L}) = n \log(\alpha) + n\alpha \log(x_m) - (\alpha + 1) \sum_{i=1}^{n} \log(X_i).$$

It is not hard to see that this function is concave with respect to α. Using this fact, we can maximize with respect to α by simply setting its derivative to zero. In other words, $\hat{\alpha}_{\mathrm{mle}}$ satisfies

$$\frac{n}{\alpha} + n \log(x_m) - \sum_{i=1}^{n} \log(X_i) = 0.$$

It follows that

$$\hat{\alpha}_{\mathrm{mle}} = \frac{n}{\sum_{i=1}^{n} \log\left(\frac{X_i}{x_m}\right)},$$

which matches the form of the estimator in the theorem statement. $\qquad\square$

Given the form for $\hat{\alpha}_{\mathrm{mle}}$ in Theorem 8.1, we can immediately apply it to the running example we considered when discussing the use of linear regression – the MLE corresponding to the 3,000 Pareto-distributed (with tail index 2) samples we have been using works out to $\hat{\alpha}_{\mathrm{mle}} = 1.94$. We give more empirical examples showing the use of $\hat{\alpha}_{\mathrm{mle}}$ (and some issues with it) later in this chapter. For now, note that, while there is a modest improvement in the accuracy of the estimate using MLE compared to that obtained from regression on the log-log scaled rank plot, the real power of MLE lies in its statistical guarantees, which we discuss next.

8.3 Properties of the Maximum Likelihood Estimator

The form of the $\hat{\alpha}_{\mathrm{mle}}$ is perhaps a bit mysterious at first sight. However, there is strong intuition for why this form should yield an accurate estimator. In particular, $\hat{\alpha}_{\mathrm{mle}}$ can be viewed as the reciprocal of the sample average of n random variables $E_i = \log(X_i/x_m)$.

To understand the power of this viewpoint, recall the following useful property, which we have used a couple of times already in this book: E_i is exponentially distributed with parameter α. To see why, observe that

$$\Pr\left(\log\left(\frac{X_i}{x_m}\right) > t\right) = \Pr\left(X_i > x_m e^t\right) = \left(\frac{x_m}{x_m e^t}\right)^{\alpha} = e^{-\alpha t}. \qquad (8.6)$$

This observation implies that $\mathbb{E}[E_i] = 1/\alpha$. Thus, a very natural way to estimate α is to use the inverse of the sample average of $E_i = \log(X_i/x_m)$ – which is exactly what $\hat{\alpha}_{\mathrm{mle}}$ does. That is,

$$\hat{\alpha}_{\mathrm{mle}} = \frac{1}{S_n/n}, \quad \text{where } S_n = \sum_{i=1}^{n} E_i.$$

This form not only provides intuition for $\hat{\alpha}_{\mathrm{mle}}$; it also immediately yields an important property of $\hat{\alpha}_{\mathrm{mle}}$. Since $\hat{\alpha}_{\mathrm{mle}}$ is the sample average of E_i, which are i.i.d., we can apply the strong law of large numbers to conclude that

$$\hat{\alpha}_{\mathrm{mle}} = \frac{1}{S_n/n} \xrightarrow{a.s.} \frac{1}{\mathbb{E}[E_i]} = \alpha.$$

Thus, with enough data, $\hat{\alpha}_{\mathrm{mle}}$ will yield an accurate estimate of α with probability 1. This property is termed "strong consistency" and is an important first-order check on the quality of an estimator.

Theorem 8.2   *Assuming that $\{X_i\}_{1 \le i \le n}$ are drawn from a Pareto(x_m, α) distribution, the maximum likelihood estimator $\hat{\alpha}_{\text{mle}}$ of α is strongly consistent (i.e., $\hat{\alpha}_{\text{mle}} \overset{a.s.}{\to} \alpha$ as $n \to \infty$).*

However, it is important to note that strong consistency is a statement about the limiting behavior of $\hat{\alpha}_{\text{mle}}$ only. In fact, despite being accurate in the limit, $\hat{\alpha}_{\text{mle}}$ is biased on finite samples. To see this, note that S_n is the sum of Exponential random variables, and thus follows a Gamma distribution. In particular, we can write

$$\mathbb{E}\left[\hat{\alpha}_{\text{mle}}\right] = \mathbb{E}\left[n/S_n\right]$$

$$= n \int_0^\infty \frac{1}{t} \frac{\alpha^n}{(n-1)!} t^{n-1} e^{-\alpha t} dt$$

$$= \alpha \left(\frac{n}{n-1}\right) \int_0^\infty \frac{\alpha^{n-1}}{(n-2)!} t^{n-2} e^{-\alpha t} dt$$

$$= \alpha \left(\frac{n}{n-1}\right).$$

This calculation shows that $\hat{\alpha}_{\text{mle}}$ has a small and disappearing bias, $n/(n-1)$. Clearly, this bias can be corrected for, if desired, by simply using the estimator $\hat{\alpha}^*_{\text{mle}} = (n-1)/S_n$ instead of $\hat{\alpha}_{\text{mle}}$.

The observation that S_n follows a Gamma distribution also lets us easily calculate the variance of $\hat{\alpha}_{\text{mle}}$ and $\hat{\alpha}^*_{\text{mle}}$. We leave the details of the calculation as Exercise 3 and simply state the results here:

$$\text{Var}\left[\hat{\alpha}_{\text{mle}}\right] = \alpha^2 \left(\frac{n^2}{(n-1)^2(n-2)}\right) \sim \frac{\alpha^2}{n}, \tag{8.7}$$

$$\text{Var}\left[\hat{\alpha}^*_{\text{mle}}\right] = \frac{\alpha^2}{(n-2)} \sim \frac{\alpha^2}{n}. \tag{8.8}$$

Without a comparison, it is difficult to say whether this variance is large or small; however, it turns out that this is actually the smallest variance that is possible for any unbiased estimator. More specifically, the following lemma, which follows from a Cramér–Rao inequality (the interested reader is referred to [43, 178]), provides a lower bound on the variance of any unbiased estimator.

Lemma 8.3   *Assuming that $\{X_i\}_{1 \le i \le n}$ are drawn from a Pareto(x_m, α) distribution, the variance of any unbiased estimator $\hat{\alpha}$ of α is bounded from below as $\text{Var}(\hat{\alpha}) \ge \frac{\alpha^2}{n}$.*

This lemma implies that the variance of $\hat{\alpha}_{\text{mle}}$ and $\hat{\alpha}^*_{\text{mle}}$ asymptotically matches the lower bound; this property is termed *asymptotic efficiency*.

Theorem 8.4   *Assuming that $\{X_i\}_{1 \le i \le n}$ are drawn from a Pareto(x_m, α) distribution, the maximum likelihood estimators $\hat{\alpha}_{\text{mle}}$ and $\hat{\alpha}^*_{\text{mle}}$ of α are asymptotically efficient.*

So far, we have seen that $\hat{\alpha}_{\text{mle}}$ and $\hat{\alpha}^*_{\text{mle}}$ are asymptotically consistent and asymptotically efficient, but we have not provided any characterization of the estimation errors. The interpretation of $\hat{\alpha}_{\text{mle}}$ as the sample average of E_i gives us an easy way to provide such a characterization. In particular, if we apply the central limit theorem to S_n in the same way we earlier applied the law of large numbers, we can see that $\hat{\alpha}_{\text{mle}}$ is *asymptotically normal*, that is, $\sqrt{n}(\hat{\alpha}_{\text{mle}} - \alpha) \xrightarrow{d} \alpha Z$, where Z is the standard Gaussian. This is powerful, not just because it shows that the error is approximately Gaussian, but also because it reveals that the estimation errors decay as $O(1/\sqrt{n})$, which is consistent with the calculation of the variance in (8.7).

Theorem 8.5 *Assuming that $\{X_i\}_{1 \leq i \leq n}$ are drawn from a Pareto(x_m, α) distribution, the maximum likelihood estimator $\hat{\alpha}_{\text{mle}}$ of α is asymptotically Gaussian. Formally, $\sqrt{n}(\hat{\alpha}_{\text{mle}} - \alpha) \xrightarrow{d} \alpha Z$, where Z is the standard Gaussian.*

Asymptotic normality of maximum likelihood estimators can be established under fairly general conditions (see [26]), so it is not surprising that it holds here. However, this case provides an elegant illustration for why this property holds.

Proof To start, recall that, since E_i is exponentially distributed with parameter α, we have that $\mathbb{E}[E_i] = 1/\alpha$, $\text{Var}(E_i) = 1/\alpha^2$. As we discussed earlier, the key idea of the proof is to apply the central limit theorem to $S_n = \sum_{i=1}^{n} E_i$. In particular, the central limit theorem implies that as $n \to \infty$,

$$\frac{S_n - \frac{n}{\alpha}}{\sqrt{n}} \xrightarrow{d} \frac{Z}{\alpha},$$

where N is the standard Gaussian.

To use this to analyze $\hat{\alpha}_{\text{mle}}$, let $Y_n := \frac{S_n - \frac{n}{\alpha}}{\sqrt{n}}$. The estimator $\hat{\alpha}_{\text{mle}}$ can now be represented in terms of Y_n as follows:

$$Y_n \sqrt{n} = S_n - \frac{n}{\alpha}$$

$$\Rightarrow \quad \alpha Y_n \frac{n}{S_n} = \sqrt{n}(\alpha - \frac{n}{S_n}) = \sqrt{n}(\alpha - \hat{\alpha}_{\text{mle}}).$$

Now, since $\alpha Y_n \xrightarrow{d} Z$ and $\frac{n}{S_n} \xrightarrow{a.s.} \alpha$ (by the strong law of large numbers), it follows that

$$\sqrt{n}(\alpha - \hat{\alpha}_{\text{mle}}) \xrightarrow{d} \alpha Z,$$

which completes the proof (since the Gaussian is symmetric, $\sqrt{n}(\alpha - \hat{\alpha}_{\text{mle}}) \xrightarrow{d} \alpha Z$ implies $\sqrt{n}(\hat{\alpha}_{\text{mle}} - \alpha) \xrightarrow{d} \alpha Z$). Note that the last equation follows from the more general fact that, for any two sequences of scalar random variables Y_n and Z_n, if $Y_n \xrightarrow{d} Y$, where Y is a continuous random variable, and $Z_n \xrightarrow{a.s.} c$, where $c \neq 0$ is a constant, then $Y_n Z_n \xrightarrow{d} cY$. (See Exercise 7.) $\qquad\square$

8.4 Visualizing the MLE via Regression

To this point we have discussed a number of important results highlighting the quality of the maximum likelihood estimator $\hat{\alpha}_{\text{mle}}$. Together with the numerical results we have seen so far, our discussion has painted a very positive picture of the MLE. However, the MLE fails at some of the things that made linear regression so appealing. Regression allowed a visual exploration of the data that could both identify if the power-law hypothesis was appropriate and provide insight into the quality of the estimator. In contrast, maximum likelihood estimation does not have a way to accomplish either of these tasks. In particular, the MLE is simply a number and so, without further exploration, it is unclear how much confidence one should place in this estimate or whether the Pareto distribution hypothesis was appropriate in the first place.

While these drawbacks can be mitigated to some extent via goodness-of-fit tests (as we discuss in Section 8.5), the lack of a visual representation of the MLE is a serious drawback when one is performing exploratory analysis of the data to understand if heavy-tailed phenomena are present. For example, consider our running examples from Figure 8.7 of a Pareto distribution ($\alpha = 2$, $x_m = 0.5$) and an Exponential distribution ($\mu = 1$). In both cases we can compute $\hat{\alpha}_{\mathrm{mle}}$ and obtain an estimate $-$ 1.94 and 7.81 respectively. However, these calculations do not give any indication about whether the data is truly from a Pareto distribution or not. In fact, they provide no distinction between the two cases. Thus, maximum likelihood estimation cannot be used for exploratory analysis and must be paired with other statistical tools (goodness-of-fit tests) in order to ensure that it is only applied in appropriate settings.

It turns out that there is a way around this concern too. In particular, there is a little-known connection between the MLE and linear regression that provides a way to visualize the fit of the MLE. This may be surprising since, so far, we have presented maximum likelihood estimation as a completely different approach from linear regression. However, it is possible to reconcile these two approaches for estimation into a unified framework. In particular, by choosing weights carefully, one can view maximum likelihood estimation as a form of *weighted* least squares regression. This view of maximum likelihood estimation is quite powerful because it both (i) provides a visual interpretation of maximum likelihood estimation, allowing the MLE to be used for exploratory analysis, and (ii) highlights why the estimates provided by (unweighted) linear regression are inaccurate.

To state the relationship between linear regression and maximum likelihood estimation more formally, let us begin with a bit of notation. As previously, we index the samples in sorted, nondecreasing order (i.e., $X_{(1)} \leq X_{(2)} \leq \cdots \leq X_{(n)}$) so that the value of the empirical c.c.d.f. (aka, the rank plot defined in (8.3)) at $X_{(i)}$ is given by $\bar{F}_n(X_{(i)}) = \frac{n+1-i}{n}$.

Using this notation, the weighted least-squares (WLS) estimate $\hat{\alpha}_{wls}(w)$ of α given nonnegative weights $w = (w_i, 1 = 1, \ldots, n)$ is a minimizer of the weighted sum of the squared error between the rank plot and the c.c.d.f. at the sample points on a logarithmic scale. That is,

$$\hat{\alpha}_{wls}(w) := \arg\min_{\alpha} \sum_{i=1}^{n} w_i \left(\log \bar{F}_n(X_{(i)}) - \log \bar{F}(X_{(i)}; \alpha)\right)^2$$

$$= \arg\min_{\alpha} \sum_{i=1}^{n} w_i \left(\log\left(\frac{n+1-i}{n}\right) + \alpha \log\left(\frac{X_{(i)}}{x_m}\right)\right)^2. \qquad (8.9)$$

From this formulation it is not hard to derive a closed form expression for $\hat{\alpha}_{wls}(w)$. In particular, the objective is convex; thus, we can obtain its minimizer by setting its derivative w.r.t. α to zero, that is,

$$0 = \sum_{i=1}^{n} 2w_i \left[\log\left(\frac{n+1-i}{n}\right) + \alpha \log\left(\frac{X_{(i)}}{x_m}\right) \right] \log\left(\frac{X_{(i)}}{x_m}\right)$$

$$= 2 \sum_{i=1}^{n} w_i \log\left(\frac{n+1-i}{n}\right) \log\left(\frac{X_{(i)}}{x_m}\right) + 2\alpha \sum_{i=1}^{n} w_i \log^2\left(\frac{X_{(i)}}{x_m}\right).$$

This gives us the following expression for the WLS estimate:

$$\hat{\alpha}_{wls}(w) = \frac{-\sum_{i=1}^{n} w_i \log\left(\frac{n+1-i}{n}\right) \log\left(\frac{X_{(i)}}{x_m}\right)}{\sum_{i=1}^{n} w_i \log^2\left(\frac{X_{(i)}}{x_m}\right)}.$$

It is not immediately clear how this expression for $\hat{\alpha}_{wls}(w)$ relates to $\hat{\alpha}_{mle}$; however, the theorem that follows shows that, if the weights are chosen as

$$w_i = \left[\log\left(\frac{X_{(i)}}{x_m}\right) \right]^{-1},$$

then $\hat{\alpha}_{mle}$ can be recovered from $\hat{\alpha}_{wls}(w)$. Formally, we have the following result.

Theorem 8.6 *Suppose the weights w for $\hat{\alpha}_{wls}(w)$ are defined as $w_i = \left(\log\left(\frac{X_{(i)}}{x_m}\right) \right)^{-1}$ for $1 \le i \le n$. Then, as $n \to \infty$, $\hat{\alpha}_{wls}(w) \sim \hat{\alpha}_{mle}$, that is, the $\hat{\alpha}_{wls}(w)$ is asymptotically equivalent to $\hat{\alpha}_{mle}$.*

The choice of weights in this theorem is interesting because it conveys how to "correct" the estimator given by linear regression. In particular, it highlights the source of the errors in the regression estimator: too much weight given to the data in the *tail* and too little weight given to the data in the *body*. This makes sense intuitively because the data in the tail is much more noisy than data in the body. Given that a parametric distribution is being estimated, fitting the body implies a fit of the tail, and so placing a higher weight on the body, where the noise is smaller, is appropriate.

The contrast between the weighted and unweighted regression estimators can be quite striking visually, as shown in Figure 8.9. It is easy to see the impact of the additional weight on data from the body of the distribution in this case.

We now prove the connection between regression and the MLE estimators formally.

Proof of Theorem 8.6 Given our choice of weights, the expression for $\hat{\alpha}_{wls}(w)$ simplifies to

$$\hat{\alpha}_{wls}(w) = \frac{-\sum_{i=1}^{n} \log\left(\frac{n+1-i}{n}\right)}{\sum_{i=1}^{n} \log\left(\frac{X_{(i)}}{x_m}\right)}.$$

Recalling the expression for $\hat{\alpha}_{mle}$ from Theorem 8.1, it suffices to prove that, as $n \to \infty$,

$$-\sum_{i=1}^{n} \log\left(\frac{n+1-i}{n}\right) \sim n.$$

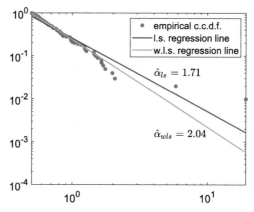

Figure 8.9 *Rank plot corresponding to 100 samples from a Pareto distribution with* $x_m = 0.5$ *and* $\alpha = 2$, *alongside the least-squares and weighted least-squares regression lines. Note that the least-squares regression line is "swayed" by the considerable tail noise in the rank plot. Clearly, the weighted least squares estimate of the tail index is more accurate.*

This follows from Stirling's approximation as follows:[2]

$$-\sum_{i=1}^{n} \log\left(\frac{n+1-i}{n}\right) = \log\left(\frac{n^n}{n!}\right)$$

$$\sim \log\left(\frac{n^n}{\sqrt{2\pi n}(n/e)^n}\right)$$

$$= n - \frac{1}{2}\log(2\pi n) \sim n.$$

□

8.5 A Recipe for Parametric Estimation of Power-Law Distributions

The connection to regression we proved in the previous section serves to address one of the biggest concerns about the MLE and to allow exploratory visualization of data that is statistically grounded. Combined with the results in Section 8.3 showing that the MLE is a minimal variance unbiased estimator, it should be clear by now that the MLE is a powerful tool for parametric estimation of power-law distributions. However, it is important to realize that the MLE should not be used in isolation. It is not enough to perform estimation with the MLE; one must also evaluate the quality of the fit and contrast the fit with what is possible using other distributional models.

While our goal in this chapter is to focus on estimation tools, we would be remiss if we did not end with a brief discussion of how the estimation tools we present fit into the broader statistical pipeline. To that end, we provide a recipe for the whole pipeline in what follows. Note that we make particular choices in this recipe about goodness-of-fit tests and other ingredients, and there are certainly alternatives that are possible. We make these choices to align with the presentation of [37, 48] and to provide a concrete approach that is reliable in

---

[2] Stirling's approximation provides an asymptotic approximation of the factorial function: $n! \sim \sqrt{2\pi n}\left(\frac{n}{e}\right)^n$.

general settings. However, we also discuss some potential alternatives to this recipe in the additional notes at the end of the chapter.

1. *Estimate the parameters x_m and α.* An estimate of α can be computed using the tools presented in this chapter. To compute an estimate of x_m, it is typically sufficient to use the minimal observed sample in the data. However, a more sophisticated approach is to estimate x_m using Kolmogorov–Smirnov (KS) minimization [48]. In this approach, x_m is selected so as to minimize the maximum difference between the cumulative empirical distribution on $x > x_m$ and the c.d.f. of the best fitting Pareto distribution for those observations. We will discuss this technique, also referred to as PLFIT (short for Power-Law FIT), in greater detail in Section 9.6 of Chapter 9, where we address the question "Where does the power law tail begin?" in a more general semi-parametric setting.

2. *Calculate the goodness-of-fit between the data and the Pareto distribution using the Kolmogorov–Smirnov (KS) statistic.* To do this, we first compute the KS statistic for the fit of the estimated parameters to the data. Then, a large number of synthetic datasets with parameters matching those estimated from the data is generated. Each is individually fit to a Pareto distribution, using the tools in this chapter, and then the KS statistic for the fit is computed. Finally, a p-value is calculated by computing the fraction of the time that the KS statistic is larger than the value for the empirical data.

3. *Compare the fit of the Pareto distribution to alternative distributions using a likelihood ratio test.* To contrast the fit of the Pareto distribution with other alternatives (e.g., the Exponential, Weibull, or LogNormal distributions) the procedures in steps 1 and 2 are repeated with the alternative distributions. In each case, the log-likelihood of the best fit is computed and compared to that of the Pareto fit. If the log-likelihood of the Pareto is higher, then the Pareto is a better fit than the alternatives. However, because the log-likelihoods are random quantities, one must calculate a p-value to test the null hypothesis that the log-likelihoods are the same. Details on how to compute the p-values can be found in [48].

This is a very standard recipe that is used across a variety of fields, and there is public code available to run through the whole sequence (e.g., see [47]). It has been used successfully in a wide range of network science settings. However, the recipe provides, perhaps, an undue sense of certainty. It is important to remember that this recipe is only appropriate for *parametric* estimation when the data comes from a precise power-law (Pareto) distribution. If this is not the case (e.g., if only the *tail* comes from an *approximate* power-law distribution), then one should be cautious with this approach.

To emphasize the importance of this point, let us highlight an example from the field of network science. As we discussed in Chapter 6, a celebrated result in network science is that many complex networks in the world around us exhibit scale-free, power-law degree distributions. In Section 6.3 we study one prominent, simple mechanism that has been used to explain this phenomenon – preferential attachment – and prove that this mechanism yields power-law degree distributions. However, when one applies the recipe in order to estimate the tail of the degree distribution for a preferential attachment graph from data, it is likely to reject the hypothesis of a power-law degree distribution; see [37, 212] for a discussion of an example where this has happened. This is surprising and troubling, but the reason behind it is simple: the recipe just presented only leads to a consistent estimator when applied to samples from pure Pareto distributions and, potentially, a limited class of regularly varying distributions. The degree distribution of preferential attachment graphs is more complex. Given this, it is actually *correct* to reject the conclusion that preferential attachment yields a *pure* power law degree distribution. However, that is not a reason to conclude that the *tail* of the degree distribution under preferential attachment is not a power law, which we have proven is true.

This example highlights how crucial it is to ensure that the appropriate statistical tool is used, and so we end this chapter with the following important disclaimer:

Disclaimer. *It is important to remember that all the results in this section assume the data is from a precise, parametric Pareto distribution. The MLE is well-suited for this sort of parametric estimation. If instead, just the tail of the distribution is power-law, the MLE is no longer appropriate and the techniques discussed in Chapter 9 should be applied.*

8.6 Additional Notes

We have focused our discussion in this chapter on understanding two of the most common techniques used for estimating the parameters of Pareto distributions from data. We have provided an illustration of the general approaches, their properties, and their limitations, but there are many details of the procedures that have been studied carefully in the literature.

For example, we have not focused in depth on the wide variety of approaches for binning data when forming histograms of the empirical p.d.f. This is because we do not recommend using approaches that require binning the data. Instead, using the rank plot of the empirical c.c.d.f. makes much better use of the data. However, there is a large literature studying the impact of different approaches for determining bin sizes that can be used to help guide choices if one needs to look directly at the empirical p.d.f.; see [214] and the references therein.

To conclude the chapter, we commented briefly about how the estimation procedures we discuss fit into the broader statistical pipeline for studying heavy-tailed phenomena. In particular, any estimate must be validated using a goodness-of-fit test and then contrasted with alternative hypotheses using a likelihood ratio test. Without these steps one cannot be confident in a fit to a Pareto distribution. We have suggested a concrete approach for this based on the procedure in [37, 48]; however, there are a variety of alternative techniques related to evaluating goodness of fit. For example, interested readers can explore [115] for a detailed discussion of the topic.

Finally, we have focused this chapter entirely on estimating parameters of the Pareto distribution. The intuition and tools we have presented can be applied much more broadly. To illustrate this, Exercises 4, 5, and 6 task the reader with deriving the MLE for some common heavy-tailed distributions. When exploring other heavy-tailed distributions using graphical tools, one should use the Quantile-Quantile (QQ) plot instead of the log-log plot. For a distribution, F, the quantiles of the distribution can be determined by the inverse F^{-1}, which is often referred to as the quantile function. Then, the QQ plot is constructed by plotting the quantiles of the distribution versus the quantiles of the data points. Specifically, letting $X_{(i)}$ be the ith largest of n samples, the QQ plot for F and the data is generated by plotting

$$\left(X_{(i)}, F^{-1}\left(\frac{n - i + 1}{n + 1} \right) \right), \text{ for } i = 1, \ldots, n. \tag{8.10}$$

If the data matches the distribution perfectly, its QQ plot is a line that follows the 45 degree line $y = x$. This provides a valuable exploratory tool, and then estimation can be done using the MLE for the distribution of interest. Exercise 11 tasks the reader with exploring the use of QQ plots for heavy-tailed distributions. The interested reader can refer to [183] for an introduction to statistical tools related to estimation and exploration via QQ-plots.

8.7 Exercises

1. Consider a histogram obtained from i.i.d. Pareto-distributed samples, with tail index α.
   (a) Show that, when using uniform bin sizes, the expected number of samples in any bin is eventually monotonically decreasing.
   (b) Show that the monotonicity of the expected number of samples per bin in part (a) also holds under logarithmic binning, assuming $\alpha > 1$.
   Note: This shows that the tail of the histogram is necessarily "noisier" than the body, even under logarithmic binning.

2. Consider a histogram obtained from i.i.d. Pareto-distributed samples with tail index α, under logarithmic binning. Taking the bin centers to be the geometric means of the bin boundaries, show that the slope of the histogram on a log-log plot provides an estimate of $-\alpha$.

3. Let $X_1, X_2, \ldots, X_n$ denote i.i.d. Pareto random variables with known minimum value x_m and unknown tail index α. Let $\hat{\alpha}_{\text{mle}}$ denote the MLE of α, and let $\hat{\alpha}^*_{\text{mle}} = \frac{n-1}{n}\hat{\alpha}_{\text{mle}}$ denote its unbiased version. Prove that

$$\text{Var}\left[\hat{\alpha}_{\text{mle}}\right] = \alpha^2 \left(\frac{n^2}{(n-1)^2(n-2)} \right) \sim \frac{\alpha^2}{n},$$

$$\text{Var}\left[\hat{\alpha}^*_{\text{mle}}\right] = \frac{\alpha^2}{(n-2)} \sim \frac{\alpha^2}{n}.$$

4. Suppose that $X_1, X_2, \ldots, X_n$ denote i.i.d. samples from a $\text{LogNormal}(\mu, \sigma^2)$ distribution, where both μ and σ^2 are unknown. Derive the MLE for μ and σ^2 based on the available samples.

5. Suppose that $X_1, X_2, \ldots, X_n$ denote i.i.d. samples from a $\text{Weibull}(\alpha, \beta)$ distribution, where both α and β are unknown. Characterize the MLE of α and β based on the available samples.
   Note: Unlike in the case of the LogNormal, you will not get a closed form for the MLE in this case. However, you should be able to express $\hat{\beta}_{\text{mle}}$ as a function of $\hat{\alpha}_{\text{mle}}$, which is in turn the unique positive solution of a certain equation that can be solved numerically.

6. Suppose that $X_1, X_2, \ldots, X_n$ denote i.i.d. samples from a heavy-tailed distribution having support $[0, \infty)$, with a c.c.d.f. given by $\bar{F}(x) = (1 + x)^{-\alpha}$ for $x \geq 0$. (Note that this distribution is just a special case of the family of Burr distributions we discussed in Chapter 1.) Derive the MLE of the tail index α based on the available samples.

7. Consider two sequences of scalar random variables Y_n and Z_n. As $n \to \infty$, if $Y_n \overset{d}{\to} Y$, where Y is a continuous random variable, and $Z_n \overset{a.s.}{\to} c$, where $c \neq 0$ is a constant, then $Y_n Z_n \overset{d}{\to} cY$.

8. In this exercise, we analyze an alternative moment-based estimator for the tail index of the Pareto distribution. Recall that the mean of a Pareto(x_m, α) random variable X is given by $\mathbb{E}[X] = \frac{\alpha x_m}{\alpha - 1}$ for $\alpha > 1$. This relation can be rewritten as $\alpha = \frac{\mathbb{E}[X]}{\mathbb{E}[X] - x_m}$. Replacing the expectation with an empirical average yields an estimator of α. Formally, suppose that $X_1, X_2, \ldots, X_n$ are i.i.d. samples from a Pareto(x_m, α) distribution, where x_m is known, but α is unknown. The moment estimator $\hat{\alpha}_{\text{mom}}$ of α is given by

$$\hat{\alpha}_{\text{mom}} = \frac{\frac{1}{n} \sum_{i=1}^{n} X_i}{\frac{1}{n} \sum_{i=1}^{n} X_i - x_m}.$$

   Note that this estimator is only meaningful if $\alpha > 1$.

   (a) Assuming $\alpha > 1$, prove that $\hat{\alpha}_{\text{mom}}$ is strongly consistent.

   (b) Assuming $\alpha > 2$, prove that $\sqrt{n}(\hat{\alpha}_{\text{mom}} - \alpha) \overset{d}{\to} \sqrt{\alpha^2 + \frac{\alpha}{\alpha - 2}} Z$, where Z is a standard Gaussian.

   (c) Compare the asymptotic variance of $\hat{\alpha}_{\text{mom}}$ to that of $\hat{\alpha}_{\text{mle}}$.

9. Simulate 1,000 samples from a Pareto distribution with $x_m = 1$, $\alpha = 3$. Use the following different techniques to estimate the tail parameter α from this data (assuming the value of x_m is known to the estimator if needed):
   - Least-squares regression on the histogram (on a log-log scale) using uniform bin sizes.
   - Least-squares regression on the histogram (on a log-log scale) using logarithmic bin sizes.
   - Least-squares regression on the rank plot (on a log-log scale).
   - Weighted least-squares regression on the rank plot (on a log-log scale) using the weights specified in Section 8.4.
   - Maximum likelihood estimation.
   - The moment-based method described in Exercise 8.

   Compare the accuracy of these methods by computing their mean square error averaged over several simulation runs.

10. Simulate 2,000 samples from a Pareto distribution with $x_m = 1$, $\alpha = 2$. Follow the recipe outlined in Section 8.5 to estimate the minimum value x_m and the tail index α from this data. For Step 3, compare the fit of your Pareto hypothesis to the best LogNormal, Weibull, and Exponential fits.

11. Visualize the goodness of fit between the data and your different hypotheses in Exercise 10 (the best of Pareto, LogNormal, Weibull, and Exponential fits) using a QQ plot.

12. Repeat Exercise 10, but start with 2,000 data points sampled from a LogNormal distribution having unit mean and $\sigma = 4$.

Estimating Power-Law Tails: Let the Tail Do the Talking

When developing and applying statistical tools to estimate heavy-tailed phenomena, it is appealing to hope that you have a situation with i.i.d. samples from a precise parametric distribution, like the Pareto. As we saw in the previous chapter, in such situations it is possible to apply classical tools to explore the data, estimate the distribution, and even test the quality of the fit obtained. Though there are some potential pitfalls that are important to avoid (like applying unweighted regression to perform the estimation), there is a clear recipe that can be used to obtain reliable estimates.

However, if we step back and think carefully about the situations when we want to apply these tools, we quickly realize that the assumptions that underlie parametric tools are naive. In particular, it is almost never the case that a real dataset follows a precise, parametric Pareto distribution over the full range of the data. Parametric distributions are simple approximations of what we expect to see, but there are too many complexities in real data for us to believe they hold precisely. Instead, if we are honest with ourselves, what we should expect to see is an *approximate* version of power-law distributions, and likely an approximation that holds *only in the tail*, not for the whole distribution.

If the data only has a tail that approximately follows a power-law distribution, it means that the statistical regime we consider must change. Instead of the parametric modeling assumption that the data are i.i.d. samples from a Pareto distribution, we should assume that the data are i.i.d. samples from a *regularly varying* distribution. Under this semi-parametric assumption, our goal is still to estimate the shape parameter α, only now we cannot use a precise distributional form to do the analysis. We must make use of the properties we have derived in Parts I and II of this book to understand how the asymptotic behavior of these distributions can allow estimation of the tail from data.

These two competing perspectives – *parametric* versus *semi-parametric* – represent fundamentally different views on how to estimate heavy-tailed phenomena. One may hope that small changes to how the classical parametric tools are applied might allow them to be adapted to the semi-parametric setting. Unfortunately, statistical estimation of heavy-tailed phenomena must be approached carefully and, like in the parametric setting, intuitive approaches to estimation produce misleading conclusions in the semi-parametric setting. For example, it is natural to think that "censoring" the data by applying the classical tools (e.g., the MLE) to only the data above a certain size threshold might provide reliable estimation of the tail. However, we illustrate in this chapter that such an approach is potentially unreliable, and censoring must be done in a careful, data-driven way. This is because it is difficult to accurately assess where the tail "begins" and, further, the tail may only be approximately power-law, in which case the parametric estimators themselves may be unreliable.

197

As a result, statistically rigorous estimation in the semi-parametric regime requires the development of new tools. In particular, in the parametric setting, classical estimators "listen to the body," since tail behavior can be inferred from the body and estimates of the body are less noisy. In contrast, to estimate the tail directly in the semi-parametric setting, statistical tools must "let the tail do the talking." What this means in practice is that it is necessary to look at only the largest samples in the dataset, the outliers, and to infer properties of the tail from the properties of these extremes. This leads to what are referred to as *extremal estimators*, which make use of the properties of extremal processes that we discuss in Chapter 7.

The most prominent of these extremal estimators is the Hill estimator. While the Hill estimator is the most popular of the extremal estimators, it is still used far less frequently than classical parametric tools such as regression and maximum likelihood estimation. This is unfortunate, and is likely the result of it being less intuitive and seemingly more mysterious than these simple parametric tools. However, there is strong intuition behind the Hill estimator, and it can be viewed as an extremal version of the MLE, as we discuss in Section 9.2. Further, it is easy to apply and avoids many of the potential pitfalls associated with the MLE, since it can be used to both estimate the shape parameter of the tail, α, as well as *where* the tail begins and ends. Additionally, the so-called *Hill plot* provides a visualization that can be used to explore the data and develop confidence in both the quality of the estimate and the appropriateness of the heavy-tailed model in the first place.

However, a common theme in our chapters on estimation of heavy-tailed phenomena is that no estimator is perfect; and the Hill estimator is not immune from this critique. In particular, the Hill estimator can be misleading, especially in cases where the slowly varying component of a regularly varying distribution does not behave like a constant. In such cases, *Hill horror plots* provide examples where one may be tempted to have confidence in estimators that are flawed. To address such situations, there is a large literature that has developed and analyzed variations of the Hill estimator and the Hill plot. We discuss several such variations in Section 9.4.

While the shift from the classical statistical tools associated with parametric estimation of power-law distributions to the extremal tools associated with semi-parametric models provides increased statistical power, it also brings with it added analytic complexity. It is our hope that the background provided in Parts I and II of this book helps make the material easily digestible. One powerful consequence of the shift toward extremal estimators and the analytic tools that come with them is that it is possible to move beyond power-law distributions to more general classes of heavy-tailed (and light-tailed) distributions.

Specifically, note that the Hill estimator is only applicable for regularly varying data, that is, when the data is sampled from a distribution that lies in the maximal domain of attraction (MDA) of the Fréchet distribution. However, tools from exteme value theory enable the development of more general estimators that are applicable when the data is sampled from the MDA of *any* extreme value distribution. We study three such generalized extremal estimators, including the *moments estimator* (aka, the *Dekkers–Einmahl–de Haan estimator*), the *Pickands estimator*, and an estimator based on the *peaks over thresholds* approach. These tools not only provide an estimate of the tail index when the data is sampled from a regularly varying distribution, but crucially also help determine whether or not the data is regularly varying in the first place. This enables a statistically rigorous disambiguation between regularly varying data and data sampled from other heavy-tailed distributions, including the LogNormal and the Weibull.

Finally, despite the generality of the extremal estimators we discuss, they all suffer from one fundamental challenge: applying them in practice requires first determining where the tail begins in order to determine which subset of the data on which to apply the estimator. Tools such as the Hill plot and the corresponding plots for other extremal estimators provide ad hoc tools for making a choice, but they can be challenging to apply in practice and can sometimes lead to misleading results. Ideally, this choice should be made in a data-driven statistically rigorous manner, and we conclude the chapter by presenting two such approaches that have provable guarantees: PLFIT (Power-Law FIT), which is provably consistent in the case of regularly varying distributions, and the double bootstrap method, which provably minimizes the mean squared error in the case of regularly varying distributions with a certain second-order condition.

We hope that the material in this chapter and the previous one leaves you with a set of tools for the identification and estimation of heavy-tailed phenomena: parametric and semi-parametric; power-law and beyond. However, our goal is not to leave you with a precise recipe that can be followed in every circumstance. Rather, the discussion in these chapters demonstrates that all estimators can be misleading if they are not applied carefully. Thus, it is crucial to obtain confidence in conclusions about heavy-tailed phenomena through the use of complementary approaches. While we cannot give a specific general recipe for how to estimate heavy-tailed phenomena, we do end this chapter with guidelines that should be used when exploring data (Section 9.7). We hope that these guidelines can help to avoid future mistaken identifications and thus ease some of the controversy that surrounds heavy tails in many fields.

9.1 The Failure of Parametric Estimation

We begin the technical part of this chapter where we left off the previous chapter – with parametric estimation. Except, now we are no longer in a parametric setting. That is, instead of considering the parametric setting where $X_1, \ldots, X_n$ are sampled i.i.d. from a Pareto distribution, we consider the semi-parametric setting where $X_1, \ldots, X_n$ are sampled i.i.d. from a regularly varying distribution with index $-\alpha$. One can argue that this situation is much more relevant in most practical settings since it is unlikely that the full distribution precisely follows a power law; instead, it is likely that the *tail* of the distribution *approximately* follows a power law.

Clearly, the data in the semi-parametric setting no longer satisfies the assumptions that the parametric estimators were derived under; however it is natural to wonder if they can still be "saved." That is,

Can parametric estimators be adapted for use in settings where only the tail follows a power law?

In short, the answer to this question is "no." However, in practice, many people still attempt to use parametric estimators in such situations. They do so by "censoring" the data, that is, fixing a threshold where the tails "begins" and then throwing away (censoring) all data from below the threshold. Given the censored data, they then apply a parametric estimator to what remains (i.e., the data above the threshold).

While this simple approach is natural, it is also very fragile. Attempts to apply parametric estimators to censored versions of the data often lead to unreliable, sometimes misleading, conclusions. For example, consider the task of determining a lower bound, $b_{\min}$, above which a distribution has a power-law behavior. It is often too optimistic to hope for a threshold where there is a stark transition in behavior. Instead, the distributional behavior below $b_{\min}$ typically blends smoothly into the behavior above $b_{\min}$. In such situations it is impossible to determine a clear cutoff visually, and the specific choice used severely impacts the estimate obtained. Further, and perhaps more importantly, it is often too optimistic to hope that the behavior above $b_{\min}$ is precisely a Pareto distribution. Rather, there is a mixture of different behaviors that are only approximately power-law. Unfortunately, in both of these cases a censored version of a parametric estimator is unlikely to be appropriate. Formally, the application of parametric estimators in such situations leads to estimators that are no longer consistent.

The remainder of this section drives this point home using a simple, illustrative example. In particular, the example highlights how sensitive censored parametric approaches are to the choice of $b_{\min}$, while also leading us toward a more robust approach.

Specifically, we consider the following scenario in this section. Suppose the observations $X_1, \ldots, X_n$ are sampled from a mixture of shifted Pareto distributions defined as follows:

$$\bar{F}(x) = q\bar{F}_1(x) + (1-q)\bar{F}_2(x) = q(1+x)^{-\alpha_1} + (1-q)(1+x)^{-\alpha_2}. \qquad (9.1)$$

This distribution is the mixture of two power laws F_1 and F_2; thus, the tail of the distribution is regularly varying with parameter $-\min\{\alpha_1, \alpha_2\}$ (see Section 2.4). Note that we are considering shifted Pareto distributions with $x_m = 0$ (more precisely, Burr distributions) here in order to simplify the exposition and reduce the complexity of the equations.

A Censored Maximum Likelihood Estimator. The simple form of F in our example means that one could optimistically hope that, by choosing a suitable cutoff $b_{\min}$, it would be possible to estimate the tail of $\bar{F}$ using a censored version of the maximum likelihood estimator. We have derived $\hat{\alpha}_{\mathrm{mle}}$ for the (unshifted) Pareto distribution in the previous chapter (see Theorem 8.1). A parallel argument (see Exercise 6 in Chapter 8) yields the following maximum likelihood estimator for shifted Pareto distributions such as F_1 and F_2:

$$\hat{\alpha}_{\mathrm{mle}} = \frac{1}{\frac{1}{n}\sum_{i=1}^{n}\log(1+X_i)}.$$

To apply $\hat{\alpha}_{\mathrm{mle}}$ with a cutoff $b_{\min}$, we simply censor the observations used by $\hat{\alpha}_{\mathrm{mle}}$ to include only those larger than $b_{\min}$. Let k be the number of observations larger than $b_{\min}$ and recall that we use the following notation to index the samples in sorted decreasing order $X_{(1)} \geq \cdots \geq X_{(k)}$. Then, the censored estimate is

$$\hat{\alpha}_{\mathrm{mle},k} = \frac{1}{\frac{1}{k}\sum_{i=1}^{k}\log\left(\frac{1+X_{(i)}}{1+b_{\min}}\right)}.$$

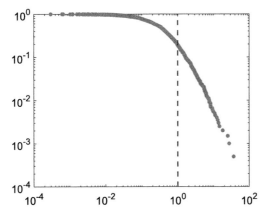

Figure 9.1 Example illustrating the issues with censoring: The figure shows the rank plot on a log-log scale obtained from 2,000 samples from the distribution defined in (9.1), with $q = 1/2$, $\alpha_1 = 2$, $\alpha_2 = 3$. Note that the rank plot "looks" linear beyond $b_{\min} = 1$; censoring the data less than this value results in a reasonable estimate, $\hat{\alpha} = 2.12$, as opposed to the uncensored estimate, $\hat{\alpha}_{\mathrm{mle}} = 2.38$.

Figure 9.1 illustrates what happens when this estimate is applied using a value of $b_{\min}$ chosen by "inspecting" the rank plot. The censored estimator $\hat{\alpha}_{\mathrm{mle},k}$ is certainly an improvement on the uncensored estimator, but it still leaves much to be desired.

A Problem with Consistency

While the performance of the censored estimator is "okay," we would certainly hope to do better. To do so, it is useful to understand what is going wrong with the censored MLE $\hat{\alpha}_{\mathrm{mle},k}$. The issue is actually quite simple: $\hat{\alpha}_{\mathrm{mle},k}$ is not a consistent estimator for F.

It is not too hard to understand why $\hat{\alpha}_{\mathrm{mle},k}$ is not consistent. To do so, let $k = k_1 + k_2$, where k_i is the number of observations larger than $b_{\min}$ drawn from F_i. Now, observe that

$$
\frac{1}{\hat{\alpha}_{\mathrm{mle},k}} = \frac{1}{k} \sum_{i=1}^{k} \log\left(\frac{1 + X_{(i)}}{1 + b_{\min}}\right)
$$

$$
= \frac{k_1}{k} \frac{1}{k_1} \sum_{i=1}^{k_1} \log\left(\frac{1 + X_{1,i}}{1 + b_{\min}}\right) + \frac{k_2}{k} \frac{1}{k_2} \sum_{i=1}^{k_2} \log\left(\frac{1 + X_{2,i}}{1 + b_{\min}}\right), \tag{9.2}
$$

where we have split the sums between samples taken from F_1 and F_2 and relabeled data as $X_{j,i}$ to make the distinction explicit. This form reveals that each term contains a censored maximum likelihood estimate of $1/\alpha_i$. Thus, since the maximum likelihood estimator is consistent,

$$
\frac{1}{k_j} \sum_{i=1}^{k_j} \log\left(\frac{1 + X_{j,i}}{1 + b_{\min}}\right) \to \frac{1}{\alpha_j}.
$$

Further, by the law of large numbers, as the sample size $n \to \infty$, $\frac{k_j}{n} \to q_j \bar{F}_j(b_{\min})$, where $q_1 = q$, $q_2 = 1 - q$. So,

$$
k_j/k = \frac{k_j/n}{k/n} \to q_j \bar{F}_j(b_{\min})/\bar{F}(b_{\min}).
$$

Combining these two observations, we see that as $n \to \infty$,

$$\frac{1}{\hat{\alpha}_{\mathrm{mle},k}} \to \frac{q\bar{F}_1(b_{\min})}{\bar{F}(b_{\min})}\frac{1}{\alpha_1} + \frac{(1-q)\bar{F}_2(b_{\min})}{\bar{F}(b_{\min})}\frac{1}{\alpha_2} \tag{9.3}$$
$$< 1/\min\{\alpha_1, \alpha_2\} \text{ for all } b_{\min} < \infty.$$

Consequently, $\hat{\alpha}_{\mathrm{mle},k}$ is not a consistent estimator of $\min\{\alpha_1, \alpha_2\}$ – it is asymptotically biased for any fixed value of $b_{\min}$. But, of course, it is possible to control the asymptotic bias of $\hat{\alpha}_{\mathrm{mle},k}$ by picking $b_{\min}$ sufficiently large. This fact gives hope that it is still possible to design a consistent estimator in this setting.

A Consistent Estimator

We have just seen that "fixing" a parametric estimator through the use of a static lower bound to censor the data is dangerous and can lead to estimators that are inconsistent. However, with a bit more care, the approach can become successful. The key idea that is missing in the preceding discussion is that the lower bound used must *depend on* n, not be set as a constant. That is, we should use $b_{\min}(n)$ instead of $b_{\min}$.

To understand why this works, let us return to the derivation presented earlier and ask:

What properties must $b_{\min}(n)$ satisfy in order for $\hat{\alpha}_{\mathrm{mle},k(n)}$ to be a consistent estimator?

It turns out that there are two key properties that we need to enforce on $b_{\min}(n)$ in order to guarantee that the resulting estimator $\alpha_{\mathrm{mle},k(n)}$ is consistent. The first follows from (9.2), copied here:

$$\frac{1}{\hat{\alpha}_{\mathrm{mle},k(n)}} = \frac{1}{k(n)} \sum_{i=1}^{k(n)} \log\left(\frac{1+X_{(i)}}{1+b_{\min}(n)}\right)$$
$$= \frac{k_1(n)}{k(n)}\frac{1}{k_1(n)}\sum_{i=1}^{k_1(n)}\log\left(\frac{1+X_{1,i}}{1+b_{\min}(n)}\right) + \frac{k_2(n)}{k(n)}\frac{1}{k_2(n)}\sum_{i=1}^{k_2(n)}\log\left(\frac{1+X_{2,i}}{1+b_{\min}(n)}\right). \tag{9.4}$$

Let us assume, without loss of generality, that $\alpha_1 < \alpha_2$, so that consistency implies $\alpha_{\mathrm{mle},k(n)} \to \alpha_1$ as $n \to \infty$. For this to hold, we require the censored maximum likelihood estimate in the first term to be consistent, that is,

$$\frac{1}{k_1(n)}\sum_{i=1}^{k_1(n)}\log\left(\frac{1+X_{1,i}}{1+b_{\min}(n)}\right) \to \frac{1}{\alpha_1}.$$

Importantly, this step requires that $k_1(n) \to \infty$ as $n \to \infty$. This occurs whenever

$$k(n) \to \infty \text{ as } n \to \infty,$$

which means that $b_{\min}$ *must grow slowly enough to ensure that enough large samples are present.*

The second property required for $\alpha_{\mathrm{mle},k(n)}$ to be consistent is that the second term in (9.4), which contributes to the bias of the estimator, diminishes as $n \to \infty$. In particular, we require $k_2(n)/k(n) \to 0$ as $n \to \infty$. Intuitively, this means we need $b_{\min}(n) \to \infty$, since (i) for any static choice of $b_{\min}$, we have $\frac{k_2(n)}{k(n)} \to (1-q)\bar{F}_2(b_{\min})/\bar{F}(b_{\min})$, and (ii) $\bar{F}_2(x)/\bar{F}(x) \to 0$ as $x \to \infty$.[1] The condition $b_{\min}(n) \to \infty$ is in turn implied by the condition that the number of retained samples grows *sublinearly* in n (see Exercise 2), that is,

$$k(n)/n \to 0 \text{ as } n \to \infty.$$

Thus, to ensure that $\hat{\alpha}_{\text{mle},k(n)}$ is consistent, *a small, vanishing fraction of the data must be used.*

The combination of the two properties highlights an interesting trade-off: we must keep enough data for the parametric estimators to be useful ($k(n) \to \infty$) but only a vanishing fraction of all the observations ($k(n)/n \to 0$) in order to ensure that the estimator remains unbiased. Of course, there is an additional trade-off if one considers the variance of the estimator – a larger $k(n)$ will reduce the variance at the expense of an increased bias.

9.2 The Hill Estimator

The example in the previous section highlights a promising approach for adapting parametric estimators for power law distributions to settings where only the tail follows a power law distribution: use censored parametric estimators that focus only on the largest few observations. In fact, we have chosen the example because it provides strong intuition for the form of one of the most popular nonparametric estimators for power law distributions: *the Hill estimator.*

The Hill estimator, $\hat{\alpha}_H(X_1, \ldots, X_n) = \hat{\alpha}_{k,n}^H$, has a simple definition that is very reminiscent of the definition of the MLE. In particular,

$$\hat{\alpha}_H = \hat{\alpha}_{k,n}^H = \frac{1}{\frac{1}{k}\sum_{i=1}^{k} \log\left(\frac{X_{(i)}}{X_{(k)}}\right)}, \tag{9.5}$$

where we have again indexed the samples in sorted decreasing order using $X_{(1)} \geq \cdots \geq X_{(n)}$.

Perhaps the most natural way to interpret the Hill estimator is as a censored maximum likelihood estimator that uses only the k largest samples from the data. Thus, it "lets the tail do the talking." While it looks similar to the estimators we discussed, and rejected, in the previous section, there is a subtle but important difference: it uses a *random* threshold (determined by the samples) rather than the *deterministic* one discussed in the previous section.

Already, the example of shifted Pareto distributions in the previous section highlights connections between the Hill estimator and the MLE, but it is not hard to see that the connection extends more broadly. To do so, let us consider a situation where $X_1, \ldots, X_n$ come from a Pareto distribution. We have already seen that the MLE of α in this case is

---

[1] At this point, we do not attempt to formalize this statement; the goal of this discussion is simply to motivate the two conditions required to make our censored estimator consistent. That the presented approach does result in a consistent estimator will be shown formally in Section 9.3 as part of our discussion of the Hill estimator.

$$\hat{\alpha}_{\mathrm{mle}} = \frac{1}{\frac{1}{n} \sum_{i=1}^{n} \log\left(\frac{X_i}{x_m}\right)},$$

which matches the form of the Hill estimator where x_m is replaced by $X_{(k)}$ and only the k largest samples are considered.

To make the connection even more explicit, instead of taking the maximum likelihood estimate of α given $X_1, \ldots, X_n$, let us compute the maximum likelihood estimator of α given only the largest k samples, $X_{(1)}, \ldots, X_{(k)}$. To do this, we first need to compute the joint distribution of $X_{(1)}, \ldots, X_{(k)}$, and then we can find the α which maximizes the likelihood of these samples. To this end, the p.d.f. of the largest k of n samples can be computed as follows:

$$f_{k,n}(x_1, \ldots, x_k; \alpha) = \frac{n!}{(n-k)!}(F(x_k))^{n-k} \prod_{i=1}^{k} f(x_i),$$

where the multiplier in front follows from noting that the k largest samples $X_{(1)}, \ldots, X_{(k)}$ can be chosen in $\binom{n}{k}$ ways from the n samples and then arranged in $k!$ ways. So, every specific ordered collection could have come from $n!/(n-k)!$ different realizations.

Using the preceding calculation of the density, we can write the maximum likelihood estimator of the censored data, $\hat{\alpha}_{k,n}^{\mathrm{mle}}$, as

$$\hat{\alpha}_{k,n}^{\mathrm{mle}} = \arg\max_{\alpha} f_{k,n}(X_{(1)}, \ldots, X_{(k)}; \alpha).$$

From here, a specialization to the Pareto distribution follows from a straightforward calculation that yields

$$\hat{\alpha}_{k,n}^{\mathrm{mle}} = \frac{1}{\frac{1}{k} \sum_{i=1}^{k} \log\left(\frac{X_{(i)}}{X_{(k)}}\right)} = \hat{\alpha}_{k,n}^{H}. \tag{9.6}$$

Note that the connection between weighted regression and the MLE shown in Theorem 8.6 gives another useful view of the Hill estimator as the least squares fit of the k largest samples. In fact, by noting that the weights used in the equivalence between the MLE and weighted regression are "nearly" the same for the largest samples, we can expect that unweighted regression of the k largest samples would lead to an unbiased, minimal variance estimator as well. This intuition turns out to be correct, as shown in [132].

Though the connections to the MLE of the Pareto distribution provide helpful intuition, they do not truly expose the generality of the Hill estimator. The key power of the estimator is that it applies even when the distribution is not a precise Pareto. To highlight the generality of the Hill estimator in this dimension, we can return to the semi-parametric setting where the observations $X_1, \ldots, X_n$ come from a regularly varying distribution F. Thus, we no longer have a parametric form for the distribution; instead, we simply know that $\bar{F}$ satisfies

$$\lim_{x \to \infty} \frac{\bar{F}(xy)}{\bar{F}(x)} = y^{-\alpha}.$$

In this situation one cannot derive an estimator via the maximum likelihood approach. However, we can still provide intuition for the form of the Hill estimator by taking advantage of Karamata's theorem (Theorem 2.10). In particular, we use the following (equivalent) modification of Karamata's theorem:

$$\int_t^\infty \frac{\bar{F}(x)}{x}dx \sim \frac{\bar{F}(t)}{\alpha}, \text{ as } t \to \infty.$$

From this version of Karamata's theorem, it is possible to construct a simple relationship that can be used to estimate α as follows. Taking X to be a generic random variable having distribution F,

$$\frac{1}{\bar{F}(b)}\mathbb{E}\left[(\log(X/b))_+\right] = \frac{1}{\bar{F}(b)}\int_b^\infty (\log x - \log b)dF(x) \tag{9.7}$$

$$= \frac{1}{\bar{F}(b)}\int_b^\infty \frac{\bar{F}(x)}{x}dx \to \frac{1}{\alpha}, \text{ as } b \to \infty,$$

where the second equality follows from partial integration, and the limit follows from Karamata's theorem. The form of (9.7) shows that we can obtain an estimator of α using an empirical estimate of the left-hand side of (9.7). To do this, we need to make two decisions: (i) we need to decide how to estimate $\bar{F}$ and (ii) we need to choose b. The decision for (i) is natural – use the empirical c.c.d.f. $\bar{F}_n(x)$ (see (8.3) for the definition). For (ii) the choice is less clear, but it is clear that the choice should depend on n, that is, $b = b(n)$, with $b(n) \to \infty$. Setting $b(n) = X_{(k)}$, that is, the kth largest sample, and noting that $\bar{F}_n(X_{(k)}) = k/n$, we arrive at an estimator satisfying

$$\frac{1}{\hat{\alpha}} = \frac{1}{k}\sum_{i=1}^k \log\left(\frac{X_{(i)}}{X_{(k)}}\right),$$

which of course exactly matches the Hill estimator.

Interestingly, Karamata's theorem can be used in other ways to derive the Hill estimator as well. One of the most natural is to focus on the mean residual life instead of the c.c.d.f., which we explore in Exercise 1. This interpretation has a strong connection to the Peaks over Thresholds estimator we discuss in Section 9.5.3. However, for now, we move on to understanding the properties of the Hill estimator.

9.3 Properties of the Hill Estimator

By this point, hopefully the form of the Hill estimator is no longer mysterious. In fact, given the connection to the MLE, you might even *expect* that the Hill estimator inherits many of the strong properties of the MLE. However, we show in this section that the devil is in the details for the Hill estimator. If k is set appropriately, then the Hill estimator inherits many of the nice properties of the MLE; but if k is not set appropriately, then it is a very unreliable estimator.

In fact, we have already seen in the example of a shifted Pareto that if one considers a deterministic scaling of $k(n)$, then $k(n)$ needs to grow unboundedly with n, that is, $k(n) \to \infty$, to ensure that there is "enough" data being used by the Hill estimator, but it also needs to grow sublinearly, that is, $k(n)/n \to 0$, so that the estimator remains unbiased. While these requirements were only argued for the special case of a mixture of shifted Pareto distributions, it turns out that they are broadly applicable.

This is perhaps easiest to see in the case of one of the most fundamental properties that one seeks in statistical estimators: consistency. Specifically, exactly the conditions on k that show up in the example of shifted Pareto distributions are what is needed in order for the Hill estimator to be *weakly consistent*, that is, to satisfy $\hat\alpha_H = \hat\alpha^H_{k,n} \xrightarrow{p} \alpha$ as $n \to \infty$.

Theorem 9.1 (Weak Consistency)   *Consider X_n sampled i.i.d. from a regularly varying distribution with index $-\alpha$. If $k(n) \to \infty$ and $k(n)/n \to 0$ as $n \to \infty$, then $\hat\alpha_H = \hat\alpha^H_{k(n),n} \xrightarrow{p} \alpha$ as $n \to \infty$.*

Though this theorem is a good start, if you recall, a stronger property holds for the MLE in the parametric setting considered in Chapter 8. In that setting, the MLE guarantees *strong consistency*, where the convergence in probability just presented is replaced by almost sure convergence. The Hill estimator also satisfies strong consistency; however, it requires a slightly more restrictive assumption on the choice of k.

Theorem 9.2 (Strong Consistency)   *Consider X_n sampled i.i.d. from a regularly varying distribution with index $-\alpha$. If $k(n)/\log\log n \to \infty$ and $k(n)/n \to 0$ as $n \to \infty$, then $\hat\alpha_H = \hat\alpha^H_{k(n),n} \xrightarrow{a.s.} \alpha$ as $n \to \infty$.*

The extra condition that k is increasing faster than $\log\log n$ is needed to prove strong consistency. It is hard to provide intuition for this, but readers familiar with the law of the iterated logarithm will appreciate the need for this condition.

These results about consistency should be treated as a natural extension of the analysis of the MLE in the parametric setting. In the previous section, we gave a heuristic argument for the consistency of the Hill estimator for the special case of a mixture of two shifted Pareto distributions. We now provide a rigorous proof of weak consistency for the general case of regularly varying samples by combining several results from Chapter 2. We omit the proof of strong consistency of the Hill estimator here; the interested reader is referred to [57].

Proof of Theorem 9.1   Throughout the proof we assume that $k(n), n \geq 1$ is a fixed and nondecreasing sequence of natural numbers. Set $S_n = 1/\hat\alpha_{k(n),n}$.

We first consider the special case where F is Pareto with $x_m = 1$, that is, $F(x) = 1 - x^{-\alpha}$, before extending the proof to regularly varying distributions. To begin, we expand S_n as follows

$$S_n = \frac{1}{k(n)} \sum_{i=1}^{k(n)-1} \left(\log X_{(i)} - \log X_{(k(n))}\right) = \frac{1}{k(n)} \sum_{i=1}^{k(n)-1} \left(E_{(i)} - E_{(k(n))}\right),$$

where $E_i := \log X_i$, and $(E_{(1)}, \ldots, E_{(n)})$ denotes the order statistics of $(E_1, \ldots, E_n)$. Since $(E_1, \ldots, E_n)$ is a vector of i.i.d. exponentially distributed random variables (with rate α), the memorylessness property implies that the joint distribution of the order statistics can be expressed as follows (see Exercise 3):

$$\left(E_{(1)}, \ldots, E_{(n)}\right) \stackrel{d}{=} \left(E'_1, \ldots, E'_n\right),$$

where

$$E'_i = \sum_{j=i}^{n} E_j/j, \quad i = 1, \ldots, n.$$

Thus,

$$S_n \stackrel{d}{=} \frac{1}{k(n)} \sum_{i=1}^{k(n)-1} (E_i' - E_{k(n)}')$$

$$\stackrel{d}{=} \frac{1}{k(n)} \sum_{i=1}^{k(n)-1} \left(\sum_{j=i}^{n} \frac{E_j}{j} - \sum_{j=k(n)}^{n} \frac{E_j}{j} \right)$$

$$= \frac{1}{k(n)} \sum_{i=1}^{k(n)-1} \sum_{j=i}^{k(n)-1} \frac{E_j}{j} = \frac{1}{k(n)} \sum_{j=1}^{k(n)-1} \frac{E_j}{j} \sum_{i=1}^{j} 1 = \frac{1}{k(n)} \sum_{j=1}^{k(n)-1} E_j.$$

As the $E_j, j \geq 1$, are independent and exponentially distributed with rate α, and since $k(n) \to \infty$ as $n \to \infty$, we can apply the weak law of large numbers to conclude that $S_n \to 1/\alpha$ in probability. This concludes the proof for the Pareto case.

We now turn to the general case. For this, we need additional notation. Define the generalized inverse $F^{-1}(u) = \inf\{x : F(x) \geq u\}$ and let $U_i, i = 1, \ldots, n$ be n independent and identically distributed uniform random variables. Observe that we can use the representation $X_i = F^{-1}(U_i)$, and $X_{(i)} = F^{-1}(U_{(i)})$, $i = 1, \ldots, n$.

We now invoke the following result on inverses of regularly varying functions:

$$F^{-1}(1 - y) = y^{-1/\alpha} \ell(1/y) \text{ as } y \downarrow 0,$$

for a slowly varying function $\ell(\cdot)$. A formal justification of this result is beyond the scope of this book (see Section 2.2 of [31] for a proof), though the reader should verify that this result holds under the special case where F is a precise power-law distribution.

Using the preceding representation of $F^{-1}(\cdot)$, we can express S_n as follows:

$$S_n = \frac{1}{k(n)} \sum_{i=1}^{k(n)-1} \left[\log F^{-1}(U_{(i)}) - \log F^{-1}(U_{(k(n))}) \right]$$

$$= \frac{1}{k(n)} \sum_{i=1}^{k(n)-1} \left[\frac{1}{\alpha} \log \left(\frac{1}{1 - U_{(i)}} \right) + \log \ell \left(\frac{1}{1 - U_{(i)}} \right) \right.$$

$$\left. - \frac{1}{\alpha} \log \left(\frac{1}{1 - U_{(k(n))}} \right) - \log \ell \left(\frac{1}{1 - U_{(k(n))}} \right) \right].$$

Defining $E_i := \frac{1}{1-U_i}$ and $E_{(i)} := \frac{1}{1-U_{(i)}}$, we have

$$S_n = \frac{1}{k(n)} \sum_{i=1}^{k(n)-1} \frac{1}{\alpha} \left[\log E_{(i)} - \log E_{(k(n))} \right] + \frac{1}{k(n)} \sum_{i=1}^{k(n)-1} \left[\log \ell \left(E_{(i)} \right) - \log \ell \left(E_{(k(n))} \right) \right]$$

$$=: T_1 + T_2.$$

Noting now that $\{\log E_i\}$ are i.i.d. Exponential random variables with mean 1, it follows from our analysis of the Pareto case that $T_1 \stackrel{P}{\to} \frac{1}{\alpha}$ as $n \to \infty$. It therefore remains to show that $T_2 \stackrel{P}{\to} 0$ as $n \to \infty$. For this, we use Karamata's representation theorem (see Theorem 2.12) to write

$$\ell(x) = a(x) \exp\{ \int_1^x b(u) u^{-1} du \}, \tag{9.8}$$

where $a(x) \to a \in (0, \infty)$ and $b(x) \to 0$ as $x \to \infty$. Thus, T_2 can be expressed as follows:

$$T_2 = \frac{1}{k(n)} \sum_{i=1}^{k(n)-1} \log\left(\frac{a(E_{(i)})}{a(E_{(k(n))})}\right) + \frac{1}{k(n)} \sum_{i=1}^{k(n)-1} \int_{E_{(k(n))}}^{E_{(i)}} \frac{b(u)}{u} du$$

$$=: T_{2a} + T_{2b}.$$

To see that $T_{2a} \xrightarrow{P} 0$ as $n \to \infty$, set

$$M(x) := \sup_{y \ge x} |\log(a(y)/a(x))| \tag{9.9}$$

and note that $M(x) \to 0$ since $a(x) \to a \in (0, \infty)$ as $x \to \infty$. Now, observe that

$$|T_{2a}| \le M(E_{(k(n))}). \tag{9.10}$$

Since $\lim_{n\to\infty} k(n)/n = 0$, $\lim_{n\to\infty} E_{(k(n))} \xrightarrow{P} \infty$ (see Exercise 2). This means that the upper bound on T_{2a} in (9.10) tends to zero as $n \to \infty$, which in turn shows that $T_{2a} \xrightarrow{P} 0$.
Finally, we can bound T_{2b} as follows:

$$|T_{2b}| \le \alpha |T_1| \sup_{u \ge E_{(k(n))}} |\frac{b(u)}{u}|.$$

Once again, since $\lim_{n\to\infty} E_{(k(n))} \to \infty$ and $\lim_{x\to\infty} b(x) = 0$, we conclude that $T_{2b} \to 0$.
□

Beyond consistency, another important property that the Hill estimator inherits from the MLE is *asymptotic normality*. Asymptotical normality is a crucial condition, since it provides a distributional understanding of the estimation errors that occur when using the Hill estimator, showing that they are not too large, that is, they decay as $O(\sqrt{n})$. However, the conditions that must be imposed to obtain asymptotic normality are stronger than those needed for consistency. In particular, it is necessary to impose a second-order regular variation assumption.

Theorem 9.3 (Asymptotic normality)   *Consider X_n sampled i.i.d. from a regularly varying distribution with index $-\alpha$, such that the second-order condition (2.20) holds. If $k \to \infty$ and $k/n \to 0$ as $n \to \infty$, then $\sqrt{k}(\alpha_{k,n}^H - \alpha) \xrightarrow{d} N(0, \alpha^2)$ as $n \to \infty$.*

As you can guess from the need for the second-order regular variation assumptions, the proof of asymptotic normality is technical and beyond the scope of this book. A proof revealing the essence of the second-order condition is given in [97]. For a textbook treatment, see Chapter 9.1 of [183].
At this point, we have discussed consistency and asymptotic normality. These are two of the three properties we studied in the previous chapter when presenting the MLE. We now turn to the third one, *asymptotic efficiency*. In this case, the story is different. The Hill estimator does not inherit the properties of the MLE regarding variance. In fact, you may have already anticipated this based on the example of a shifted Pareto we discussed earlier in the chapter.

The issue with asymptotic efficiency in the case of the Hill estimator is easy to understand intuitively. In particular, Theorem 9.3 shows that the asymptotic variance of the Hill estimator decreases as k grows. Thus, if one seeks to minimize variance under the Hill estimator, one should choose large k; incorporating more data into the estimator and reducing variance. However, doing so may lead to bias, as the results on consistency highlight and as we show using the example of the shifted Pareto distribution. Consequently, there is a fundamental trade-off between bias and variance in the Hill estimator. We explore this trade-off qualitatively in the next section as we discuss how to use the so-called Hill plot in order to select k in practice. But, interested readers should know that it is also possible to explore the trade-off analytically in the case of restricted classes of regularly varying distributions (e.g., see [75], p. 341).

9.4 The Hill Plot

Modulo some assumptions, the Hill estimator maintains the statistical properties we look for in estimators. However, the properties we have highlighted in the previous section should not be interpreted as saying that the Hill estimator is infallible. In particular, quite the opposite is true. The results make it clear that the Hill estimator's properties are very sensitive to the number of order statistics used (i.e., the choice of k). Further (and unfortunately) while the theorems about consistency and asymptotic normality give guidance on choosing k, this guidance is not helpful when choosing k for a particular dataset in practice. Thus, the key question we are left with at this point is:

How should the number of order statistics, k, be chosen in practice when using the Hill estimator?

Unfortunately, there is no perfect answer to this question. There are a variety of approaches that are very successful in practice; but all come with drawbacks, and care needs to be used when applying them. The key to many of these approaches is a visualization known as the *Hill plot*.

The Hill plot illustrates the Hill estimator $\hat{\alpha}_{k,n}^H$ as k changes, thus showing how the estimator is impacted by the number of order statistics included. In particular, the Hill plot shows $\hat{\alpha}_{k,n}^H$ as a function of k for $1 \leq k \leq n$. One hopes that the Hill plot is stable, that is, $\hat{\alpha}_{k,n}^H$ does not change too much with k over a "large" region. If this is the case, the plot provides confidence in the estimate $\hat{\alpha}_{k,n}^H$ given by k in that range.

Illustrations with Synthetic Data

To get a feeling for how the Hill plot works, it is useful to start with some ideal examples. For the first example, we return to the parametric case that was our focus in Chapter 8. Specifically, Figure 9.2 shows the rank and Hill plots for 5,000 samples from a Pareto distribution with $\alpha = 2$ and $x_m = 1$. As you would expect, both approaches work well in this case. The rank plot shows a nearly straight line yielding a WLS estimate of 1.97. Similiarly, in the case of the Hill plot, while there is noise when only a few order statistics are used (k is small), the plot shows a nearly stable line. This means that the Hill estimator is giving a consistent value regardless of k, which gives a lot of confidence in the estimate it provides.

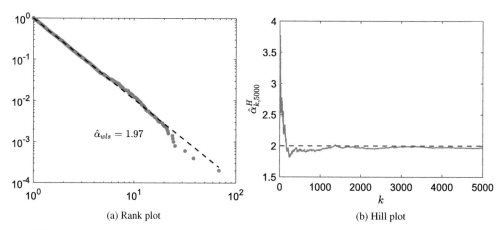

(a) Rank plot   (b) Hill plot

Figure 9.2 Contrasting the rank and Hill plots using 5,000 samples from a Pareto distribution with $x_m = 1$ and $\alpha = 2$. The dotted line on the Rank plot shows the weighted least square fit, and the dotted line on the Hill plot shows the correct value of the tail index.

The case of a pure Pareto gives an idyllic illustration of how to use the Hill plot, but the power of the Hill plot is that it is useful outside of parametric settings. As a first illustration of this, Figure 9.3 shows the rank and Hill plots of 5,000 samples drawn from a distribution with an Exponential body and a Pareto tail. Specifically, the c.c.d.f. of the distribution is given by $\frac{e^{-x/5} - e^{-2}}{2(1 - e^{-2})} + \frac{1}{2}$ for $x \in [0, 10]$ and $50x^{-2}$ for $x > 10$. In this case, the rank plot is problematic, since one needs to estimate the lower threshold before trying to estimate the tail index, which, as we have seen, can bias the estimator. In contrast, the Hill plot shows a clear stable region for small-to-moderate order statistics. In this region, the Hill estimator is stable across choices of k, and these estimators yield an accurate estimate of the tail index of the distribution. This example highlights a key feature of the Hill estimator: it automatically determines where the tail behavior is power-law, rather than relying on the user to determine the right threshold for where the tail "begins."

To drive this last point home, let us return to the example of a mixture of shifted Pareto distributions that we discussed as the beginning of this chapter. Specifically, each sample is drawn from a Pareto$(1, 2)$ (i.e., $\alpha = 2$) or a Pareto$(1, 4)$ (i.e., $\alpha = 4$) distribution with equal probability. This is shown in Figure 9.4. Here a weighted least-squares regression on the rank plot gives a biased estimate of the tail (i.e., 2.67). In contrast, the Hill plot, for small values of k, does take a value very close to the tail index. This is even more apparent when we zoom in to that part of the Hill plot, as shown in the inset on Figure 9.4(b).[2]

Illustrations with Real-World Data

It is one thing to apply the Hill plot to data where you know the correct answer, and another to see if it can provide confidence in estimators in situations where the answer is unknown. To highlight how the Hill plot can be used in real-world scenarios, we consider two examples in this section: the size of blackouts and the population of cities.

---

[2] The task of estimating the tail index in this case becomes easier when we look at smoothed and scaled versions of the Hill plot later in this section.

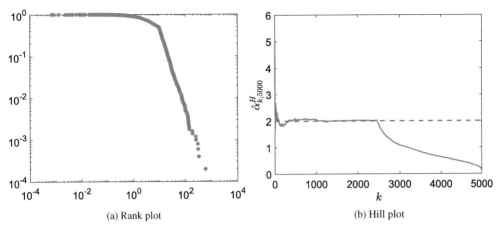

(a) Rank plot
(b) Hill plot

Figure 9.3 Contrasting the rank and Hill plots using 5,000 samples drawn from a distribution with an exponential body and a Pareto tail with $\alpha = 2$. The dotted line on the Hill plot shows the correct value of the tail index.

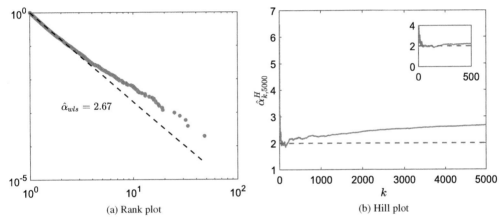

(a) Rank plot
(b) Hill plot

Figure 9.4 Contrasting the rank and Hill plots using 5,000 samples drawn from a mixture of two Pareto distributions with $\alpha = 2, 4$. The dotted line on the Rank plot shows the weighted least squares fit, and the dotted line on the Hill plot shows the correct value of the tail index.

The first example we consider is blackouts. Specifically, we consider a dataset of the sizes, in terms of number of households affected, of US blackouts from 2002–2018. For details behind the dataset, we refer to [161]. Figure 9.5 shows the rank and Hill plots for the blackout data. We see that the rank plot is not definitive. It could be interpreted as supporting a power-law hypothesis or not depending on how a lower cutoff on the data is set. In contrast, the Hill plot provides significant evidence of power law behavior. In particular, it contains a significant flat portion that yields an estimate slightly in excess of 1. Thus, the Hill plot reveals that, while there is not likely a pure power law in this data, there is evidence of a power-law tail. To highlight the danger of relying on the rank plot, note that if one were to fit a pure Pareto distribution using all the data, a much lower index (0.2) would be estimated. This would lead to the incorrect conclusion that blackout sizes have an infinite mean, which

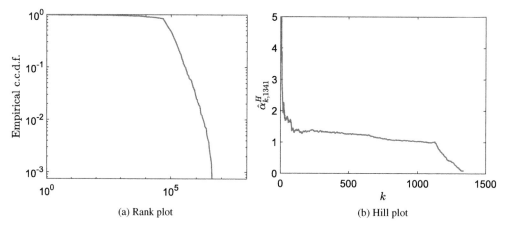

(a) Rank plot   (b) Hill plot

Figure 9.5 Contrasting the rank and Hill plots using data about the the number of customers affected by power blackouts in the US between 2002 and 2018.

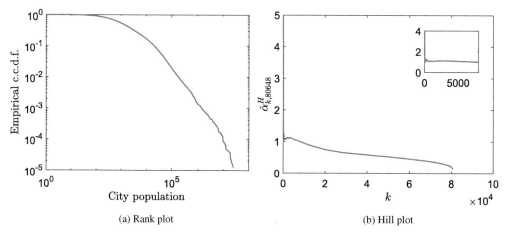

(a) Rank plot   (b) Hill plot

Figure 9.6 Contrasting the rank and Hill plots corresponding to the population of US cities in the 2010 census.

would suggest different approaches for mitigating their impact. Such procedures have been used in the power systems community to draw potentially incorrect conclusions about the behavior of blackouts.

The second example we consider is city populations. Here, we study the population of US cities as per the 2010 census [2]. The rank and Hill plots for this data are shown in Figure 9.6. Again, the evidence from the rank plot is not definitive. Depending on how one determines a lower cutoff for the power law tail, it would be possible to accept or reject the power-law hypothesis based on this plot. The Hill plot is also less convincing here than in the case of blackouts. The data certainly identifies some power-law like phenomena, since the Hill estimator does not quickly go to zero as k grows, but there is not a large flat region evident. However, zooming in, one sees behavior much like the example of a mixture of Pareto distributions in Figure 9.4. For a moderately sized range of k, $k = 0 - 8000$, there is a flat, reliable estimate of the tail index to be obtained (see the inset on Figure 9.6(b)). This shows that the extreme tail does indeed seem to follow a power law, even though the body

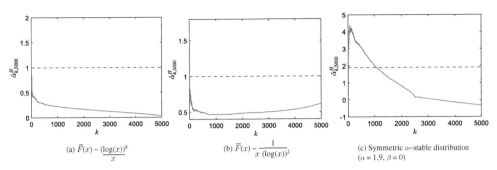

Figure 9.7 Hill horror plots. In each case, the dotted red line on the Hill plot shows the correct value of the tail index.

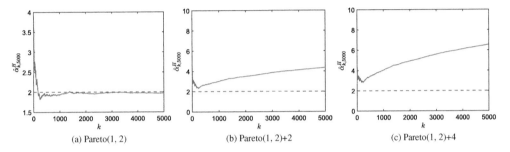

Figure 9.8 Impact of data translation on the Hill estimator.

seems to be a more complex mixture of distributions. However, this example highlights that it is sometimes delicate to apply the Hill plot in practice. Unlike the case of Figure 9.4, we cannot be sure of our conclusion and must search for other ways to validate our estimate.

Issues with the Hill Plot

While the examples we have considered so far paint a somewhat rosy picture of how the Hill plot can guide the choice of k, things do not always work out so nicely. In particular, there are many examples of so-called *Hill horror plots*, where it is nearly impossible to infer a reliable choice of k from observing the shape of the Hill plot.

Some examples of Hill horror plots are shown in Figure 9.7. In Figures 9.7(a) and 9.7(b), we consider distributions with a tail that decays as x^{-1}, but multiplied/divided by a poly-logarithmic slowly varying function. This simple transformation badly distorts the Hill plots, making it impossible to deduce the true tail index by looking at them. Another example is provided in Figure 9.7(c), where we take samples from a symmetric α-stable distribution with tail index 1.9.

Things get even worse when one realizes that the Hill estimator is not location invariant. So, the same example of a pure Pareto distribution we discuss in Figure 9.2 can become a horror plot of its own simply by shifting the distribution; see Figure 9.8.

These Hill horror plots all highlight the difficulty, even impossibility, of determining a reliable choice of k from the Hill plot. Further, because of the bias that can creep into the estimator for large k and the lack of scale invariance, the Hill plot can sometimes be deceptive even when it seems to provide a reliable estimate. However, this does not mean that one

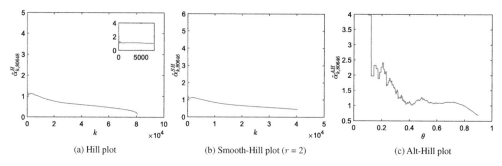

(a) Hill plot (b) Smooth-Hill plot ($r = 2$) (c) Alt-Hill plot

Figure 9.9 US city sizes: Smooth- and Alt-Hill plots.

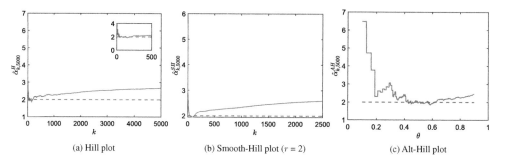

(a) Hill plot (b) Smooth-Hill plot ($r = 2$) (c) Alt-Hill plot

Figure 9.10 Mixture of Pareto distributions: Smooth-Hill and Alt-Hill plots.

should ignore the possibility of making a Hill plot. Many of the examples that lead to Hill horror plots would also be problematic for other estimators and, as we have already seen, when the Hill plot works, it works remarkably well. Further, there are simple adjustments that can be made to the Hill plot to mitigate (some of) the issues associated with using it in practice.

Variations of the Hill Plot

Two of the most problematic issues surrounding the Hill plot are (i) the volatility in the Hill estimates using small k and (ii) the potential bias in estimates using large k. Many variations of the Hill plot have been proposed with the goal of addressing these issues via "simple" adjustments. For example, attempts have been made to reduce volatility by smoothing the Hill estimator, and to reduce the impact of the bias for large k by rescaling the Hill plot.

We begin by introducing the following procedure for "smoothing" the Hill estimator, due to Resnick and Stărică [184]. For smoothing parameter $u > 1$, set

$$\hat{\alpha}_{k,n}^{SH} = \frac{1}{(u-1)k} \sum_{j=k+1}^{uk} \hat{\alpha}_{j,n}^{H}.$$

Importantly, this smoothes the Hill estimator, reducing the volatility, while still guaranteeing consistency (see [184]).[3]

---

[3] There is another approach for smoothing the Hill estimator, via the use of a kernel function [54]. This approach combines *all* the order statistics to define the tail index estimator, but with a higher weight to the higher-order statistics; we discuss kernel-based estimators in Section 9.8.

A similarly simple change in scaling can reduce the dangers associated with the potential bias in the Hill estimator for large values of k. One prominent version of this rescaling is the following, where for $\theta \in (0, 1]$ we set

$$\hat{\alpha}_{\theta,n}^{AH} = \hat{\alpha}_{\lceil n^\theta \rceil,n}^{H}$$

and then plot $\hat{\alpha}_{\theta,n}^{AH}$ against θ. This places emphasis on estimates using fewer order statistics, and thus avoids the potential bias that comes from using the Hill estimator with a large k. Of course, the same rescaling can be done for $\hat{\alpha}_{k,n}^{SH}$ as well.

Figures 9.9 and 9.10 illustrate the impact of using these variations of the Hill plot (called the Smooth-Hill and Alt-Hill plots) in two of our running examples: city sizes and a mixture of Pareto distributions. In both cases, they provide the desired benefits: the smoothed Hill plot succeeds at reducing the volatility, and the Alt-Hill plot ensures the focus is on estimates associated with small k. However, in neither case do the plots result in visualizations that provide estimates that instill confidence. For example, it is still difficult to interpret any of the variations of the Hill plots for the mixture of Pareto distributions in a way that leads to the correct estimate of the tail. Thus, while Figures 9.9 and 9.10 demonstrate that these variations of the Hill plot avoid some of the dangers associated with choosing the right k to use for the Hill estimator, it is clear that issues remain. In fact, it is important to realize the reason that so many variations exist in the literature is that no simple variations have been able to address all the issues identified by the Hill horror plots. Thus, while the Hill plot is a powerful estimation tool, it needs to be used with care; and complementary approaches should be used to build confidence in the estimates it provides. Providing such complementary approaches is the goal of the next section.

9.5 Beyond Hill and Regular Variation

By this point in this chapter we have seen that the Hill estimator is a powerful tool, with strong theoretical properties, but that it also has some drawbacks that come up in practical applications related to how to choose the number of order statistics, k, to use in the estimator. While the Hill plot provides an approach for choosing k that works remarkably well in many cases, it can sometimes be fooled into giving misleading estimates.

It is initially tempting to try to fix the issues with the Hill plot via small changes to the Hill estimator and the Hill plot, and we discussed two such approaches in the previous section. However, one quickly realizes that more fundamental changes need to be made in order to truly address the issues with the Hill estimator. Concretely, the Hill horror plots highlight issues with volatility for small k and bias for large k. Further, they show that the Hill estimator becomes problematic when the distribution is less like a pure Pareto and that the estimator is not location invariant. These issues cannot be addressed with simple changes to the Hill plot and require more fundamental changes to the estimator itself.

To address these more serious issues one must go back and think about how to extract other features from the data that can be used to improve estimation. As you might expect, there is an enormous literature developing potential extremal estimators (see, for example, [75, 183, 212]). Our goal in this section is not to cover all the competing estimators in this literature, but rather to highlight the features of the data that underlie these estimators and how they differ from what is used in the Hill estimator. Thus, we choose to focus on three prominent estimators that each use a different feature of the data, particularly of the extremes of the data, in order to estimate the tail. While the Hill estimator uses the empirical mean of the log of the order statistics to estimate α, the methods we discuss here make use of a variety of other properties (e.g., the second moment of the order statistics, the gaps between the order statistics, and the size and frequency with which samples exceed a certain level). These different features each lead to different estimators, specifically the moments estimator (aka the Dekkers–Einmahl–De Haan estimator), the Pickands estimator, and the peaks over thresholds estimator, respectively.

An important benefit of the contrasting approaches underlying these estimators is that each provides an estimator that can be used for a broader class of distributions than the Hill estimator. In particular, the Hill estimator is built upon properties of extreme value theory but focuses only on regularly varying distributions (i.e., distributions within the domain of attraction of the Fréchet distribution). By making use of different features of the data, this section presents more general estimators that allow estimation of distributions from the domains of attraction of the Fréchet, Gumbel, as well as the Weibull distribution. Recall from Chapter 7 in (7.2) that we can include all three extreme value distributions in one parametric form via

$$H(x) = \exp\left\{ -\left(1 + \xi\left(\frac{x - \mu}{\psi} \right) \right)^{-1/\xi} \right\}, \tag{9.11}$$

which is referred to as the *generalized extreme value distribution*. Here, $\xi \in \mathbb{R}$ is the shape parameter, $\psi > 0$ is the scale parameter, and $\mu \in \mathbb{R}$ is the location parameter. If $\xi = 0$, this corresponds to the Gumbel distribution, $\xi > 0$ corresponds to the Fréchet distribution (i.e., the regularly varying case, with $\xi = 1/\alpha$), while $\xi < 0$ corresponds to the Weibull distribution. The three approaches we introduce in this section all estimate the ξ parameter corresponding to that extreme value distribution whose maximal domain of attraction (MDA) contains the distribution of the data samples.

The ability of the estimators in this section to be used for distributions beyond power laws is crucial for the task of validating the power-law hypothesis as well as for ensuring the ability to estimate more complex heavy-tailed phenomena. Specifically, estimating ξ from data allows one to decide if power laws or regular variation are worthy of consideration at all – if the estimator of ξ turns out to be in $(-\infty, 0]$, then it may make sense to abandon the hypothesis of a power law in favor of a different family of distributions.

Before moving to our first estimator, there is one final important point we would like to emphasize. You may hope that this section ends with the presentation of a single, "best" estimator for the semi-parametric setting. Unfortunately, this is not the case. In fact, even among the estimators we present, which are perhaps the most prominent in the literature, none uniformly dominates the others in terms of variance (see [75, Section 6.4.2]). This serves to emphasize a point we have repeated multiple times already: when estimating the tail, there is no single recipe. Confidence must be built by seeking alignment between multiple estimators.

9.5.1 The Moments Estimator

The moments estimator, sometimes referred to as the Dekkers–Einmahl–De–Haan estimator, is perhaps the most natural extension of the Hill estimator. Recall that we interpreted the Hill

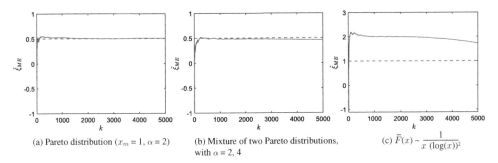

(a) Pareto distribution ($x_m = 1$, $\alpha = 2$)   (b) Mixture of two Pareto distributions, with $\alpha = 2, 4$ (c) $\bar{F}(x) \sim \dfrac{1}{x\,(\log(x))^2}$

Figure 9.11 Plots of the moments estimator $\hat{\xi}_{ME}$ versus k for data sampled from different regularly varying distributions. In each case, the dotted red line shows the correct value of the parameter ξ.

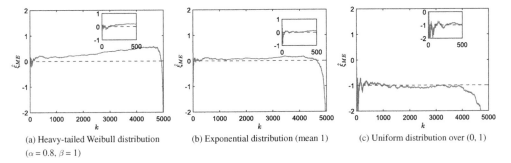

(a) Heavy-tailed Weibull distribution ($\alpha = 0.8$, $\beta = 1$)   (b) Exponential distribution (mean 1)   (c) Uniform distribution over (0, 1)

Figure 9.12 Plots of the moments estimator $\hat{\xi}_{ME}$ versus k for data sampled from different distributions having tails lighter than power law. In each case, the dotted red line shows the correct value of the parameter ξ.

estimator as (the reciprocal of) an empirical estimator of $\mathbb{E}\left[\log(X/b)_+\right]/\bar{F}(b)$ for large b. The moments estimator incorporates an additional empirical estimator of the normalized second moment of the same quantity (i.e., of $\mathbb{E}\left[\log^2(X/b)_+\right]/\bar{F}(b)$, for large b). Indeed, Dekkers, Einmahl, and De Haan showed that incorporating this second moment provides a consistent estimator of ξ not only for distributions in the domain of attraction of the Fréchet distribution, but also for distributions in the domains of attraction for the Weibull and Gumbel distributions [59].

To define the moments estimator formally, we consider n i.i.d. samples $X_1, \ldots, X_n$ from a distribution in the domain of attraction of a generalized extreme value distribution with index ξ. We denote the order statistics of the samples by $X_{(1)}, \ldots, X_{(n)}$. The moments estimator is defined as follows:

$$\hat{\xi}_{ME} = \hat{\xi}_{k,n}^{ME} = \underbrace{\hat{\xi}_{k,n}^{H,1}}_{T_1} + \underbrace{1 - \frac{1}{2}\left(1 - \frac{\left(\hat{\xi}_{k,n}^{H,1}\right)^2}{\hat{\xi}_{k,n}^{H,2}}\right)^{-1}}_{T_2}, \qquad (9.12)$$

where

$$\hat{\xi}_{k,n}^{H,1} = \frac{1}{k}\sum_{i=1}^{k}\log X_{(i)} - \log X_{(k+1)},$$

$$\hat{\xi}_{k,n}^{H,2} = \frac{1}{k}\sum_{i=1}^{k}(\log X_{(i)} - \log X_{(k+1)})^2.$$

Notice that $\hat{\xi}_{k,n}^{H,1}$ is essentially the reciprocal of the Hill estimator $\alpha_{k,n}^{H}$ (modulo the insignificant replacement of $X_{(k)}$ by $X_{(k+1)}$). On the other hand, $\hat{\xi}_{k,n}^{H,2}$ represents the empirical second moment discussed earlier.

To build intuition for the form of the moments estimator, we note that the term T_1 of the estimator, which is simply the reciprocal of the Hill estimator, converges to $\max(\xi, 0)$. On the other hand, the second term T_2 converges to $\min(\xi, 0)$. In other words, the first term ensures the consistency of the estimator when the data is regularly varying,[4] whereas the second ensures consistency when the data distribution lies in the MDA of the Weibull. Of course, both terms converge to zero when the data is sampled from a distribution in the MDA of the Gumbel.

The intuition that the moments estimator is an extension of the Hill estimator means that we should expect the moments estimator to maintain the same statistical properties as the Hill estimator. Indeed this is the case. The moments estimator guarantees consistency and asymptotic normality under the same conditions as the Hill estimator; see [59]. But, as we have already mentioned, a key benefit that comes from the use of higher moments is that the moments estimator maintains these properties for a broader class of distributions than the Hill estimator. We summarize the conditions needed for consistency in the following theorem.

Theorem 9.4 (Consistency)   *Consider i.i.d. samples $X_1, \ldots, X_n$ from a distribution in the domain of attraction of the generalized extreme value distribution having parameter ξ that satisfies the condition $\sup\{x | \Pr(X_i \geq x) < 1\} > 0$. If $k(n)/n \to 0$ and $k(n) \to \infty$ as $n \to \infty$, then $\hat{\xi}_{k,n}^{ME} \xrightarrow{p} \xi$ as $n \to \infty$. If additionally $k(n)/(\log n)^\delta \to \infty$ as $n \to \infty$ for some $\delta > 0$, then $\hat{\xi}_{k,n}^{ME} \xrightarrow{a.s.} \xi$ as $n \to \infty$.*

Note that the condition $\sup\{x | \Pr(X_i \geq x) < 1\} > 0$, which means that the upper endpoint of the support is positive, can be satisfied via a simple translation of the data if needed. Like the Hill estimator, the moments estimator can sometimes work beautifully and other times fail dramatically. The challenge, again, is in choosing the correct k and a plot parallel to the Hill plot is crucial for this. We illustrate this in Figure 9.11 for the same examples we consider when illustrating the Hill plot. In these plots we show the moments estimator $\hat{\alpha}_{ME} = \hat{\alpha}_{k,n}^{ME}$ as a function of k. Note that the estimator works well when the samples are drawn from a Pareto distribution, or a mixture of Pareto distributions. However, significant deviations from power laws can result in *horror plots* similar to those we encountered with the Hill estimator (see Figure 9.11(c) for an example).

---

[4] If the samples are regularly varying with index $-\alpha$, it can be shown that $\hat{\xi}_{k,n}^{H,2} \to \frac{2}{\alpha^2}$. Substituting this limit into (9.12), it is easy to see that the second term T_2 tends to zero when $\gamma > 0$.

In general, plots of the moments estimator are less volatile than the Hill estimator. But, this is not the reason to use the moments estimator over the Hill estimator. Instead, the most important contrast is that it can provide more reliable estimation than the Hill estimator when the distribution deviates from a power-law. We illustrate this fact in Figure 9.12, where we consider samples from a heavy-tailed Weibull distribution, the Exponential distribution, as well as the uniform distribution. The nonpositive estimate of ξ suggested by the moments plots provides a strong signal that the data is *not* power law.

To summarize, the moments estimator provides a natural generalization of the Hill estimator to distributions within the maximal domain of attraction of any extremal distribution. This yields a useful means of rejecting the hypothesis that the data follows a power law. However, the moments estimator also inherits some of the issues with the Hill estimator: it is not location invariant, and there are instances (horror plots) where it is nearly impossible to find a reliable choice of k.

9.5.2 The Pickands Estimator

The moments estimator extends the Hill estimator by using higher moments of the order statistics and is successful in reducing volatility and extending the class of distributions that can be estimated. However, it still lacks location invariance – something that is particularly troubling when applying the estimator in practice. In this section, we present an alternative estimator, the Pickands estimator, that achieves both location and scale invariance, though at the cost of larger volatility.

The Pickands estimator (proposed by James Pickands III [174]) is built fundamentally differently than the Hill and moments estimators we have considered so far. Instead of using the order statistics to infer an estimate of ξ using empirical moments, the Pickands estimator focuses on the gaps between the order statistics. Specifically, the gaps between logarithmically spaced order statistics:

$$\hat{\xi}_P = \hat{\xi}_{k,n}^P = \frac{1}{\log 2} \log \left(\frac{X_{(k)} - X_{(2k)}}{X_{(2k)} - X_{(4k)}} \right). \tag{9.13}$$

Upon first glance, the form of the Pickands estimator is quite curious. The estimators we have considered so far seemingly make use of much more detailed information in the data since they use $k \to \infty$ order statistics, whereas the Pickands estimator only considers the spacing between *three* order statistics: $X_{(k)}$, $X_{(2k)}$, and $X_{(4k)}$. This begs the question:

Are three order statistics enough information to allow accurate estimation of the tail?

Indeed they are. The spacing between the order statistics conveys enough information to allow the Pickands estimator to be weakly consistent and asymptotically Gaussian under similar conditions to the moments estimator, and more generally than the Hill estimator. We summarize these results in the following theorem from [58].

Theorem 9.5 (Consistency)   *Consider n i.i.d. samples from a distribution in the domain of attraction of the generalized extreme value distribution having parameter ξ. If $k(n)/n \to 0$ and $k(n) \to \infty$ as $n \to \infty$, then $\hat{\xi}_{k,n}^P \overset{p}{\to} \xi$ as $n \to \infty$. If additionally $k(n)/\log\log n \to \infty$ as $n \to \infty$, then $\hat{\xi}_{k,n}^P \overset{a.s.}{\to} \xi$ as $n \to \infty$.*

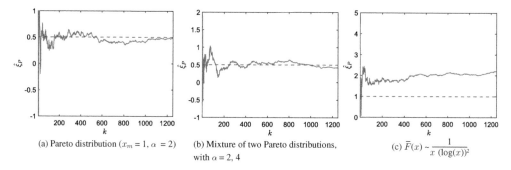

Figure 9.13 Plots of the Pickands estimator $\hat{\xi}_P$ versus k for data sampled from different regularly varying distributions. In each case, the dotted red line shows the correct value of the parameter ξ.

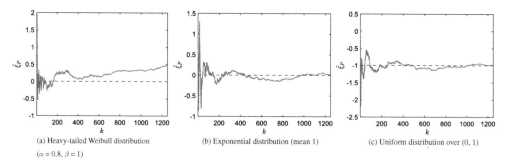

Figure 9.14 Plots of the Pickands estimator $\hat{\xi}_P$ versus k for data sampled from different distributions having tails lighter than power law. In each case, the dotted red line shows the correct value of the parameter ξ.

While we do not want to dwell too much on the technicalities underlying this result, it turns out that intuition for the form of the Pickands estimator can be gained from the simple derivation of weak consistency for the special case of the Fréchet MDA (i.e., assuming the samples are drawn from a regularly varying distribution). In particular, the proof illustrates the sufficiency of just three order statistics for the consistency of the Pickands estimator.

Proof of weak consistency in Theorem 9.5 for regularly varying distributions. Suppose that $\{X_i\}_{i=1}^n$ are i.i.d. and regularly varying with index $-\alpha$. Our goal is to prove that in this case,

$$\hat{\xi}_{k,n}^P \xrightarrow{p} \frac{1}{\alpha}.$$

We reuse the representation from the proof of Theorem 9.1, that is,

$$X_{(i)} = F^{-1}(U_{(i)}) = \left(\frac{1}{1-U_{(i)}}\right)^{1/\alpha} \ell\left(\frac{1}{1-U_{(i)}}\right),$$

where $\{U_i\}_{i=1}^n$ are i.i.d. and uniformly distributed random variables in $[0,1]$. Define $V_{(ak)} := \frac{k}{n(1-U_{(ak)})}$ for $a \in \mathbb{N}$. Under our hypotheses ($k \to \infty$, $k/n \to 0$), it can be shown that $V_{(ak)} \xrightarrow{p} 1/a$ (see Exercise 4). Now,

$$
\frac{X_{(ak)}}{F^{-1}\left(1-k/n\right)} = \frac{\left(\frac{1}{1-U_{(ak)}}\right)^{1/\alpha}\ell\left(\frac{1}{1-U_{(ak)}}\right)}{(n/k)^{1/\alpha}\ell(n/k)}
$$

$$
= \frac{(n/k)^{1/\alpha}V_{(ak)}^{1/\alpha}\,\ell((n/k)V_{(ak)})}{(n/k)^{1/\alpha}\ell(n/k)}
$$

$$
\xrightarrow{p} (1/a)^{1/\alpha}.
$$

Here, we have used the uniform convergence property of slowly varying functions (see Section 2.6). We are now ready to analyze the Pickands estimator:

$$
\hat{\xi}_{k,n}^{P} = \frac{1}{\log 2}\log\left(\frac{X_{(k)}/F^{-1}\left(1-k/n\right) - X_{(2k)}/F^{-1}\left(1-k/n\right)}{X_{(2k)}/F^{-1}\left(1-k/n\right) - X_{(4k)}/F^{-1}\left(1-k/n\right)}\right)
$$

$$
\xrightarrow{d} \frac{1}{\log 2}\log\left(\frac{1-(1/2)^{1/\alpha}}{(1/2)^{1/\alpha}-(1/4)^{1/\alpha}}\right)
$$

$$
= \frac{1}{\alpha}.
$$

This completes the proof. □

Like the other extremal estimators we consider in this chapter, the challenge when applying the Pickands estimator in practice is choosing k. Doing so relies on a plot that parallels the Hill plot, which shows $\hat{\alpha}_{k}^{P}$ as a function of k. We illustrate the so-called *Pickands plot* for our running examples in Figures 9.13 and 9.14.

Like the Hill plots, these Pickands plots demonstrate that the Pickands estimator can be quite effective in many situations, but can be useless or even deceptive in others (see the horror plot in Figure 9.13(c)). Compared to the Hill and moments estimators, volatility is often more of a problem for the Pickands estimator since it uses only three order statistics. However, like the moments estimator, the Pickands estimator can be used to rule out the power-law hypothesis, which is something the Hill plot cannot be relied on to accomplish. And, of course, the Pickands estimator is location- *and* scale-invariant, a property that neither the Hill estimator nor the moments estimator possess.

There have been many attempts to develop variations of the Pickands estimator that reduce volatility (e.g., [12, 70, 79, 192, 208, 225]). These variations tend to still use linear combinations of logarithmically separated order statistics. The variation in [208] is of particular interest because it has been proven to be consistent (see [69]) when used in the double bootstrap method we discuss in Section 9.6.2.

9.5.3 Peaks over Threshold

All of the extremal estimators we have considered so far make use of order statistics to perform estimation. Our next estimator is different. It focuses instead on the distribution of *residual life*, which forms the basis of the definition of the class of long-tailed distributions in Chapter 4 and also played an important role when studying extremal processes in Chapter 7.

Recall that the residual life distribution associated with the distribution F at *age* x is defined as

$$\bar{R}_x(t) = \frac{\bar{F}(x+t)}{\bar{F}(x)} \quad (t \geq 0).$$

Indeed, if the random variable X has distribution F, then $\bar{R}_x(t) = \Pr\left(X > x + t \mid X > x\right)$. Additionally, the *mean residual life* (MRL) is defined as

$$m(x) = E[X - x | X > x] = \int_0^\infty \frac{\bar{F}(x+t)}{\bar{F}(x)} dt.$$

While we have referred to $X - x$ as the residual life in this book, it has other names as well. In the statistical literature it is often referred to via the language of exceedances. Specifically, given a threshold x, an *exceedance* occurs if a random variable is larger than x, and the size of the exceedance $X - x$ is referred to as the *excess*. Thus, $m(x)$ is sometimes referred to as the *mean excess function*.

Like the density function and the moment generating function, the MRL function $m(x)$ uniquely determines the distribution. In fact, it can be quite useful for distinguishing heavy-tailed and light-tailed behavior. Recall that many heavy-tailed distributions have an increasing mean residual life (IMRL), while many light-tailed distributions have a decreasing mean residual life (DMRL), but we should not make the mistake of linking IMRL/DMRL too tightly with heavy-tailed/light-tailed distributions (as we have discussed in Chapter 4).

However, it is possible to identify a large class of heavy-tailed distributions via properties of the mean residual life function. A particularly useful result of this type in Chapter 4 is that, under some regularity conditions, a distribution is long-tailed if and only if $m(x) \to \infty$ as $x \to \infty$ (see Theorem 4.9). The fact that $m(x)$ grows unboundedly under exactly the class of long-tailed distributions (which includes subexponential and regularly varying distributions) already provides a useful connection for exploratory data analysis, but we can go further for the case of regularly varying distributions. In the case of regularly varying distributions, $m(x)$ grows (asymptotically) linearly. To see this, we can use an application of Karamata's theorem and mimic the derivation of $m(x)$ for the Pareto distribution in (4.5). Consider a distribution F that is regularly varying with index $-\alpha$. Then expressing $\bar{F}(x) = x^{-\alpha}L(x)$, where $L(\cdot)$ is slowly varying,

$$\begin{aligned}
\lim_{x \to \infty} \frac{m(x)}{x} &= \lim_{x \to \infty} \frac{1}{x} \int_0^\infty \frac{\bar{F}(x+t)}{\bar{F}(x)} dt \\
&= \lim_{x \to \infty} \frac{1}{x\bar{F}(x)} \int_x^\infty L(y)y^{-\alpha} dy \\
&= \frac{1}{\alpha - 1}.
\end{aligned}$$

Thus, for regularly varying distributions the mean excess function is asymptotically linear with slope $1/(\alpha - 1)$. At first glance, this appears like a useful observation for statistical estimation, in the same spirit as the regression-based tail estimation described in Chapter 8. Specifically, one might hope to fit a regression line to the empirical mean residual life function to estimate α. However, as we illustrate in what follows, this approach can be quite unreliable in practice.

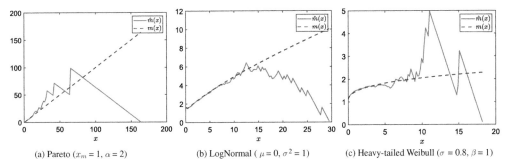

(a) Pareto ($x_m = 1$, $\alpha = 2$)   (b) LogNormal ($\mu = 0$, $\sigma^2 = 1$)   (c) Heavy-tailed Weibull ($\sigma = 0.8$, $\beta = 1$)

Figure 9.15 Plots of the empirical mean residual life corresponding to 5,000 data points sampled from different heavy-tailed distributions.

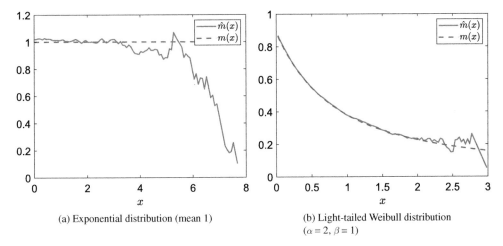

(a) Exponential distribution (mean 1)   (b) Light-tailed Weibull distribution ($\alpha = 2$, $\beta = 1$)

Figure 9.16 Plots of the empirical mean residual life corresponding to 5,000 data points sampled from different light-tailed distributions. In each plot, the corresponding mean residual life function is plotted in red.

Using the Empirical Mean Residual Life Function for Estimation:

We saw earlier that the mean residual life distribution is asymptotically linear for regularly varying distributions. In fact, the mean residual life distribution grows unboundedly for (most) heavy-tailed distributions. As a result it is natural to try to use the empirical mean residual life function to conclude whether or not the underlying data is sampled from a heavy-tailed distribution, or specifically, from a regularly varying one.

Concretely, the empirical mean residual life (EMRL) is defined as follows:

$$m_n(x) = \frac{\sum_{i=1}^n (X_i - x) I(X_i > x)}{\sum_{i=1}^n I(X_i > x)}.$$

Observe that $m_n(x)$ approaches zero as x approaches the largest sample in the dataset. In other words, the EMRL is necessarily bounded for *any* distribution. This is an important contrast to the true mean residual life function $m(x)$. Moreover, one should expect that the EMRL would be highly variable around the largest samples in our dataset. This means that asymptotic properties of the mean residual life function must be translated very carefully to its empirical variant.

Figures 9.15 and 9.16 show some examples of the empirical mean residual life for data sampled from Pareto, LogNormal, Weibull (heavy-tailed and light-tailed), and Exponential distributions. In the heavy-tailed cases, note that $m_n(x)$ tends to increase at first (as expected), but eventually decays to zero once x approaches the largest of our (finitely many) data samples. However, can one conclude from looking just at the EMRL that the corresponding MRL grows unboundedly?

In the Pareto example (Figure 9.15(a)), note the nearly linear growth of $m_n(x)$ for moderate x. However, estimating the tail index by computing the slope of a regression line fit to the EMRL would be problematic, for the same reasons discussed in Chapter 8 in the context of the rank plot and the histogram. Moreover, the estimate of the slope would depend critically on the interval over which this estimation is performed (i.e., how much of the "head" and "tail" portions of the EMRL are disregarded). Finally, a "near linear" behavior of the EMRL for moderate values of x is also seen in the LogNormal case (Figure 9.15(b)), further cautioning against the use of the EMRL for statistical estimation.

Thus, one should be very careful when using the mean residual life for exploratory analysis. It is not a statistical estimation tool, and we do not recommend attempting to estimate the tail index using linear regression of the empirical mean excess. A more robust approach is to exploit the limiting behavior of the residual life *distribution* as $x \to \infty$, as we do next. Importantly, this approach also generalizes beyond the maximal domain of attraction of the Fréchet, to the union of the maximal domains of attraction of all extreme value distributions.

Statistical Estimation

We now present a statistically sound approach for using the residual life distribution to estimate the ξ parameter of the extremal distribution whose domain of attractions contains the distribution of our data. This approach, which has been used widely in the field of hydrology, is called *peaks over threshold* (POT).

The key observation that underlies the use of the residual life distribution as a tool for semi-parametric estimation is that the threshold for determining exceedances can be chosen in such a way that the exceedances are relatively unaffected by the body of the distribution. That is, if a large enough threshold is chosen, then the behavior of the tail will dominate the behavior of the residual life distribution, and so estimates that use the residual life distribution are effective even without parametric assumptions on the body of the distribution.

More concretely, the peaks over thresholds estimator (aka the POT estimator) relies on a deep but simple-to-state observation: for distributions lying in the domain of attraction of an extreme value distribution, the sizes of the residual life (i.e., "peaks") beyond a (large) threshold are i.i.d. and well approximated by a scaled version of a generalized Pareto distribution. This means that the tail of the sampled distribution can be parametrically estimated by fitting the residual life values to a generalized Pareto distribution.

Before explaining why this is true, let us recall the definition of a generalized Pareto distribution. We define it in terms of its c.c.d.f.:

$$\text{The generalized Pareto distribution: } \bar{G}_{\xi,\beta}(x) = (1 + \xi x/\beta)^{-1/\xi}. \qquad (9.14)$$

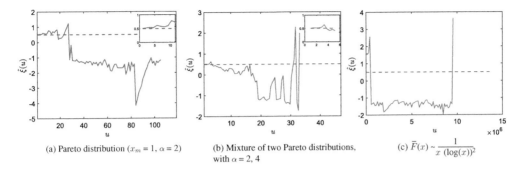

(a) Pareto distribution ($x_m = 1$, $\alpha = 2$)   (b) Mixture of two Pareto distributions, with $\alpha = 2, 4$ (c) $\bar{F}(x) \sim \dfrac{1}{x\,(\log(x))^2}$

Figure 9.17 Plots of the POT estimator $\hat{\xi}$ versus threshold u for data sampled from different regularly varying distributions. In each case, the dotted red line shows the correct value of the parameter ξ.

Note that this should be interpreted as $e^{-x/\beta}$ if $\xi = 0$. The following result is the heart of the peaks over thresholds estimator.

Lemma 9.6   *Let x^F denote the right endpoint of F and suppose that F lies in the maximum domain of attraction of an extreme value distribution. Then there exists a positive function $\beta(u)$ such that*

$$\lim_{u \to x^F} \sup_{y < x^F - u} \left| \frac{\bar{F}(u+y)}{\bar{F}(u)} - \bar{G}_{\xi,\beta(u)}(y) \right| = 0.$$

With Lemma 9.6 in hand (see [75] for a proof), we have now reduced the task of semi-parametric estimation of the tail to a parametric estimation task. The goal at this point is simply to fit a generalized Pareto distribution to the exceedences of the samples above a suitably large threshold u. Given our discussion in Chapter 8, it is perhaps not surprising that we recommend using maximum likelihood estimation for this task. The maximum likelihood estimators of ξ and β can be derived in a straightforward manner (see Exercise 6) and are simple to use in practical settings. In particular, let $k = k(u)$ be the number of values of the sample exceeding u, and then the maximum likelihood estimate of (ξ, β), denoted $(\hat{\xi}_k, \hat{\beta}_k)$, can be computed by making the substitution $\tau = -\xi/\beta$ and solving the following system of equations for $(\hat{\xi}_k, \hat{\tau}_k)$:

$$\hat{\xi}_k = \frac{1}{k} \sum_{i=1}^{k} \log(1 - \hat{\tau}_k(X_{(i)} - u)), \tag{9.15}$$

$$\frac{1}{\hat{\tau}_k} + \frac{1}{k}\left(\frac{1}{\hat{\xi}_k} + 1\right) \sum_{i=1}^{k} \frac{X_{(i)} - u}{1 - \hat{\tau}_k(X_{(i)} - u)} = 0. \tag{9.16}$$

Finally, we have $\hat{\beta}_k = -\hat{\xi}_k/\hat{\tau}_k$.

Using these estimators for the generalized Pareto distribution, the POT estimator achieves basically the same statistical guarantees we have seen for the other extremal estimators we have presented. Interestingly, the same conditions are required for k, the number of samples

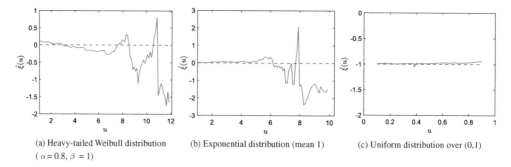

(a) Heavy-tailed Weibull distribution
($\alpha = 0.8$, $\beta = 1$)

(b) Exponential distribution (mean 1)

(c) Uniform distribution over (0,1)

Figure 9.18 Plots of the POT estimator $\hat{\xi}$ versus threshold u for data sampled from different distributions having tails lighter than power law. In each case, the dotted red line shows the correct value of the parameter ξ.

larger than the threshold u, as are required for the number of order statistics in the Hill and moments estimators.

Theorem 9.7 *Consider n i.i.d. samples from a distribution in the domain of attraction of the generalized extreme value distribution having parameter ξ. If $k(n)/n \to 0$ and $k(n) \to \infty$ as $n \to \infty$, then $\hat{\xi}_k \overset{p}{\to} \xi$ as $n \to \infty$.*

We refer the interested reader to [75] for a proof of Theorem 9.7. The form of the theorem highlights that, again, we are left with a delicate decision when using the peaks over thresholds estimator. As with all of the extremal estimators we have discussed, there is a decision about how much data to use. Here, that decision boils down to choosing a threshold u, which ensures that enough, but not too many, exceedances occur. If u is too large, then there are too few samples used in the estimator and the variance may be high; but if u is too small, then the resulting estimator may be biased.

When making this choice, it is often effective to visualize the estimator. There are a few common options for approaching this visualization. One is to make plots similar to the Hill plot and Pickands plot, which show the estimator $\hat{\xi}_k$ as a function of k (as we have done in Figures 9.17 and 9.18). Another approach is to pick u such that the EMRL $m_n(x)$ looks linear for x larger than u. This can be effective as long as there is reason to believe that $\xi > 0$; however, as our experiments with the EMRL reveal, it is often optimistic to think that a clear threshold will be visible.

We illustrate the use of the peaks over thresholds estimator on our running examples in Figures 9.17 and 9.18. Note that for suitable choices of u, a reasonably accurate estimator is obtained in most examples. This enables, as with the moments and the Pickands estimators, a means of rejecting the power law hypothesis entirely. However, as was also the case with former estimators, *horror plots* are possible (see Figure 9.17(c)), where the estimator is outright deceptive.

9.6 Where Does the Tail Begin?

We started this chapter with the question of where does the tail begin, and now we return to it with more tools in hand. In the meantime, we have shown sufficient conditions for consistency and asymptotic normality of the Hill estimator and related estimators. We have also seen that desirable large sample properties can be achieved by letting the number of order statistics k used by the estimator depend on the size of the dataset n in such a way that $k = k(n) \to \infty$ but $k(n)/n \to 0$. This is a useful guide but is difficult to apply in practice, since it does not prescribe a specific choice of k for a given (finite) dataset. Instead, for finite n, we have shown some exploratory graphical tools that can help in selecting k – the Hill plot, the Pickands plot, and so on. However, we have illustrated that the Hill plot and its variations can be challenging to apply when data are not generated by a pure power law. Further, the use of the Hill plot is necessarily ad hoc, which can sometimes lead to misleading results (e.g., in the case of the Hill horror plots). For this reason, it is desirable to have additional tools available that allow *data driven, statistically rigorous, selection of k.* In this section, we introduce two of such methods: PLFIT and the double bootstrap method.

For simplicity of presentation, in this section we assume once more that the data consists of i.i.d. samples drawn from a regularly varying distribution with (unknown) index $-\alpha$. That is, we revert to the setting of Sections 9.1–9.4. The first technique we introduce, PLFIT, is designed specifically for this setting (where the data is assumed to be drawn from a distribution that lies in the MDA of the Fréchet); however, there are variants of the second technique we introduce, the double bootstrap method, that work in the more general setting of Section 9.5 (where the data is assumed to be drawn from a distribution that lies in the MDA of *any* extremal distribution); see, for example, [69, 105].

9.6.1 PLFIT

Perhaps the most prominent approach for choosing the cutoff for when the tail begins (i.e., k_n) is PLFIT (Power-Law FIT), proposed by Clauset et al. [48]. The approach is simple, popular, and intuitive; and it is the basis for the guidelines proposed in Section 8.5.

Specifically, in PLFIT, the cutoff k_n is chosen to be the value that minimizes the Kolmorogorov–Smirnov distance between the fitted power-law and the empirical conditional distribution function associated with the k_n largest observations, under the assumption that the latter distribution is a pure Pareto distribution. More formally, define

$$D_{n,k} := \sup_{y \geq 1} \left| \frac{1}{k-1} \sum_{i=1}^{k-1} I\left(\frac{X_{(i)}}{X_{(k)}} > y \right) - y^{-\hat{\alpha}_{n,k}^H} \right|, \qquad (9.17)$$

which is the Kolmogorov distance between the empirical distribution of the sequence $(X_{(i)}/X_{(k)})_{i=1}^{k-1}$ and a pure Pareto distribution with index $\hat{\alpha}_{n,k}^H$ on $[1, \infty)$.[5] The PLFIT method consists of choosing the cutoff as

$$\kappa_n^\star := \underset{1 \leq k \leq n-1}{\arg\min} D_{n,k}.$$

Finally, the estimate of the tail index under this approach is $\hat{\alpha}_{n,\kappa_n^\star}^H$.

---

[5] Note that if $X_{(1)}, X_{(2)}, \ldots, X_{(k-1)}$ are assumed to be drawn from a Pareto distribution with minimum value $X_{(k)}$ and tail index $\hat{\alpha}_{n,k}^H$, then the normalized values $(X_{(i)}/X_{(k)})_{i=1}^{k-1}$ are drawn from a Pareto distribution with minimum value 1 and the same tail index $\hat{\alpha}_{n,k}^H$.

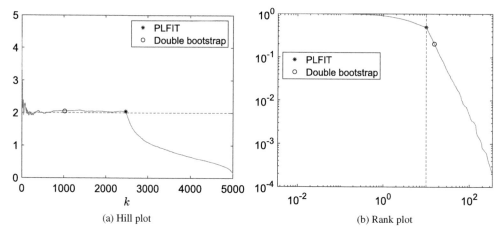

(a) Hill plot   (b) Rank plot

Figure 9.19 Rank plot and Hill plot corresponding to data sampled from distribution with Exponential body, Pareto tail. The left panel depicts the number of order statistics recommended by the Hill double bootstrap method and PLFIT; the right panel shows the corresponding cutoffs on the rank plot. Tail index estimates: 2.045 (PLFIT) and 2.048 (double bootstrap).

The simplicity and intuitive nature of the PLFIT estimator has meant that it has become popular across a broad range of academic disciplines, as exemplified by the large number of citations to [48]. Nevertheless, until recently, it had not been known whether this approach leads to a consistent estimator. A reason behind the lack of mathematical results is that PLFIT is an example of a *minimum distance estimation* (MDE) procedure, in which a criterion function is minimized over the parameter space. The criterion function is a functional measuring the distance between the empirical distribution and a parametric family of distributions. Popular choices of criterion functions include the Cramér–von Mises, the Kolmogorov–Smirnov (KS), and the Anderson–Darling criterion. MDE is a prominent approach in the statistics literature due to the robustness it provides, that is, MDE estimates are not affected much by small departures of the population distribution from the parametric family. However, it is typically hard to prove formal statistical guarantees for MDE estimators. It is only recently that the (weak) consistency of the tail index estimator based on PLFIT was established by [28]. A formal statement of the consistency guarantee for PLFIT is the following.

Theorem 9.8   *Consider n i.i.d. samples from a distribution F that is regularly varying with index $-\alpha$. Assuming that F and its generalized inverse F^{-1} are continuous, $\hat{\alpha}^H_{n,\kappa_n} \xrightarrow{p} \alpha$ as $n \to \infty$.*

We refer the interested reader to [28] for a proof of Theorem 9.8. Intuitively, the proof boils down to showing that the random sequence $\{\kappa^\star_n\}$ satisfies $\kappa^\star_n \xrightarrow{p} \infty$ and $\frac{\kappa^\star_n}{n} \xrightarrow{p} 0$, and that these conditions suffice to ensure weak convergence of $\hat{\alpha}^H_{n,\kappa^\star_n}$.

While we now know that PLFIT leads to a consistent estimator, many other properties of PLFIT remain to be understood. The literature studying asymptotic properties of minimal distance estimates based on the Kolmogorov–Smirnov distance is very limited, even under parametric assumptions, due to technical challenges (e.g., see [123, Section 3.5] for a discussion on the difficulties). Despite the challenges, a richer picture of the properties of the

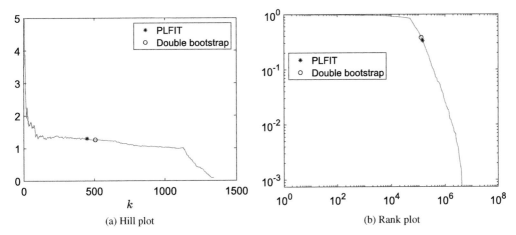

(a) Hill plot   (b) Rank plot

Figure 9.20 Rank plot and Hill plot corresponding to data on number of customers affected by power blackouts in the US between 2002 and 2018. The left panel depicts the number of order statistics recommended by the Hill double bootstrap method and PLFIT; the right panel shows the corresponding cut-offs on the rank plot. Tail index estimates: 1.305 (PLFIT) and 1.266 (double bootstrap).

PLFIT estimator is now beginning to emerge. For example, it has been shown that the PLFIT estimator is not asymptotically normal, and tends to choose a smaller k than is asymptotically optimal, resulting in higher variance of the index estimate (see [71]). Given this, it is important to keep in mind that PLFIT should be used in combination with other approaches to build confidence.

Despite its limitations, PLFIT is quite popular and is the workhorse of heavy-tailed statistical estimation across a wide variety of fields. We illustrate its application using our running examples in Figures 9.19–9.21. Figure 9.19 illustrates the most simple case, when PLFIT is applied to data that comes from a distribution with an Exponential body and a Pareto tail, with $\alpha = 2$. In this case, Figure 9.19(a) shows the Hill Plot and we see that PLFIT identifies the right endpoint of the flat region, which is indeed the place where the tail begins. The red dotted line highlights the estimate corresponding to the k selected by PLFIT. The estimate provided by PLFIT in this case is 2.045, which is very close to the true value of 2. Figure 9.19(b) uses the rank plot to highlight the estimate of "where the tail begins" that PLFIT provides. Here, PLFIT correctly identifies the transition point between the Exponential body and the Pareto tail. Figures 9.20 and 9.21 show the application of PLFIT to the US power blackouts and US city sizes datasets. In these cases, we do not know the true values, but we can see that PLFIT seems to identify the flat region in the Hill plot and that the point identified as the "beginning of the tail" in the rank plot seems reasonable upon a visual inspection.

9.6.2 The Double Bootstrap Method

While PLFIT provides a data-driven, statistically consistent approach for selecting the number of order statistics, the resulting estimate of the tail index is known to be suboptimal from

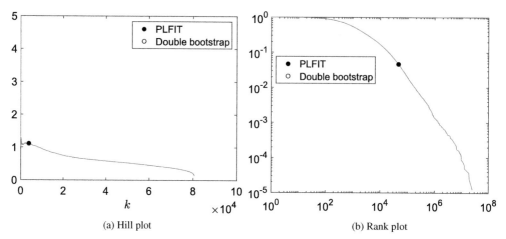

(a) Hill plot   (b) Rank plot

Figure 9.21 Rank plot and Hill plot corresponding to data on populations of US cities as per the 2010 census. *(a)* depicts the number of order statistics recommended by the Hill double bootstrap method and PLFIT; *(b)* shows the corresponding cutoffs on the rank plot. Tail index estimates: 1.105 (PLFIT) and 1.115 (double bootstrap).

the standpoint of the (asymptotic) mean square error (see [71]). The second approach we present seeks to provide an optimal mean square error via a double bootstrap method. Interestingly, as the name suggests, this involves combining two different consistent estimators of the tail index.

More specifically, given a specific consistent estimator $\hat{\xi}_{n,k}^{(1)}$ of $\xi = 1/\alpha$ (like the Hill/moments/Pickands estimator) that utilizes the k largest order statistics out of the n available data samples, double bootstrap methods seek to select the *optimal* number of order statistics in order to minimize the mean square error, that is, they seek to estimate

$$\kappa_0^\star(n) = \arg\min_k \mathbb{E}\left[\left(\hat{\xi}_{n,k}^{(1)} - \xi\right)^2\right].$$

An exact calculation of $\kappa_0^\star(n)$ is infeasible since that would require prior knowledge of the precise distribution from which the data is sampled. Instead, double bootstrap methods produce $\hat{\kappa}^\star(n)$ satisfying $\frac{\hat{\kappa}^\star(n)}{\kappa_0^\star(n)} \xrightarrow{p} 1$ as $n \to \infty$. This is achieved by combining the originally chosen estimator $\hat{\xi}^{(1)}$ with a second consistent estimator $\hat{\xi}^{(2)}$.

There is not a single double bootstrap method; instead, there is a family of such methods in the literature. For example, the *Hill double bootstrap* approach, which we describe here, uses $\hat{\xi}^{(1)}$ as the Hill estimator, and therefore seeks to obtain the asymptotically optimal choice of k for the Hill estimator, using a related moment-based estimator for $\hat{\xi}^{(2)}$. Similarly, the *moments double bootstrap* approach sets $\hat{\xi}^{(1)}$ as the moments estimator of Section 9.5.1, with an analogous third moment estimator used as $\hat{\xi}^{(2)}$ (see [69]). A double bootstrap approach has also been developed for the Pickands estimator of Section 9.5.2 (see [69]).

These approaches all differ, but there is a general structure that is typically followed. Typically, double bootstrap methods follow the generic approach outlined in what follows.

Double Bootstrap (Generic)

1. Select n_1 samples from the n data points randomly, independently, and with replacement (such a sampling from the empirical distribution of the data is referred to as a *bootstrap*), and compute $\hat{\xi}^{(1)}_{n_1,k}$ and $\hat{\xi}^{(2)}_{n_1,k}$ for different k. Typically, $n_1 = O(n^{1-\epsilon})$ for $\epsilon \in (0, 1/2)$.
2. Repeat Step 1 for r independent bootstrap samples, denoting the estimators corresponding to the jth bootstrap by $\hat{\xi}^{(\ell)}_{n_1,k}(j)$ for $j = 1, 2, \ldots, r$ and $\ell = 1, 2$.
3. Pick $\hat{\kappa}^{\star}_1$ as that value of k that maximizes the *empirical agreement* between $\hat{\xi}^{(1)}$ and $\hat{\xi}^{(2)}$. Intuitively, one may think of $\hat{\kappa}^{\star}_1$ as follows:

$$\hat{\kappa}^{\star}_1 = \arg\min_k \frac{1}{r} \sum_{j=1}^{r} \left[\hat{\xi}^{(1)}_{n_1,k}(j) - \hat{\xi}^{(2)}_{n_1,k}(j)\right]^2.$$

   However, this expression is only provided as an illustration of how one might capture agreement between $\hat{\xi}^{(1)}$ and $\hat{\xi}^{(2)}$; the precise definition $\hat{\kappa}^{\star}_1$ depends on the specific double bootstrap method under consideration.
4. Repeat Steps 1–3 for a smaller bootstrap of size $n_2 = \frac{n_1^2}{n}$, and compute $\hat{\kappa}^{\star}_2$ analogously. (The use of this second bootstrap explains why these methods are called "double bootstrap.")
5. The output $\hat{\kappa}^{\star}$ is now defined as

$$\kappa^{\star} = \frac{(\kappa^{\star}_1)^2}{\kappa^{\star}_2} A(\kappa^{\star}_1, n_1, n),$$

   where the function A depends on the specific double bootstrap method under consideration.

To give a concrete example of this generic approach, we focus on the most prominent example: the Hill double bootstrap method, which was proposed in [55].

The Hill Double Bootstrap Method

As the name suggests, the Hill double bootstrap method seeks to select the number of order statistics so as to optimize the asymptotic mean square error of the Hill estimator. Accordingly, it uses the Hill estimator as the first estimator:

$$\hat{\xi}^{(1)}_{n,k} := \hat{\xi}^{H}_{n,k} = \frac{1}{k} \sum_{i=1}^{k} \log\left(\frac{X_{(i)}}{X_{(k)}}\right). \tag{9.18}$$

The second estimator used for the double bootstrap is the following moment-based estimator:

$$\hat{\xi}^{(2)}_{n,k} = \frac{M_{n,k}}{2\hat{\xi}^{H}_{n,k}}, \tag{9.19}$$

where

$$M_{n,k} := \frac{1}{k} \sum_{i=1}^{k} \left(\log \left(\frac{X_{(i)}}{X_{(k)}} \right) \right)^2. \tag{9.20}$$

When $\xi > 0$, note that the consistency of this estimator was alluded to in our discussion of the moments estimator in Section 9.5.1.

Given the choice of these two estimators, we can characterize their difference as

$$|\hat{\xi}_{n,k}^{(1)} - \hat{\xi}_{n,k}^{(2)}| = \frac{|M_{n,k} - 2\left(\hat{\xi}_{n,k}^{H}\right)^2|}{2|\hat{\xi}_{n,k}^{H}|}.$$

As in Step 3 of the generic template, the Hill bootstrap approach selects $\hat{\kappa}_1^\star$ so as to minimize the empirical mean square of the numerator. More specifically, this is accomplished as follows:

1. Select r independent bootstrap samples of size n_1 from the given data. Compute, for each bootstrap sample j, $\hat{\xi}_{n_1,k}^{(1)}(j)$ (using (9.18)) and $M_{n_1,k}(j)$ (using (9.20)) for different values of k. Compute

$$\hat{\kappa}_1^\star := \arg\min_k \frac{1}{r} \sum_{j=1}^{r} \left[M_{n_1,k}(j) - 2\left(\hat{\xi}_{n_1,k}^{H}(j)\right)^2 \right]^2.$$

2. Repeat the previous step with a smaller bootstrap of size $n_2 = \frac{n_1^2}{n}$, to compute $\hat{\kappa}_2^\star$ analogously.

3. Finally, output

$$\hat{\kappa}^\star = \frac{(\hat{\kappa}_1^\star)^2}{\hat{\kappa}_2^\star} \left(\frac{\log \hat{\kappa}_1^\star}{2\log n_1 - \log \hat{\kappa}_1^\star} \right)^{\frac{2(\log n_1 - \log \hat{\kappa}_1^\star)}{\log n_1}}.$$

The power of the approach in the Hill double bootstrap is that it can be proven that it optimizes the asymptotic mean square error. Recall that a characterization of the asymptotic mean square error (and also asymptotic normality) of extremal estimators like the Hill estimator assumes a second-order condition on the distribution the data is sampled from. Since double bootstrap methods seek to optimize the asymptotic mean square error, it is therefore not surprising that the same second-order condition plays a crucial role in the design and analysis of these methods.[6] Specifically, the following is a semiformal statement summarizing the performance of the Hill double bootstrap. A formal statement, as well as its proof, can be found in [55].

---

[6] Formally, the second-order condition is stated as follows. Recall from our proof of Theorem 9.1 that if F is a regularly varying distribution with index $-\alpha$, then $U(x) = F^{-1}(1 - 1/x) = x^\gamma \ell(x)$ for slowly varying function $\ell(\cdot)$. In other words, $\lim_{t\to\infty} \frac{U(tx)}{U(t)} = x^\gamma$. The distribution F satisfies the second-order condition if there exists a function A, which does not change sign near infinity, such that for $x > 0$,

$$\lim_{t\to\infty} \frac{\frac{U(tx)}{U(t)} - x^\gamma}{A(t)} = x^\gamma \frac{x^\rho - 1}{\rho}$$

for some $\rho < 0$.

Theorem 9.9 *Suppose that $X_1, X_2, \ldots, X_n$ are i.i.d. samples from a regularly varying distribution F satisfying a certain second-order condition. Taking $n_1 = O(n^{1-\epsilon})$ for $\epsilon \in (0, 1/2)$, under the Hill double bootstrap method, $\frac{\hat{\kappa}^\star}{\kappa_0^\star} \xrightarrow{P} 1$ as $n \to \infty$. Moreover, $\hat{\xi}_{n,\hat{\kappa}^\star}^{(1)}$ has the same asymptotic efficiency as $\hat{\xi}_{n,\kappa_0^\star}^{(1)}$, meaning that $\sqrt{\kappa_0^\star}(\hat{\xi}_{n,\kappa_0^\star}^{(1)} - \xi)$ and $\sqrt{\hat{\kappa}^\star}(\hat{\xi}_{n,\hat{\kappa}^\star}^{(1)} - \gamma)$ share the same (normal) limit in distribution as $n \to \infty$.*

While we do not present the full details of this result and its proof here, it is worth noting that the need for the second bootstrap in double bootstrap methods actually stems from not knowing the value of the parameter associated with the second-order condition. The interested reader can refer to [55, 181] for a detailed discussion. Additionally, note that the requirement of the second-order condition also implies that, while double bootstrap methods have provable asymptotic optimality guarantees, these guarantees hold for a narrower class of distributions than one may hope. For example, a "pure power law" Pareto distribution does not satisfy the second-order condition. That said, empirical experience suggests that double bootstrap methods perform well in practice, even when the underlying distribution does not satisfy the second-order condition (see [212]).

It is important to emphasize that double bootstrapping is the first approach we discuss in this chapter that guarantees an (asymptotically) optimal choice of k, albeit subject to a second-order condition on the distribution the data is sampled from. In contrast, PLFIT guarantees consistency (but not optimality), without the need for the second-order condition. Additionally, while the Hill double bootstrap method described here only works when $\xi > 0$ (i.e., the data is drawn from a regularly varying distributions), there are other double bootstrap techniques that work for all values of ξ (i.e., when the data distribution lies in the MDA of any extremal distribution (see [69])). As a result, double bootstrap methods are particularly appealing for use in practice. In fact, the approach is growing in popularity. At this point, open source code has been made public [211] and has recently been used in the context of network science and beyond [212].

We illustrate the application of the Hill double bootstrap using our running examples in Figures 9.19–9.21. Figure 9.19 illustrates the most simple case, when the Hill double bootstrap is applied to data that comes from a distribution with an Exponential body and a Pareto tail, with $\alpha = 2$. In this case, Figure 9.19(a) shows the Hill Plot and we see that the Hill double bootstrap identifies a point near the center of the flat region, which yields an estimate of the tail index of 2.045, which is very close to the true value of 2. Additionally, the figure highlights how close the estimates of the Hill double bootstrap and PLFIT are in this case. Figure 9.19(b) shows that the estimate of "where the tail begins" provided by the Hill double bootstrap is again very close to that of PLFIT. Figures 9.20 and 9.21 show the application of the Hill double bootstrap to the US power blackouts and US city sizes datasets. In these cases, we do not know the true values, but we can see that the Hill double bootstrap and PLFIT arrive at nearly the same estimate, which adds confidence to the estimates each provides.

9.7 Guidelines for Estimating Heavy-Tailed Phenomena

We have covered a lot of ground in this chapter and the previous one. These two chapters on estimation present a wide variety of parametric and semi-parametric tools that are appropriate for identifying and estimating power-laws and broader heavy-tailed phenomena. Given the variety of estimators presented in these chapters, it is important to end this chapter by stepping back and thinking about how and when to use each of these tools when exploring data.

Perhaps the most important message that these chapters convey is that it is crucial to think carefully about whether parametric or semi-parametric analysis is appropriate for the data in question. Doing this involves asking the following question:

> *Is there reason to believe the whole **distribution** follows a power-law,*
> *or does just the **tail** follow a power-law?*

Answering this question often requires thinking deeply about the source of the data, but exploratory analysis can be helpful as well. Investigating the rank plot and the mean residual life plot can be very helpful in understanding if a power-law is hiding in the data. Another important tool for exploratory analysis that we have discussed only briefly in these chapters is the QQ plot. Like the mean residual life plot, the QQ plot can be very effective at distinguishing power-law behavior from other forms of heavy-tailed distributions, and we definitely recommend its use.

While many tools are available to help answer the question of whether the parametric or semi-parametric approach is more appropriate, often it is the case that one is still left with some uncertainty about which approach to apply. In such situations it is important to remember that using parametric tools when they are not appropriate can lead to untrustworthy conclusions, as we discussed in Section 8.5. The example discussed there of parametric tools being used to reject the power-law hypothesis when applied to data from graphs generated via preferential attachment is a cautionary tale that highlights a crucial lesson: if there is uncertainty about whether the parametric tools are appropriate, then it is safest to use semi-parametric techniques.

Once exploratory analysis has led to a conclusion about whether parametric or semi-parametric analysis is appropriate, statistical estimation can begin. If parametric estimation is appropriate, the material in Chapter 8 shows that the MLE is a reliable and effective estimator, and the fit it provides can be visualized using a form of weighted least squares regression. Of course, the fit provided by the MLE should also be evaluated for statistical significance. For this purpose, it is standard to perform a goodness-of-fit test using the Kolmogorov–Smirnov (KS) statistic. Additionally, it is important to test the hypothesis of a power-law by contrasting the fit to what could be achieved via a different parametric assumption. For this, a standard approach is to use the likelihood ratio test, as described in Section 8.5, though there are also many other options including cross-validation (see [201]), and Bayesian approaches (see [124]).

Things are more complex if parametric analysis is not appropriate. In cases where only the tail of the distribution is power-law, it may be tempting to still use parametric tools like the MLE by censoring the data. The material in Section 9.1 makes clear that this can be disastrous. Instead, extremal estimators should be used in the semi-parametric setting. To that end, this chapter presents four extremal estimators that each rely on different features of the data: the Hill estimator, the moments estimator, the Pickands estimator, and peaks over thresholds. Interestingly, none of these dominates the others in terms of the variance of the estimator, which highlights the important lesson that no extremal estimator is uniformly the "best." Further, while each works beautifully in many situations, there are situations where each can lead to misleading estimation. As a result, the best option in practice is to derive confidence in any estimate by looking for alignment between different estimators. We recommend using these approaches together (including PLFIT and the different versions of the double bootstrapping approach) in order to build confidence in estimates in the semi-parametric setting.

The approach just outlined may seem complicated; however, the identification of heavy-tailed phenomena is a continual source of controversy in many fields as a result of the failure of simple, intuitive approaches. Thus, it is crucial to take care to obtain validation from multiple approaches in order to have confidence in the resulting conclusions.

9.8 Additional Notes

In this chapter, we have given an approachable overview of the state-of-the-art tools for heavy-tail estimation in the semi-parametric setting, covering both exploratory tools and estimation procedures. The discussion necessarily omits some important approaches. Two of particular interest are the QQ plot, which is most useful in the parametric setting, and regression estimators. As we have discussed, regression based on the full dataset results in biased estimates. However, it is possible to show that applying regression to the top k data points results in a consistent estimator as the dataset n grows large, and $k = k(n) \to \infty$, $k/n \to 0$. We refer to [183] for more background on QQ plots; the consistency of regression applied to the top k order statistics in the semi-parametric setting is established in [132].

Another class of estimators we have not discussed in this chapter is the class of *kernel estimators*. Unlike in the case of the Hill/moments/Pickands estimators, where a discrete number k of order statistics are used to estimate the shape parameter ξ, kernel approaches use the entire dataset, but with higher weights applied to the higher order statistics. The weights are specified by a certain kernel function (this gives the approach its name) and a certain bandwidth parameter. This results in a "smoother" estimate, less sensitive to small changes in user-specified parameters. While a kernel variant of the Hill estimator was first introduced for $\xi > 0$ (i.e., for regularly varying data) in [54], the approach has been generalized for all $\xi \in \mathbb{R}$ in [105]. Kernel-based estimators are also amenable to a double bootstrap technique to optimize the choice of the bandwidth parameter (see [105]).

To ensure the chapter is accessible, we have omitted proofs of some important results. The interested reader may find a detailed technical treatment of the results presented this chapter in more technical books such as [27, 75, 183]. Additionally, a complementary view of statistical analysis of heavy tails that avoids asymptotic methods can be found in [205].

Finally, in this chapter (and throughout the book), we have focused attention on the case of univariate heavy-tailed distributions. The theory of multivariate heavy-tailed distributions is more technically demanding, but many of the same insights and techniques apply. In particular, one of the essential ideas is to reduce the case of multivariate distributions to a univariate setting. To illustrate how this is accomplished, consider a random vector $(X_1, \ldots, X_k)$. To reduce it to a univariate setting, we consider a polar coordinate transform, and we say that the random vector $(X_1, \ldots, X_k)$ is regularly varying when its radius R is regularly varying and its angle Θ converges to a limiting distribution conditional upon

R being large. For an introduction to this deep and important area, we refer to the book by Resnick [183]. An application of this concept to a network science application can be found in [213].

9.9 Exercises

1. Suppose that X is regularly varying with index $-\alpha$, and $Y = \log(X)$. Prove, using Karamata's theorem, that the mean residual life function m_Y associated with Y satisfies

$$\lim_{x \to \infty} m_Y(x) = \frac{1}{\alpha}.$$

   Interpret the Hill estimator in light of this expression.

2. Consider an i.i.d. sequence of random variables $\{X_i\}_{i \geq 1}$, having a distribution F satisfying $\bar{F}(x) > 0$ for all $x > 0$. Let $\{k(n)\}$ denote a deterministic sequence of natural numbers satisfying (i) $k(n) \leq n$, (ii) $\lim_{n \to \infty} k(n) = \infty$, and (iii) $\lim_{n \to \infty} \frac{k(n)}{n} = 0$. For $n \in \mathbb{N}$, define $b_{\min}(n)$ to be the value of the $k(n)$th largest sample from among $\{X_1, X_2, \ldots, X_n\}$. Prove that $b_{\min}(n) \to \infty$ as $n \to \infty$ almost surely.

3. Let $(E_1, \ldots, E_n)$ denote a vector of i.i.d. exponentially distributed random variables with rate λ, and $(E_{(1)}, \ldots, E_{(n)})$ denote the associated order statistics (i.e., $E_{(i)}$ is the ith largest value among $E_1, \ldots, E_n$). Prove that

$$(E_{(1)}, \ldots, E_{(n)}) \overset{d}{=} (E_1', \ldots, E_n'),$$

   where

$$E_i' = \sum_{j=i}^{n} E_j/j, \quad i = 1, \ldots, n.$$

4. Let $(U_1, U_2, \ldots, U_n)$ denote a vector of i.i.d. uniform random variables on $[0, 1]$, and $(U_{(1)}, \ldots, U_{(n)})$ denote the associated order statistics. Show that $n(1 - U_{(k)})/k \to 1$ in probability if $k = k(n) \to \infty$ and $k/n \to 0$ as $n \to \infty$.

   Hint: Begin by expressing the order statistics of uniform random variables in term of a ratio of sums of Exponential random variables.

5. Consider the following generalization of the Pickands estimator. Suppose that we are given i.i.d. observations $X_1, X_2, \ldots, X_n$ drawn from a distribution belonging to the MDA of a generalized extreme value distribution with index ξ. Define, for $\theta \in \mathbb{N}, \theta \geq 2$,

$$\hat{\xi}_{k,n}^{GP} = \frac{1}{\log(\theta)} \log \left(\frac{X_{(k)} - X_{(\theta k)}}{X_{(\theta k)} - X_{(\theta^2 k)}} \right).$$

   Note that this estimator reduces to the Pickands estimator when $k = 2$.
   Prove that this estimator is weakly consistent under the standard conditions on k. Specifically, show that if $k = k(n) \to \infty$ and $k(n)/0 \to 0$ as $n \to \infty$, then $\hat{\xi}_{k,n}^{GP} \overset{P}{\to} \xi$ as $n \to \infty$.

6. Characterize the MLE for the generalized Pareto distribution, and verify that the first-order conditions yield Equations (9.15) and (9.16).

7. Generate 5,000 samples each from the following (regularly varying) distributions:

- Pareto(1,3).
- An equally weighted mixture of a Pareto(1,3) and a Pareto(1,2).
- The Burr$(1, 2, 1)$ distribution.
- The Fréchet$(2, 1, 0)$ distribution.
- Lévy$(0, 1)$ distribution.

For definitions of these distributions, see Section 1.2.4.

For each of the datasets, estimate the tail index using the following methods:

(a) The Hill estimator, choosing a suitable value of k by visual inspection of the Hill plot.

(b) The moments estimator, choosing a suitable value of k by similarly inspecting the plot of ξ_{ME} versus k.

(c) The Pickands estimator, choosing a suitable value of k by inspection of the Pickands plot.

(d) The POT method.[7]

(e) Linear regression applied to the rank plot on a log-log scale.

Compare the accuracies of the different methods.

8. For the datasets from Exercise 7, observe the sensitivity of the listed methods to multiplicative scaling and translation of the data.

9. For the datasets from Exercise 7, compute the estimate of the tail index using PLFIT and the Hill double bootstrap method. How close is the value of k chosen by these approaches to the one you chose via visual inspection of the Hill plot?

---

[7] Library routines for computing the MLE of the generalized Pareto distribution can be found for Python, Matlab®, etc.

Commonly Used Notation

α	Tail index of Pareto distribution; regularly varying distributions are typically assumed to have index $-\alpha$.
F	Cumulative distribution function associated with a random variable, say X, that is, $F(x) = \Pr(X \leq x)$.
$\bar{F}$	Complementary cumulative distribution function (c.c.d.f.), or tail distribution function, associated with the distribution function F, that is, $\bar{F}(x) = 1 - F(x)$.
F_e	Excess distribution function associated with a nonnegative random variable X with distribution function F and finite mean μ, that is, $\bar{F}_e(x) = \frac{1}{\mu}\int_x^\infty \bar{F}(y)dy$.
F^{n*}	n-fold convolution of a distribution function F, defined by $F^{1*} = F$, and for $n \geq 2$, $F^{n*}(t) = \int_{-\infty}^t F^{(n-1)*}(t-s)dF(s)$.
$q(\cdot)$	Hazard rate, aka the failure rate, associated with a distribution F with density f is defined as $q(x) = f(x)/\bar{F}(x)$.
$m(\cdot)$	Mean residual life associated with a nonnegative distribution F, given by $m(x) = \int_0^\infty \frac{\bar{F}(x+t)}{\bar{F}(x)}dt$.
ϕ_X	Characteristic function associated with a random variable X, that is, $\phi_X(t) := \mathbb{E}\left[e^{itX}\right]$.
ψ_X	Laplace–Stieltjes transform associated with a random variable X, that is, $\psi_X(t) := \mathbb{E}\left[e^{-tX}\right]$.
$\mathcal{RV}(\rho)$	Regularly varying function of index ρ. For $\rho \leq 0$, a distribution $F \in \mathcal{RV}(\rho)$ if $\bar{F}$ is a regularly varying function with index ρ (see Chapter 2).
$\mathcal{S}$	Set of subexponential distributions (see Chapter 3).
$\mathcal{L}$	Set of long-tailed distributions (see Chapter 4).
Φ_α	Max-stable Fréchet distribution (see Theorem 7.3).
Ψ_α	Max-stable Weibull distribution (see Theorem 7.3).
Λ	Max-stable Gumbel distribution (see Theorem 7.3).
$\Gamma(\cdot)$	Gamma function, defined by $\Gamma(z) := \int_0^\infty e^{-t}t^{z-1}dt$.
$o(\cdot)$	$f(t) = o(g(t))$ as $t \to \infty$ if $\lim_{t\to\infty} \frac{f(t)}{g(t)} = 0$.

$\xrightarrow{d}$ Convergence in distribution. Given a sequence F_n of distribution functions indexed by n and a distribution function F, $F_n \xrightarrow{d} F$ denotes that $F_n(x) \to F(x)$ at each point x where F is continuous. If X_n has distribution function F_n and X has distribution function F, we sometimes equivalently write $X_n \xrightarrow{d} X$.

$\xrightarrow{p}$ Convergence in probability. Given a sequence of random variables X_n and a random variable X, $X_n \xrightarrow{p} X$ denotes that $P(|X_n - X| > \epsilon) \to 0$ for every $\epsilon > 0$ as $n \to \infty$.

$\xrightarrow{a.s.}$ Almost sure convergence. Given a sequence of random variables X_n and a random variable X, $X_n \xrightarrow{a.s.} X$ denotes that $\Pr\left(\lim_{n\to\infty} X_n(\omega) = X(\omega)\right) = 1$.

$\overset{d}{=}$ Equality in distribution. If X and Y have the same distribution function, we write $X \overset{d}{=} Y$.

References

[1] National Centers for Environment Information. www.ngdc.noaa.gov/. Accessed: 2019-08-06.

[2] United States Census Bureau. www.census.gov/data/tables/time-series/demo/popest/2010s-total-cities-and-towns.html. Accessed: 2019-08-06.

[3] Swiss Re. Natural catastrophes and major losses in 1995. *Sigma*, 2, 1996.

[4] D. Achlioptas, A. Clauset, D. Kempe, and C. Moore. On the bias of traceroute sampling: or, power-law degree distributions in regular graphs. *Journal of the ACM (JACM)*, 56(4):21, 2009.

[5] J. Aczél. *On applications and theory of functional equations*. Academic Press, 2014.

[6] L. A. Adamic and B. A. Huberman. The nature of markets in the world wide web. *Quarterly Journal of Electronic Commerce*, 1:5–12, 2000.

[7] J. Aitchison and J. Brown. *The lognormal distribution*. Cambridge University Press, 1957.

[8] R. Albert and A.-L. Barabási. Statistical mechanics of complex networks. *Reviews of Modern Physics*, 74(1):47, 2002.

[9] D. Aldous. *Probability approximations via the Poisson clumping heuristic*. Springer, 1989.

[10] A. P. Allen, B.-L. Li, and E. L. Charnov. Population fluctuations, power laws and mixtures of lognormal distributions. *Ecology Letters*, 4(1):1–3, 2001.

[11] L. J. Allen. *An introduction to stochastic processes with applications to biology*. Chapman and Hall/CRC, 2010.

[12] M. F. Alves. Estimation of the tail parameter in the domain of attraction of an extremal distribution. *Journal of Statistical Planning and Inference*, 45(1–2):143–173, 1995.

[13] S. Amari, R. S. Lewis, and E. Anders. Interstellar grains in meteorites: I. Isolation of SiC, graphite and diamond; size distributions of SiC and graphite. *Geochimica et Cosmochimica Acta*, 58(1):459–470, 1994.

[14] C. Anderson. *The long tail: why the future of business is selling less of more*. Hachette Books, 2006.

[15] A. Araujo and E. Giné. *The central limit theorem for real and Banach valued random variables*. Wiley, 1980.

[16] S. Asmussen. *Applied probability and queues*. Springer Science & Business Media, 2008.

[17] S. Asmussen and H. Albrecher. *Ruin probabilities*. World Scientific Publishing, 2010.

[18] S. Asmussen and C. Klüppelberg. Large deviations results for subexponential tails, with applications to insurance risk. *Stochastic Processes and Their Applications*, 64(1):103–125, 1996.

[19] S. Asmussen, C. Klüppelberg, and K. Sigman. Sampling at subexponential times, with queueing applications. *Stochastic Processes and Their Applications*, 79(2):265–286, 1999.

[20] K. B. Athreya and P. Jagers. *Classical and modern branching processes*. Springer Science & Business Media, 2012.

[21] K. B. Athreya and P. E. Ney. *Branching processes*. Courier Corporation, 2004.

[22] F. Baccelli and S. Foss. Moments and tails in monotone-separable stochastic networks. *The Annals of Applied Probability*, 14(2):612–650, 2004.

[23] A.-L. Barabási. *Network science*. Cambridge University Press, 2016.

[24] A.-L. Barabási and R. Albert. Emergence of scaling in random networks. *Science*, 286(5439):509–512, 1999.

[25] A.-L. Barabási and E. Bonabeau. Scale-free networks. *Scientific American*, 288(5):60–69, 2003.

[26] O. E. Barndorff-Nielsen and D. R. Cox. *Inference and asymptotics*. Chapman & Hall, 1994.

[27] J. Beirlant, Y. Goegebeur, J. Segers, and J. L. Teugels. *Statistics of extremes: theory and applications*. John Wiley & Sons, 2006.

[28] A. Bhattacharya, B. Chen, R. van der Hofstad, and B. Zwart. Consistency of the PLFit estimator for power-law data. *arXiv preprint arXiv:2002.06870*, 2020.

[29] R. N. Bhattacharya and R. R. Rao. *Normal approximation and asymptotic expansions*. SIAM, 2010.

[30] P. Billingsley. *Probability and measure*. John Wiley & Sons, 2008.

[31] N. Bingham, C. Goldie, and J. Teugels. *Regular variation*. Cambridge University Press, 1989.

[32] O. Boxma and B. Zwart. Tails in scheduling. *Performance Evaluation Review*, 34(4):13–20, 2007.

[33] O. J. Boxma and V. Dumas. Fluid queues with long-tailed activity period distributions. *Computer Communications*, 21(17):1509–1529, 1998.

[34] S. Boyd and L. Vandenberghe. *Convex optimization*. Cambridge University Press, 2004.

[35] T. Britton. Stochastic epidemic models: a survey. *Mathematical Biosciences*, 225(1):24–35, 2010.

[36] A. Broder, R. Kumar, F. Maghoul et al. Graph structure in the web. *Computer Networks*, 33(1–6):309–320, 2000.

[37] A. D. Broido and A. Clauset. Scale-free networks are rare. *Nature Communications*, 10(1):1017, 2019.

[38] J. A. Bucklew. *Large deviation techniques in decision, simulation, and estimation*. Wiley, 1990.

[39] D. Buraczewski, J. F. Collamore, E. Damek, and J. Zienkiewicz. Large deviation estimates for exceedance times of perpetuity sequences and their dual processes. *The Annals of Probability*, 44(6):3688–3739, 2016.

[40] D. Buraczewski, E. Damek, and T. Mikosch. *Scholastic models with power-law tails*. Springer, 2016.

[41] G. Caldarelli. *Scale-free networks: complex webs in nature and technology*. Oxford University Press, 2007.

[42] J. M. Carlson and J. Doyle. Highly optimized tolerance: a mechanism for power laws in designed systems. *Physical Review E*, 60(2):1412, 1999.

[43] G. Casella and R. L. Berger. *Statistical inference* Duxbury, 2002.

[44] D. G. Champernowne. A model of income distribution. *The Economic Journal*, 63(250):318–351, 1953.

[45] H. Chen and D. D. Yao. *Fundamentals of queueing networks: performance, asymptotics, and optimization*. Springer Science & Business Media, 2013.

[46] V. Chistyakov. A theorem on sums of independent positive random variables and its applications to branching random processes. *Theory of Probability and Its Applications*, 9:640–648, 1964.

[47] A. Clauset. Plfit.m. https://aaronclauset.github.io/powerlaws/, 2010.

[48] A. Clauset, C. R. Shalizi, and M. E. Newman. Power-law distributions in empirical data. *SIAM Review*, 51(4):661–703, 2009.

[49] R. Cont. Empirical properties of asset returns: stylized facts and statistical issues. *Quantitative Finance*, 1:223–236, 2001.

[50] J. C. Cox, S. A. Ross, and M. Rubinstein. Option pricing: a simplified approach. *Journal of Financial Economics*, 7(3):229–263, 1979.

[51] H. Cramér. Sur un nouveau théoreme-limite de la théorie des probabilités. *Actualités scientifiques et industrielles*, 736(5–23):115, 1938.

[52] M. E. Crovella, M. S. Taqqu, and A. Bestavros. Heavy-tailed probability distributions in the world wide web. *A practical guide to heavy tails*, 1:3–26. Springer, 1998.

[53] E. L. Crow and K. Shimizu. *Lognormal distributions*. Marcel Dekker, 1987.

[54] S. Csorgo, P. Deheuvels, and D. Mason. Kernel estimates of the tail index of a distribution. *The Annals of Statistics*, 13(3):1050–1077, 1985.

[55] J. Danielsson, L. de Haan, L. Peng, and C. G. de Vries. Using a bootstrap method to choose the sample fraction in tail index estimation. *Journal of Multivariate Analysis*, 76(2):226–248, 2001.

[56] L. F. M. deHaan. *On regular variation and its application to the weak convergence of sample extremes*. PhD thesis, University of Amsterdam, 1970.

[57] P. Deheuvels, E. Haeusler, and D. M. Mason. Almost sure convergence of the hill estimator. *Mathematical Proceedings of the Cambridge Philosophical Society*, 104(2):371–381, 1988.

[58] A. L. Dekkers and L. De Haan. On the estimation of the extreme-value index and large quantile estimation. *The Annals of Statistics*, 17(4):1795–1832, 1989.

[59] A. L. Dekkers, J. H. Einmahl, and L. De Haan. A moment estimator for the index of an extreme-value distribution. *The Annals of Statistics*, 17(4):1833–1855, 1989.

[60] A. Dembo and O. Zeitouni. *Large deviations: techniques and applications*. Springer-Verlag, 2010.

[61] F. Den Hollander. *Large deviations*. American Mathematical Society, 2008.

[62] D. Denisov, A. B. Dieker, and V. Shneer. Large deviations for random walks under subexponentiality: the big-jump domain. *The Annals of Probability*, 36(5):1946–1991, 2008.

[63] D. Denisov and B. Zwart. On a theorem of Breiman and a class of random difference equations. *Journal of Applied Probability*, 44(4):1031–1046, 2007.

[64] J.-D. Deuschel and D. W. Stroock. *Large deviations*. Academic Press, 1989.

[65] L. Devroye. Branching processes and their applications in the analysis of tree structures and tree algorithms. In M. Habib, C. McDiarmid, J. Ramirez-Alfonsin, and B. Reed (eds.), *Probabilistic methods for algorithmic discrete mathematics*, pages 249–314. Springer, 1998.

[66] R. Doney. Local behaviour of first passage probabilities. *Probability Theory and Related Fields*, 152(3):559–588, 2012.

[67] S. N. Dorogovtsev, J. F. F. Mendes, and A. N. Samukhin. Structure of growing networks with preferential linking. *Physical Review Letters*, 85(21):4633, 2000.

[68] A. B. Downey. The structural cause of file size distributions. In *MASCOTS 2001, Proceedings Ninth International Symposium on Modeling, Analysis and Simulation of Computer and Telecommunication Systems*, pages 361–370. IEEE, 2001.

[69] G. Draisma, L. de Haan, L. Peng, and T. T. Pereira. A bootstrap-based method to achieve optimality in estimating the extreme-value index. *Extremes*, 2(4):367–404, 1999.

[70] H. Drees. Refined Pickands estimators of the extreme value index. *The Annals of Statistics*, 23(5):2059–2080, 1995.

[71] H. Drees, A. Janßen, S. I. Resnick, and T. Wang. On a minimum distance procedure for threshold selection in tail analysis. *SIAM Journal on Mathematics of Data Science*, 2(1):75–102, 2020.

[72] P. Dupuis and R. S. Ellis. *A weak convergence approach to the theory of large deviations*. John Wiley & Sons, 2011.

[73] R. Durrett. *Probability: theory and examples*. Cambridge University Press, 2010.

[74] F. Eggenberger and G. Pólya. Über die statistik verketteter vorgänge. *ZAMM-Journal of Applied Mathematics and Mechanics/Zeitschrift für Angewandte Mathematik und Mechanik*, 3(4):279–289, 1923.

[75] P. Embrechts, C. Klüppelberg, and T. Mikosch. *Modelling extremal events: for insurance and finance*. Springer-Verlag, 1997.

[76] C. Estan and G. Varghese. New directions in traffic measurement and accounting: focusing on the elephants, ignoring the mice. *ACM Transactions on Computer Systems (TOCS)*, 21(3):270–313, 2003.

[77] J.-B. Estoup. *Gammes sténographiques: méthode et exercices pour l'acquisition de la vitesse*. Institut sténographique, 1916.

[78] A. Fabrikant, E. Koutsoupias, and C. Papadimitriou. Heuristically optimized trade-offs: a new paradigm for power laws in the internet. In P. Widmayer, F. Triquero, R. Morales, M. Hennessy, S. Eidenbenz, and R. Conejo (eds.), *Lecture Notes in Computer Science*, pages 110–122. Springer-Verlag, 2002.

[79] M. Falk. Efficiency of convex combinations of Pickands estimator of the extreme value index. *Journal of Nonparametric Statistics*, 4(2):133–147, 1994.

[80] M. Faloutsos, P. Faloutsos, and C. Faloutsos. On power-law relationships of the internet topology. *ACM SIGCOMM Computer Communication Review*, 20:251–262, 1999.

[81] W. Feller. *An introduction to probability theory and its applications: Volume I*. John Wiley & Sons, 1968.

[82] W. Feller. *An introduction to probability theory and its applications: Volume II*. John Wiley & Sons, 1971.

[83] J. Feng and T. G. Kurtz. *Large deviations for stochastic processes*. American Mathematical Society, 2006.

[84] H. Fischer. *A history of the central limit theorem: from classical to modern probability theory*. Springer Science & Business Media, 2010.

[85] R. A. Fisher and L. H. C. Tippett. Limiting forms of the frequency distribution of the largest or smallest member of a sample. *Mathematical Proceedings of the Cambridge Philosophical Society*, 24(2):180–190, 1928.

[86] A. Fontanari, N. N. Taleb, and P. Cirillo. Gini estimation under infinite variance. *Physica A: Statistical Mechanics and Its Applications*, 502(15):256–269, 2018.

[87] S. Foss, T. Konstantopoulos, and S. Zachary. Discrete and continuous time modulated random walks with heavy-tailed increments. *Journal of Theoretical Probability*, 20(3):581–612, 2007.

[88] S. Foss and D. Korshunov. Heavy tails in multi-server queue. *Queueing Systems*, 52(1):31–48, 2006.

[89] S. Foss, D. Korshunov, and S. Zachary. *An introduction to heavy-tailed and subexponential distributions*. Springer, 2011.

[90] S. Foss, Z. Palmowski, and S. Zachary. The probability of exceeding a high boundary on a random time interval for a heavy-tailed random walk. *The Annals of Applied Probability*, 15(3):1936–1957, 2005.

[91] M. Fréchet. Sur la loi de probabilité de l'écart maximum. *Annales Polonici Mathematici*, 6:93–116, 1927.

[92] X. Gabaix. Zipf's law and the growth of cities. *American Economic Review*, 89(2):129–132, 1999.

[93] X. Gabaix. Zipf's law for cities: an explanation. *The Quarterly Journal of Economics*, 114(3):739–767, 1999.

[94] X. Gabaix, P. Gopikrishnan, V. Plerou, and H. E. Stanley. A theory of power-law distributions in financial market fluctuations. *Nature*, 423(6937):267, 2003.

[95] F. Galton. Problem 4001. *Educational Times*, 1, 1873.

[96] A. Ganesh, N. O'Connell, and D. Wischik. *Big Queues*. Springer, 2004.

[97] J. Geluk, L. de Haan, S. Resnick, and C. Stărică. Second-order regular variation, convolution and the central limit theorem. *Stochastic Processes and Their Applications*, 69(2):139–159, 1997.

[98] R. Gibrat. *Les Inégalités Economiques*. Sirey, Paris, 1931.

[99] B. Gnedenko. Sur la distribution limite du terme maximum d'une serie aleatoire. *Annals of Mathematics*, 44:423–453, 1943.

[100] B. Gnedenko and A. Kolmogorov. *Limit distributions for sums of independent random variables*. Addison-Wesley, 1954.

[101] C. Goldie and C. Klüppelberg. Subexponential distributions. In R. J. Adler, R. E. Feldman, and M. S. Taqqu, eds., *A practical guide to heavy tails: statistical techniques and applications*, pages 435–459. Birkhäuser, 1998.

[102] C. M. Goldie. Implicit renewal theory and tails of solutions of random equations. *The Annals of Applied Probability*, 1(1):126–166, 1991.

[103] A. Goswami and B. V. Rao. *A course in applied stochastic processes*. Springer, 2006.

[104] G. R. Grimmett and D. Stirzaker. *Probability and random processes*. Oxford University Press, 2001.

[105] P. Groeneboom, H. Lopuhaä, and P. De Wolf. Kernel-type estimators for the extreme value index. *The Annals of Statistics*, 31(6):1956–1995, 2003.

[106] F. Guess and F. Proschan. Mean residual life: theory and applications. In P.R. Krishnaiah and C.R. Rao, eds., *Quality control and reliability*, pages 215–224. Handbook of Statistics, vol. 7. Elsevier, 1988.

[107] E. J. Gumbel. *Statistics of extremes*. Columbia University Press, 1958.

[108] L. Guo and I. Matta. The war between mice and elephants. In *Proceedings of the IEEE Conference on Network Protocols*, pages 180–188, 2001.

[109] B. Gutenberg and C. F. Richter. Frequency of earthquakes in California. *Bulletin of the Seismological Society of America*, 34(4):185–188, 1944.

[110] A. P. Hackett. *70 years of best sellers: 1895–1965*. RR Bowker Company, 1967.

[111] B. Hajek. *Random processes for engineers*. Cambridge University Press, 2015.

[112] M. Harchol-Balter and A. B. Downey. Exploiting process lifetime distributions for dynamic load balancing. *ACM Transactions on Computer Systems*, 15(3):253–285, 1997.

[113] D. Heath, S. Resnick, and G. Samorodnitsky. Patterns of buffer overflow in a class of queues with long memory in the input stream. *The Annals of Applied Probability*, 7(4):1021–1057, 1997.

[114] P. Hines, K. Balasubramaniam, and E. C. Sanchez. Cascading failures in power grids. *IEEE Potentials*, 28(5):24–30, 2009.

[115] C. Huber-Carol, N. Balakrishnan, M. Nikulin, and M. Mesbah. *Goodness-of-Fit Tests and Model Validity*. Statistics for Industry and Technology. Birkhäuser, 2012.

[116] B. A. Huberman and L. A. Adamic. Internet: growth dynamics of the world-wide web. *Nature*, 401(6749):131, 1999.

[117] J. Hull. *Introduction to futures and options markets*. Prentice Hall, 1991.

[118] I. Ibragimov and Y. Linnik. *Independent and stationary sequences of random variables*. Wolters–Noordoff, 1971.

[119] R. Jain and S. Ramakumar. Stochastic dynamics modeling of the protein sequence length distribution in genomes: implications for microbial evolution. *Physica A: Statistical Mechanics and Its Applications*, 273(3-4):476–485, 1999.

[120] S. Janson. Stable distributions. *arXiv:1112.0220*, 2011.

[121] P. Jelenković and P. Momčilović. Large deviation analysis of subexponential waiting times in a processor-sharing queue. *Mathematics of Operations Research*, 28(3):587–608, 2003.

[122] P. R. Jelenković and M. Olvera-Cravioto. Implicit renewal theorem for trees with general weights. *Stochastic Processes and Their Applications*, 122(9):3209–3238, 2012.

[123] J. Jurecková and P. K. Sen. *Robust statistical procedures: asymptotics and interrelations*. John Wiley & Sons, 1996.

[124] R. E. Kass and A. E. Raftery. Bayes factors. *Journal of the American Statistical Association*, 90(430):773–795, 1995.

[125] H. Kesten. Random difference equations and renewal theory for products of random matrices. *Acta Mathematica*, 131(1):207–248, 1973.

[126] J. Kleinberg. The small-world phenomenon: an algorithmic perspective. In *Proceedings of the Thirty-Second Annual ACM Symposium on Theory of Computing*, pages 163–170, 2000.

[127] C. Klüppelberg. Subexponential distributions and integrated tails. *Journal of Applied Probability*, 25(1):132–141, 1988.

[128] C. Klüppelberg and S. Pergamenchtchikov. The tail of the stationary distribution of a random coefficient AR(q) model. *The Annals of Applied Probability*, 14(2):971–1005, 2004.

[129] L. Knopoff and D. Sornette. Earthquake death tolls. *Journal de Physique I*, 5(12):1681–1668, 1995.

[130] E. V. Koonin, Y. I. Wolf, and G. P. Karev. *Power laws, scale-free networks and genome biology*. Springer, 2006.

[131] P. L. Krapivsky, S. Redner, and F. Leyvraz. Connectivity of growing random networks. *Physical Review Letters*, 85(21), 2000.

[132] M. Kratz and S. I. Resnick. The QQ-estimator and heavy tails. *Communications in Statistics. Stochastic Models*, 12(4):699–724, 1996.

[133] M. R. Leadbetter, G. Lindgren, and H. Rootzén. *Extremes and related properties of random sequences and processes*. Springer Science & Business Media, 2012.

[134] J. Leskovec, D. Chakrabarti, J. Kleinberg, and C. Faloutsos. Realistic, mathematically tractable graph generation and evolution, using Kronecker multiplication. In *European Conference on Principles of Data Mining and Knowledge Discovery*, pages 133–145. Springer, 2005.

[135] J. Leskovec, D. Chakrabarti, J. Kleinberg, C. Faloutsos, and Z. Ghahramani. Kronecker graphs: an approach to modeling networks. *Journal of Machine Learning Research*, 11(Feb):985–1042, 2010.

[136] J. Leskovec, J. Kleinberg, and C. Faloutsos. Graphs over time: densification laws, shrinking diameters and possible explanations. In *Proceedings of the Eleventh ACM SIGKDD International Conference on Knowledge Discovery in Data Mining*, pages 177–187. ACM, 2005.

[137] J. Leskovec, J. Kleinberg, and C. Faloutsos. Graph evolution: densification and shrinking diameters. *ACM Transactions on Knowledge Discovery from Data*, 1(1):2, 2007.

[138] J. Leslie. On the non-closure under convolution of the subexponential family. *Journal of Applied Probability*, 26(1):58–66, 1989.

[139] M. Levy, S. Solomon, and G. Ram. Dynamical explanation for the emergence of power law in a stock market model. *International Journal of Modern Physics C*, 7(1):65–72, 1996.

[140] P. Lévy. Théorie des erreurs: La Loi de Gauss et Les Lois Exceptionelles. *Bulletin de la Société Mathématique de France*, 52:49–85, 1924.

[141] L. Li, D. Alderson, J. C. Doyle, and W. Willinger. Towards a theory of scale-free graphs: definition, properties, and implications. *Internet Mathematics*, 2(4):431–523, 2005.

[142] F. Lindskog, S. I. Resnick, and J. Roy. Regularly varying measures on metric spaces: hidden regular variation and hidden jumps. *Probability Surveys*, 11:270–314, 2014.

[143] M. Loeve. *Probability theory II*. Springer, 1978.

[144] E. T. Lu and R. J. Hamilton. Avalanches and the distribution of solar flares. *The Astrophysical Journal*, 380(2):L89–L92, 1991.

[145] N. M. Luscombe, J. Qian, Z. Zhang, T. Johnson, and M. Gerstein. The dominance of the population by a selected few: power-law behaviour applies to a wide variety of genomic properties. *Genome Biology*, 3(8):research0040, 2002.

[146] A. Mahanti, N. Carlsson, M. Arlitt, and C. Williamson. A tale of the tails: power-laws in internet measurements. *IEEE Network*, 27(1):59–64, 2013.

[147] B. Mandelbrot. An informational theory of the statistical structure of language. *Communication Theory*, 84:486–502, 1953.

[148] B. Mandelbrot. A note on a class of skew distribution functions: analysis and critique of a paper by HA Simon. *Information and Control*, 2(1):90–99, 1959.

[149] B. Mandelbrot. Final note on a class of skew distribution functions: analysis and critique of a model due to HA Simon. *Information and Control*, 4(2–3):198–216, 1961.

[150] B. Mandelbrot. Post scriptum to "final note." *Information and Control*, 4(2–3):300–304, 1961.

[151] B. B. Mandelbrot. *The fractal geometry of nature*. WH Freeman, 1982.

[152] A. Mazzarella and N. Diodato. The alluvial events in the last two centuries at Sarno, southern Italy: their classification and power-law time-occurrence. *Theoretical and Applied Climatology*, 72(1–2):75–84, 2002.

[153] T. Mikosch. Regular variation, subexponentiality and their applications in probability theory. *Eurandom report 1999-013*, 1998.

[154] T. Mikosch. Modeling dependence and tails of financial time series. In B. Finkenstadt and H. Rootzen, eds., *Extreme values in finance, telecommunications, and the environment*, pp. 185–286. Chapman and Hall, 2003.

[155] T. Mikosch and A. Nagaev. Large deviations of heavy-tailed sums with applications in insurance. *Extremes*, 1(1):81–110, 1998.

[156] A. Müller and D. Stoyan. *Comparison methods for stochastic models and risks*. John Wiley & Sons, 2002.

[157] R. Musson, T. Tsapanos, and C. Nakas. A power-law function for earthquake interarrival time and magnitude. *Bulletin of the Seismological Society of America*, 92(5):1783–1794, 2002.

[158] A. V. Nagaev. Integral limit theorems taking large deviations into account when Cramér's condition does not hold. I. *Theory of Probability and Its Applications*, 14(1):51–64, 1969.

[159] S. V. Nagaev. Large deviations of sums of independent random variables. *The Annals of Probability*, 7(5):745–789, 1979.

[160] T. Nakajima and A. Higurashi. A use of two-channel radiances for an aerosol characterization from space. *Geophysical Research Letters*, 25(20):3815–3818, 1998.

[161] T. Nesti, F. Sloothaak, and B. Zwart. Emergence of scale-free blackout sizes in power grids. *Physical Review Letters*, 125:058301, 2020.

[162] G. Neukum and B. Ivanov. Crater size distributions and impact probabilities on earth from lunar, terrestrial-planet, and asteroid cratering data. *Hazards due to Comets and Asteroids*, 1:359–416, 1994.

[163] M. Newman. Power laws, Pareto distributions and Zipf's law. *Contemporary Physics*, 46(5):323–351, 2005.

[164] M. A. Nowak and R. M. May. Mathematical biology of HIV infections: antigenic variation and diversity threshold. *Mathematical Biosciences*, 106(1):1–21, 1991.

[165] R. Núñez-Queija. Queues with equally heavy sojourn time and service requirement distributions. *Annals of Operations Research*, 113(1–4):101–117, 2002.

[166] M. Nuyens and B. Zwart. A large-deviations analysis of the GI/GI/1 srpt queue. *Queueing Systems*, 54(2):85–97, 2006.

[167] A. G. Pakes. On the tails of waiting-time distributions. *Journal of Applied Probability*, 12(3):555–564, 1975.

[168] F. Papadopoulos, D. Krioukov, M. Boguñá, and A. Vahdat. Greedy forwarding in dynamic scale-free networks embedded in hyperbolic metric spaces. In *Proceedings of IEEE INFOCOM*, 2010.

[169] F. Papadopoulos, D. Krioukov, M. Boguñá, and A. Vahdat. Greedy forwarding in dynamic scale-free networks embedded in hyperbolic metric spaces. In *2010 Proceedings IEEE INFOCOM*, pages 1–9. IEEE, 2010.

[170] K. Papagiannaki, N. Taft, S. Bhattacharyya, P. Thiran, K. Salamatian, and C. Diot. A pragmatic definition of elephants in internet backbone traffic. In *Proceedings of ACM SIGCOMM Workshop on Internet Measurement*, pages 175–176, 2002.

[171] V. Pareto. *Cours d'économie politique*, vol. 1. Librairie Droz, 1964.

[172] V. Petrov. *Sums of independent random variables*. Springer, 1975.

[173] V. V. Petrov. *Limit theorems of probability theory: sequences of independent random variables*. Oxford University Press, 1995.

[174] J. Pickands III. Statistical inference using extreme order statistics. *The Annals of Statistics*, 3(1):119–131, 1975.

[175] V. Pisarenko and M. Rodkin. *Heavy-tailed distributions in disaster analysis*, vol. 30. Springer Science & Business Media, 2010.

[176] E. Pitman. On the behaviour of the characteristic function of a probability distribution in the neighbourhood of the origin. *Journal of the Australian Mathematical Society*, 8(3):423–443, 1968.

[177] E. Pitman. Subexponential distribution functions. *Journal of the Australian Mathematical Society (Series A)*, 29(3):337–347, 1980.

[178] E. J. Pitman. *Some basic theory for statistical inference: monographs on applied probability and statistics*. CRC Press, 2018.

[179] D. D. S. Price. A general theory of bibliometric and other cumulative advantage processes. *Journal of the Association for Information Science and Technology*, 27(5):292–306, 1976.

[180] Y. V. Prokhorov. An extremal problem in probability theory. *Theory of Probability and Its Applications*, 4(2):201–203, 1959.

[181] Y. Qi. Bootstrap and empirical likelihood methods in extremes. *Extremes*, 11(1):81–97, 2008.

[182] C. R. Rao. *Linear statistical inference and its applications*. Wiley, 1973.

[183] S. Resnick. *Heavy-tail phenomena: probabilistic and statistical modeling*. Springer, 2006.

[184] S. Resnick and C. Stărică. Smoothing the Hill estimator. *Advances in Applied Probability*, 29(1):271–293, 1997.

[185] S. I. Resnick. *Extremal processes*. John Wiley & Sons, 1982.

[186] S. I. Resnick. *Extreme values, regular variation and point processes*. Springer, 2013.

[187] C.-H. Rhee, J. Blanchet, and B. Zwart. Sample path large deviations for Lévy processes and random walks with regularly varying increments. *The Annals of Probability*, 47(6):3551–3605, 2019.

[188] P. Rooney. Microsoft's CEO: 80-20 rule applies to bugs, not just features. *CRN*, October 3, 2002.

[189] P. Rosin and E. Rammler. The laws governing the fineness of powdered coal. *Journal of the Institute of Fuel*, 7: 29–36, 1933.

[190] G. Samorodnitsky and M. S. Taqqu. *Stable non-Gaussian random processes: stochastic models with infinite variance*. CRC Press, 1994.

[191] A. Scheller-Wolf. Necessary and sufficient conditions for delay moments in FIFO multiserver queues with an application comparing s slow servers with one fast one. *Operations Research*, 51(5):748–758, 2003.

[192] J. Segers. Generalized Pickands estimators for the extreme value index. *Journal of Statistical Planning and Inference*, 128(2):381–396, 2005.

[193] M. Shaked and J. G. Shanthikumar. *Stochastic orders*. Springer Science & Business Media, 2007.

[194] C. Shirky. Pareto principle. www.edge.org/responses/what-scientific-concept-would-improve-everybodys-cognitive-toolkit, 2011. Accessed on August 6, 2019.

[195] A. Shwartz and A. Weiss. *Large deviations for performance analysis: queues, communication and computing*. CRC Press, 1995.

[196] H. A. Simon. On a class of skew distribution functions. *Biometrika*, 42(3/4):425–440, 1955.

[197] H. A. Simon. Some further notes on a class of skew distribution functions. *Information and Control*, 3(1):80–88, 1960.

[198] H. A. Simon. Reply to Dr. Mandelbrot's post scriptum. *Information and Control*, 4(2–3):305–308, 1961.

[199] H. A. Simon. Reply to "final note" by Benoit Mandelbrot. *Information and Control*, 4(2–3):217–223, 1961.

[200] D. Sornette and R. Cont. Convergent multiplicative processes repelled from zero: power laws and truncated power laws. *Journal de Physique I*, 7(3):431–444, 1997.

[201] M. Stone. Cross-validatory choice and assessment of statistical predictions. *Journal of the Royal Statistical Society: Series B (Methodological)*, 36(2):111–133, 1974.

[202] D. Stoyan. *Comparison methods for queues and other stochastic models*. John Wiley & Sons, 1983.

[203] J. Sutton. Gibrat's legacy. *Journal of Economic Literature*, 35(1):40–59, 1997.

[204] N. N. Taleb. *The black swan: the impact of the highly improbable*. Random House, 2007.

[205] N. N. Taleb. Technical Incerto, Vol 1: Statistical consequences of fat tails, 2019. www.academia .edu/37221402/TECHNICAL_INCERTO_VOL_1_THE_STATISTICAL_CONSEQUENCES_OF _FAT_TAILS

[206] R. Tanaka, T.-M. Yi, and J. Doyle. Some protein interaction data do not exhibit power law statistics. *FEBS Letters*, 579(23):5140–5144, 2005.

[207] J. L. Teugels. The class of subexponential distributions. *The Annals of Probability*, 3(6):1000–1011, 1975.

[208] T. Themido Pereira. Second order behavior of domains of attraction and the bias of generalized Pickands' estimator. In J. Lechner, J. Galambos, and E. Simiu, eds., *Extreme Value Theory and Applications III*. Proceedings of the Gaithersburg Conference (NIST special publication), 1993.

[209] R. van der Hofstad. *Random graphs and complex networks*, vol. 1. Cambridge University Press, 2016.

[210] N. Veraverbeke. Asymptotic behaviour of Wiener-Hfopf factors of a random walk. *Stochastic Processes and Their Applications*, 5(1):27–37, 1977.

[211] I. Voitalov. Github repository. https://github.com/ivanvoitalov/tail-estimation.

[212] I. Voitalov, P. van der Hoorn, R. van der Hofstad, and D. Krioukov. Scale-free networks well done. *Physical Review Research*, 1(3):033034, 2019.

[213] Y. Volkovich, N. Litvak, and B. Zwart. Measuring extremal dependencies in web graphs. In *Proceedings of the 17th International Conference on World Wide Web*, 2008.

[214] M. Wand. Data-based choice of histogram bin width. *The American Statistician*, 51(1):59–64, 1997.

[215] H. W. Watson and F. Galton. On the probability of the extinction of families. *The Journal of the Anthropological Institute of Great Britain and Ireland*, 4:138–144, 1875.

[216] M. S. Wheatland and P. Sturrock. Avalanche models of solar flares and the distribution of active regions. *The Astrophysical Journal*, 471(2):1044, 1996.

[217] W. Whitt. *Stochastic-process limits: an introduction to stochastic-process limits and their application to queues*. Springer Science & Business Media, 2002.

[218] Wikipedia. List of earthquakes in 2010 – Wikipedia, the Free Encyclopedia, 2017. [Online; accessed on August 7, 2017].

[219] Wikipedia. List of earthquakes in 2011 – Wikipedia, the Free Encyclopedia, 2017. [Online; accessed on August 7, 2017].

[220] Wikipedia contributors. Scale invariance – Wikipedia, the Free Encyclopedia, 2018. [Online; accessed on September 7, 2018].

[221] E. Willekens. On the supremum of an infinitely divisible process. *Stochastic Processes and Their Applications*, 26:173–175, 1987.

[222] W. Willinger, D. Alderson, and J. C. Doyle. Mathematics and the internet: a source of enormous confusion and great potential. *Notices of the American Mathematical Society*, 56(5):586–599, 2009.

[223] S. J. Young and E. R. Scheinerman. Random dot product graph models for social networks. In *International Workshop on Algorithms and Models for the Web-Graph*, pages 138–149. Springer, 2007.

[224] G. U. Yule. A mathematical theory of evolution, based on the conclusions of Dr. J. C. Willis, F. R. S. *Philosophical Transactions of the Royal Society of London. Series B*, 213:21–87, 1925 doi: https://doi.org/10.1098/rstb.1925.0002.

[225] S. Yun. On a generalized Pickands estimator of the extreme value index. *Journal of Statistical Planning and Inference*, 102(2):389–409, 2002.

[226] S. Zachary. A note on Veraverbeke's theorem. *Queueing Systems*, 46(1):9–14, 2004.

[227] G. K. Zipf. *Human behavior and the principle of least effort.* Addison-Wesley Press, 1949.

[228] V. M. Zolotarev. *One-dimensional stable distributions.* American Mathematical Society, 1986.

[229] A. P. Zwart. Tail asymptotics for the busy period in the GI/G/1 queue. *Mathematics of Operations Research*, 26(3):485–493, 2001.

[230] B. Zwart. Rare events and heavy tails in stochastic systems, 2008. Lecture notes for a summer school in Wroclaw, 2008.

[231] B. Zwart, S. Borst, and K. Debicki. Subexponential asymptotics of hybrid fluid and ruin models. *The Annals of Applied Probability*, 15(1A):500–517, 2005.

[232] B. Zwart, S. Borst, and M. Mandjes. Exact asymptotics for fluid queues fed by multiple heavy-tailed on–off flows. *The Annals of Applied Probability*, 14(2):903–957, 2004.

Index